Handbook of Moiré Measurement

Series in Optics and Optoelectronics

Series Editors: **R G W Brown**, University of Nottingham, UK
E R Pike, Kings College, London, UK

Other titles in the series

Applications of Silicon-Germanium Heterostructure Devices
C K Maiti and G A Armstrong

Optical Fibre Devices
J-P Goure and I Verrier

Laser-Induced Damage of Optical Materials
R M Wood

Optical Applications of Liquid Crystals
L Vicari (Ed.)

Forthcoming titles in the series

Stimulated Brillouin Scattering
M Damzen, V I Vlad, V Babin and A Mocofanescu

High Speed Photonic Devices
N Dagli (Ed.)

Diode Lasers
D Sands

High Aperture Focussing of Electromagnetic Waves and Applications in Optical Microscopy
C J R Sheppard and P Torok

Transparent Conductive Coatings
C I Bright

Photonic Crystals
M Charlton and G Parker (Eds)

Other titles of interest

Thin-Film Optical Filters (Third Edition)
H Angus Macleod

Series in Optics and Optoelectronics

Handbook of
Moiré Measurement

Edited by

C A Walker

Department of Mechanical Engineering,
University of Strathclyde, UK

CRC Press
Taylor & Francis Group
Boca Raton London New York

CRC Press is an imprint of the
Taylor & Francis Group, an **informa** business

First published 2004 by IOP Publishing Ltd

Published 2019 by CRC Press
Taylor & Francis Group
6000 Broken Sound Parkway NW, Suite 300
Boca Raton, FL 33487-2742

First issued in paperback 2019

No claim to original U.S. Government works

ISBN 13: 978-0-367-45442-5 (pbk)
ISBN 13: 978-0-7503-0522-8 (hbk)

Visit the Taylor & Francis Web site at
http://www.taylorandfrancis.com

and the CRC Press Web site at
http://www.crcpress.com

British Library Cataloguing-in-Publication Data

A catalogue record for this book is available from the British Library.

Library of Congress Cataloging-in-Publication Data are available

Cover Design: Victoria Le Billon

Typeset in LaTeX 2_ε by Text 2 Text Limited, Torquay, Devon

Contents

1   Introduction
C A Walker   **1**

1.1   A brief history of the moiré method   2
          References   2

**CHAPTER 2: STRAIN MEASUREMENTS AT THE LIMIT—THE
MOIRÉ MICROSCOPE**   **3**

2.1   Nonlinear analysis of interferometric moiré fringes
T W Shield   **5**

2.1.1   Example fringe data reduction   12
          References   17

2.2   Microscopic moiré interferometry
Bongtae Han   **18**

2.2.1   Immersion interferometer   18
          2.2.1.1   Optical configuration   19
          2.2.1.2   Four-beam immersion interferometer   20
2.2.2   Mechanical configuration   21
2.2.3   Fringe shifting and O/DFM method   23
          References   24

**2.3   Localized thermal strains in electronic interconnections by
microscopic moiré interferometry**
Bongtae Han   **25**

2.3.1   Specimen preparation   25
2.3.2   Thin small outline package   26
2.3.3   Leadless chip carrier   27
2.3.4   Effect of underfill encapsulation on flip chip solder bump   29
2.3.5   Plated through hole   33
          References   33

2.4   Titanium in elastic tension: micromechanical deformation
Bongtae Han   **34**
2.4.1   Introduction   34
2.4.2   Specimen and loading fixture   34
2.4.3   Experimental procedure and fringe patterns   35
2.4.4   Anomalous strains along the grain boundaries   37
2.4.5   Discussion   39
         References   40

**2.5   Micromechanical thermal deformation of unidirectional
boron/aluminum composite**
Bongtae Han   **42**
2.5.1   Experimental procedure and fringe patterns   43
2.5.2   Analysis and results   44
2.5.3   Discussion   47
         References   48

CHAPTER 3: FRACTURE MECHANICS   **49**

**3.1.1 Assessment of the shape of crack-tip plastic zones as a function of
applied load**
C A Walker and Peter M MacKenzie   **51**
3.1.1.1 Introduction   51
3.1.1.2 Experimental details   52
3.1.1.3 Measurement of Von Mises yield locus   53
3.1.1.4 Discussion of results   54
3.1.1.5 Conclusions   54
         References   55

**3.1.2 Deformation around fatigue cracks from moiré fringe
measurement**
Gianni Nicoletto   **56**
3.1.2.1 Introduction   56
3.1.2.2 Basic crack-tip models   57
         3.1.2.2.1 Stationary crack under monotonic loading   57
         3.1.2.2.2 Stationary crack under cyclic loading   58
3.1.2.3 Experimental details   59
3.1.2.4 Fatigue crack-tip deformation   60
         3.1.2.4.1 Local yielding   60
         3.1.2.4.2 Non-singular stresses   61
         3.1.2.4.3 Cyclic plasticity   62
         3.1.2.4.4 Local mode-mixity   62
3.1.2.5 Summary   63
         Acknowledgment   64
         References   64

3.2.1 Applications of moiré to cellulosic (paper and wood) and concrete materials
R E Rowlands   **66**
  3.2.1.1 Introduction   66
  3.2.1.2 Analyses   67
           3.2.1.2.1 Bolted wood connections   67
           3.2.1.2.2 Fracture of linerboard   69
           3.2.1.2.3 Fracture processes in concrete   70
  3.2.1.3 Summary   72
           References   73

3.2.2 Mixed-mode stress intensity factors in finite, edge-cracked orthotropic plates
C A Walker and Jamasri   **74**
  3.2.2.1 Introduction   74
           3.2.2.1.1 Notation   75
  3.2.2.2 Theoretical analysis   75
           3.2.2.2.1 Displacement fields around the crack-tip   75
           3.2.2.2.2 Mixed-mode fracture analysis   77
  3.2.2.3 Experiments   79
           3.2.2.3.1 Specimen preparation   79
           3.2.2.3.2 Specimen testing   80
  3.2.2.4 Analysis of the results   81
           3.2.2.4.1 Moiré patterns   81
           3.2.2.4.2 Calculation of stress intensity factors   82
  3.2.2.5 Discussion and conclusions   82
           Acknowledgments   85
           Appendix: Theory of the numerical approach for solving the
           characteristic of the biharmonic equation   85
           References   87

3.2.3 A hybrid experimental/computational approach to the assessment of crack growth criteria in composite laminates
C A Walker and M Jam   **89**
           Notation   89
  3.2.3.1 Introduction   89
  3.2.3.2 Crack growth criteria—theoretical background   90
  3.2.3.3 Crack growth criterion in use—the experimental problem   91
  3.2.3.4 Outline assessment method   92
  3.2.3.5 Experimental procedure   96
  3.2.3.6 Finite element modelling   96
  3.2.3.7 Validation of the finite element model   98
           3.2.3.7.1 Stress intensity factors from the closed form
                      solution (isotropic material assumption)   98
           3.2.3.7.2 Comparison with experimental data   98

3.2.3.8 Discussion of the computational results   99
3.2.3.9 Conclusion   101
         References   101

3.3.1 Experimental and computational assessment of the T^* fracture parameter during creep relaxation
C A Walker and Peter M MacKenzie   **103**
3.3.1.1 Introduction   103
3.3.1.2 T^* integral fracture parameter   104
3.3.1.3 Integration path   104
3.3.1.4 Experimental procedure   104
3.3.1.5 Fringe analysis   106
3.3.1.6 Results   106
         References   109

3.3.2 Moiré interferometry and crack closure
Peter M MacKenzie   **110**
3.3.2.1 Introduction   110
         3.3.2.1.1 Crack closure   110
         3.3.2.1.2 Moiré interferometry   110
3.3.2.2 Example applications   111
         3.3.2.2.1 Specimens   111
         3.3.2.2.2 Controlled crack development   111
         3.3.2.2.3 Grating limitations   113
         3.3.2.2.4 Crack closure effects revealed in moiré
                    interferograms   113
3.3.2.3 Conclusions   118
         References   119

CHAPTER 4: ELECTRONIC PACKAGING   **121**

4.1 Determination of effective coefficients of thermal expansion (CTE) of electronic packaging components
Yifan Guo   **123**
4.1.1 Introduction   123
4.1.2 Definitions of effective CTE   123
4.1.3 Procedures of CTE determination   125
       4.1.3.1 High-temperature replication of specimen gratings 125
       4.1.3.2 Real-time measurements of the thermal expansion   125
4.1.4 Application I: evaluation of thin small outline packages   125
4.1.5 Application II: plastic ball grid array package   128
4.1.6 Application III: CTE as a function of temperature   130
4.1.7 Conclusions   132
       References   132

**4.2   Determinations of thermal strains in solder joints: an application
        in electric packaging**
        Yifan Guo   **134**
        4.2.1   Introduction   134
        4.2.2   CBGA packages   135
        4.2.3   Moiré experiments   136
        4.2.4   Thermal deformations in CBGA   136
        4.2.5   Thermal strain analysis   140
        4.2.6   Conclusions   141
                References   143

**4.3   Thermomechanical behaviour of two-phase solder column
        connection under highly accelerated thermal cycle**
        Bongtae Han   **145**
        4.3.1   Specimen geometry   145
        4.3.2   Experimental procedure and fringe patterns   145
        4.3.3   Analysis and discussions   150
                4.3.3.1   Deformed shape   150
                4.3.3.2   Accumulated permanent deformation   151
                References   152

**4.4   Verification of numerical models used in microelectronics product
        development**
        Bongtae Han   **153**
        4.4.1   Ceramic ball grid array package assembly   154
        4.4.2   Bleed layer on power laminate   156
        4.4.3   Discussion   158
                References   159

4.5   Analysis of residual stresses in a die-attaching adhesive
        A S Voloshin   **160**
        4.5.1   Introduction   160
        4.5.2   Experiment   161
        4.5.3   Results   164
        4.5.4   Conclusions   169
                References   169

**4.6   Characterization of thermomechanical behaviour of a flip-chip
        plastic ball grid array package assembly**
        Bongtae Han   **171**
        4.6.1   Flip-chip plastic ball grid assembly   171
        4.6.2   Grating replication for complex geometry   172
        4.6.3   Bithermal loading   172
        4.6.4   Result and implication   174
                References   175

CHAPTER 5: COMPLEX SHAPES   177

5.1   Strain gauge verification by moiré
       Peter G Ifju   179
       5.1.1   The objective   179
       5.1.2   The experiment   181
               References   184

**5.2   A high-precision calibration of electrical resistance strain gauges
       by moiré interferometry**
       Robert B Watson   185
       5.2.1   Introduction   185
       5.2.2   The challenge   186
       5.2.3   The high-sensitivity moiré solution   186
       5.2.4   Assessment of measurement system accuracy   187
       5.2.5   Experimental procedure   189
       5.2.6   Results   191
       5.2.7   Refinements   193
               References   193

**5.3   Interlaminar deformation along the cylindrical surface of a hole in
       laminated composites**
       Raymond G Boeman   195
       5.3.1   Introduction   195
       5.3.2   Experimental procedures   196
       5.3.3   Results—straight boundary free edge   198
       5.3.4   Results—cylindrical surface of the hole   199
               5.3.4.1   Tangential strains in hole at 90°   200
               5.3.4.2   Transverse strains in hole at $\theta = 90°$ 201
       5.3.5   Summary   203
               Acknowledgments   204
               References   205

**5.4   Interlaminar deformation along the cylindrical surface of a hole in
       laminated composite tensile specimens**
       David H Mollenhauer   207
       5.4.1   Introduction   207
       5.4.2   Fringe patterns   207
       5.4.3   Experimental procedure   209
       5.4.4   Results   213
               References   220

**5.5   Thick laminated composites in compression: free-edge effects on a
       ply-by-ply basis**
       Yifan Guo   221
       5.5.1   Introduction   221
       5.5.2   In-plane compression   222

5.5.3   Interlaminar compression   225
5.5.4   Laminate properties   228
5.5.5   Conclusions   228
         References   229

5.6   Loading effects in composite beam specimens
Peter G Ifju   **230**
         References   234

CHAPTER 6: ELASTIC–PLASTIC HOLE EXPANSION   **237**

**6.1   Elastic–plastic hole expansion: an experimental and theoretical
         investigation**
C A Walker, Jim McKelvie and Jim Hyzer   **239**
6.1.1   Introduction   239
6.1.2   Residual stress measurement   239
6.1.3   Moiré interferometry   239
6.1.4   Hole-drilling method   240
6.1.5   Dissection method   240
6.1.6   Experimental investigation   241
         6.1.6.1   Expansion of a hole in a quasi-infinite plate   241
         6.1.6.2   Expansion of a hole in a thick cylinder and simple
                    lug geometry   244
6.1.7   Conclusion   249
         References   249

6.2   Roller expansion of tubes in a tube-plate
C A Walker and Jim McKelvie   **250**
6.2.1   Introduction   250
6.2.2   Experimental procedure   251
         References   253

CHAPTER 7: HYBRID MECHANICS   **255**

7.1   Some applications of high sensitivity moiré interferometry
C Ruiz   **257**
7.1.1   Introduction   257
7.1.2   Interferometric study of dovetail joints   257
7.1.3   Application to the study of cracks   258
         References   262

**7.2   Hybridizing moiré with analytical and numerical techniques:
         differentiating moiré measured displacements to obtain strains**
R E Rowlands   **263**
7.2.1   Introduction   263

7.2.2   Analyses   264
      7.2.2.1   General comments   264
      7.2.2.2   Splines and regression analysis   264
      7.2.2.3   Finite-element concepts   264
7.2.3   Summary   267
      References   267

CHAPTER 8: RESIDUAL STRESSES   **269**

8.1   Automated moirè interferometry for residual stress determination in engineering objects
Małgorzata Kujawińska and Leszek Sałbut   **271**
8.1.1   Introduction   271
8.1.2   Objects of measurements   272
      8.1.2.1   Railway rail   272
      8.1.2.2   Laser beam weldment   274
8.1.3   Experimental set-up and procedure   274
8.1.4   Experimental results and discussion   278
      8.1.4.1   Residual strain determination for railway rails   278
      8.1.4.2   Residual stress determination in laser beam welds 279
8.1.5   Conclusions   284
      References   285

8.2   Determination of thermal stresses near a bimaterial interface
Judy D Wood   **287**
8.2.1   Experimental procedure   287
8.2.2   Extracting the coefficients of thermal expansion   288
      8.2.2.1   Separation of thermal strains and stress-induced strains   291
8.2.3   Results   291
      References   293

8.3   Interior strains by a hybrid 3d photoelasticity/moiré interferometry technique
C W Smith   **294**
8.3.1   Frozen stress photoelasticity   294
8.3.2   The hybrid method   295
      References   295

CHAPTER 9: HIGH TEMPERATURE EFFECTS AND THERMAL STRAINS   **297**

9.1   Moiré interferometry using grating photography and optical processing
      Gary Cloud   **299**
      9.1.1   Case study: whole-field strain analysis to 1370 °C   299
      9.1.2   Problem characteristics   299
      9.1.3   Moiré interferometry with optical Fourier processing   300
      9.1.4   Outline of the technique   301
      9.1.5   Gratings   302
      9.1.6   Grating photography   304
      9.1.7   Optical Fourier processing   306
      9.1.8   Data processing   309
      9.1.9   Closure   309
               References   310

9.2   Heat-resistant gratings for moiré interferometry
      Colin Forno   **311**
      9.2.1   Introduction   311
      9.2.2   Basic grating procedure   312
      9.2.3   Detailed preparation   313
      9.2.4   Polishing-induced stresses   314
      9.2.5   Optical system aberrations   315
      9.2.6   Annealing studies   315
      9.2.7   High-temperature studies   316
               References   317

CHAPTER 10: MICROMECHANICS AND CONTACT PROBLEMS   **319**

10.1   Grid method for damage analysis of ductile materials
      Jin Fang   **321**
      10.1.1   Introduction   321
      10.1.2   Gratings and experiments   322
               10.1.2.1   Specimens and zero-thickness gratings   322
               10.1.2.2   Quasi-static loading and impulsive loading 322
      10.1.3   Image processing of deformed gratings   325
               10.1.3.1   Determination of strain by cross grid   325
               10.1.3.2   Displacement determination by Fourier transform   327
      10.1.4   Some analyses of void damage   331
               10.1.4.1   Interaction between the neighbouring voids   331
               10.1.4.2   Flow localization and ductile failure of ligaments   337
      10.1.5   Remarks   340
               Acknowledgment   340
               References   340

10.2 A contact problem
Jin Fang   342
    10.2.1   Introduction   342
    10.2.2   The in-plane contact of two bodies   343
    10.2.3   Contact deformation measured by moiré interferometry 346
             10.2.3.1 Moiré interferometry for displacement patterns 346
             10.2.3.2 A polarized shearing method for derivative patterns 347
             10.2.3.3 An incoherent technique for gradient patterns 351
    10.2.4   The hybrid solution of the contact stresses   354
    10.2.5   Remarks   359
             References   359

10.3 Moiré interferometry for measurement of strain near mechanical fasteners in composites
Gary Cloud   361
    10.3.1   Case study: Whole-field measurement of strain
             concentrations in composites   361
    10.3.2   Approaches to design   362
    10.3.3   Design, construction and alignment of interferometer 363
    10.3.4   Specimen gratings   367
    10.3.5   Recording fringes   368
    10.3.6   Data reduction   368
    10.3.7   Closing remarks   368
             References   370

CHAPTER 11: OUT-OF-PLANE DEFORMATION, FLATNESS AND SHAPE MEASUREMENT   373

11.1 Shadow moiré interferometry applied to composite column buckling
Peter G Ifju, Xiaokai Niu and Melih Papila   375
             References   378

11.2 Shadow moiré interferometry with enhanced sensitivity by optical/digital fringe multiplication
Bongtae Han   379
    11.2.1   Optical configuration and fringe shifting   380
    11.2.2   Optical/digital fringe multiplication (O/DFM)   381
    11.2.3   Applications   381
             11.2.3.1 Warpage of electronic device   381
             11.2.3.2 Pre-buckling behaviour of an aluminium channel 384
             References   386

11.3 Shadow moiré interferometry with a phase-stepping technique applied to thermal warpage measurement of modern electronic packages
Yinyan Wang **387**
11.3.1 Introduction 387
11.3.2 Application of phase stepping to shadow moiré fringe patterns 388
11.3.3 Phase calculation 390
11.3.4 On-line measurement of thermal warpage of modern electronic packages 391
    11.3.4.1 The experimental system 391
    11.3.4.2 Normalization of the sample position 392
11.3.5 Application examples 392
    11.3.5.1 A PBGA through a heating–cooling cycle 392
    11.3.5.2 Analysis of a BGA and its seating printed wire board 394
11.3.6 Conclusion 395
    References 396

11.4 On-line system for measuring the flatness of large steel plates using projection moiré
Jussi Paakkari **397**
11.4.1 Motivation 397
11.4.2 Measurement specifications 398
11.4.3 Design of the flatness measurement system 398
    11.4.3.1 System overview 398
    11.4.3.2 Optomechanical design of the measurement unit 399
    11.4.3.3 Image analysis software 403
11.4.4 Performance 404
    11.4.4.1 Quality of the moiré signal 404
    11.4.4.2 Calibration 404
    11.4.4.3 Repeatability and depth resolution of flatness measurement 404
    11.4.4.4 Other performance values 406
11.4.5 Conclusion 406
    Acknowledgments 407
    References 407

11.5 Optical triangular profilometry: unified calibration
Anand Asundi and Zhou Wensen **408**
11.5.1 Projection ray tracing 408
11.5.2 Image ray tracing 410
11.5.3 Choice 411
11.5.4 Spot inspection 412
11.5.5 Single-line system 414
11.5.6 Projection grating system 415
    References 416

11.6 Grating diffraction method
Anand Asundi and Bing Zhao   **417**
 11.6.1 A strain microscope with grating diffraction method 419
 11.6.2 Grating diffraction strain sensor 421
    References 423

CHAPTER 12: REAL TIME AND DYNAMIC MOIRÉ **425**

12.1 Real-time video moiré topography
Joris J J Dirckx and Willem F Decraemer **427**
 12.1.1 Introduction 427
 12.1.2 Theory of fringe formation 427
    12.1.2.1 Modulation and demodulation 427
    12.1.2.2 Shape and deformation measurement 429
    12.1.2.3 Fringe characteristics and fringe enhancement 429
 12.1.3 Apparatus 430
    12.1.3.1 Optical set-up 430
    12.1.3.2 Electronic image processing 431
 12.1.4 Measurement results 432
    12.1.4.1 Adjustment and calibration 432
    12.1.4.2 Measuring resolution 433
 12.1.5 Application example: measurements on the eardrum 435
 12.1.6 Discussion and conclusions 439
    References 441

12.2 Tensile impact delamination of a cross-ply interface studied by
moiré photography
L G Melin, J C Thesken, S Nilsson and L R Benckert **442**
 12.2.1 Introduction 442
 12.2.2 Experimental methods 443
    12.2.2.1 Optical set-up 443
    12.2.2.2 Tensile split-Hopkinson bar 445
    12.2.2.3 Specimen design and manufacturing 446
    12.2.2.4 Procedure 447
    12.2.2.5 Fringe analysis 447
 12.2.3 Results 447
 12.2.4 Discussion and conclusions 451
    Acknowledgments 452
    References 453

12.3 Real-time moiré image processing
M Bruynooghe **455**
 12.3.1 Introduction 455
 12.3.2 Experimental set-up for real-time defect detection 456
 12.3.3 Moiré image data 457

12.3.3.1	Inverse moiré image data	457
12.3.3.2	Moiré data acquisition by the phase shifting technique	458
12.3.4	Fringe skeletonizing	461
12.3.4.1	Moiré image enhancement and subsampling	462
12.3.4.2	Extraction of bright fringes by image segmentation and constrained contour modelling	463
12.3.4.3	The skeletonizing of dark fringes and background	466
12.3.4.4	Extraction of dark fringes by a graph technique	468
12.3.4.5	Experimental results	468
12.3.5	Defect detection by hit/miss transform	469
12.3.5.1	Optical and digital methods for real-time processing of moiré images by morphological hit/miss transform	470
12.3.5.2	Experimental results	471
12.3.6	Defect detection by multi-dimensional statistical analysis	473
12.3.6.1	Multi-dimensional statistical analysis of the results of optical correlation experiments	473
12.3.6.2	Factorial correspondence analysis	474
12.3.6.3	Analysis of phase moiré images by correspondence analysis	475
12.3.6.4	Rules and decision areas for supervised defect detection	476
12.3.6.5	Experimental results	476
12.3.7	Conclusion	482
	References	482

A   Micro-scale strain measurement: limitations due to restrictions inherent in optical systems and phase-stepping routines
James McKelvie   **485**

A.1	Introduction	485
A.2	The fringe structure	486
A.2.1	Phase-shifting	487
A.2.2	Displacement *versus* strain sensitivity	487
A.3	The optical limits	488
A.3.1	Under ideal conditions	488
A.3.2	With non-ideal conditions	492
A.3.3	How then to proceed?	493
A.4	The recording and analysis systems	494
A.5	Conclusions	498
	References	498

Index   **499**

Chapter 1

Introduction

C A Walker
University of Strathclyde, Glasgow, UK

The aim of this *Handbook* is to illustrate the range of problems that have been tackled using moiré methods, in the hope that the sharing of experience will enrich the area as a whole, and will also serve as an introduction to the application of moiré in its diverse embodiments—mainly in strain analysis but with sections also on three-dimensional shape measurement. The form of the *Handbook* is a series of case studies of individual investigations, briefly discussed by one or more of the original authors who carried out the work. In each example, our intention has been to discuss the reasons behind the experimental design, the problems faced and the steps taken to overcome them. What the reader will find missing, however, is a detailed exposition of the theory of moiré and the intricacies of grating design and application. These topics have been described at length in specialist monographs [1, 2], and in the Society for Experimental Mechanics *Handbook on Experimental Mechanics* [3]. The one exception to this rule is the detailed consideration given to the degree of information that may be extracted from moiré fringes. This section is an expansion of ideas first enunciated by McKelvie some time ago, and further elaborated in the intervening period. Moiré fringes may be magic in the colloquial sense but they are not magic in that they do not contain infinitely divisible information within their quasi-sinusoidal intensity profile.

The individual studies have been grouped together under chapter headings that reflect a degree of common purpose. It will quickly become evident, though, that the ideas discussed, and the way that problems were approached, read across the spectrum of application.

1

1.1   A brief history of the moiré method

The notion of using a grating as a reference for a measurement system has been around for some time—Lord Rayleigh is generally credited with the first reference to moiré in a scientific publication [4]. Linear gratings were developed for metrology purposes in the 1950s and 1960s and are now a well-established technology. The use of gratings as an experimental tool to look at two-dimensional deformation patterns was analysed theoretically by Guild [5] and reduced to practice by Post and others in the 1960s [6]. Moiré interferometry, which greatly increased the sensitivity of strain analysis, was developed in the 1970s, while fringe-shifting methods appeared on the scene in the 1980s. Improvements in lasers for illumination and in solid-state detectors for imaging mean that now the technology has advanced to a point where it is well understood, flexible and, more than ever, capable of generating large amounts of data in a form that is readily understood and which may be precisely quantified.

As an editor, one's thanks are due, above all, to the contributing authors, in particular to Dan Post for coordinating his co-workers' sections and for early discussions on the form of the *Handbook*.

References

[1] Cloud G 1995 *Optical Methods of Engineering Analysis* (Cambridge: Cambridge University Press)
[2] Post D, Han B and Ifju P 1997 *High Sensitivity Moiré: Experimental Analysis for Mechanics and Materials* (New York: Springer)
[3] Kobayashi A S (ed) 1993 *Handbook on Experimental Mechanics* 2nd edn (Bethel, CT: Society for Experimental Mechanics)
[4] Lord Rayleigh 1874 On the manufacture and theory of diffraction gratings. Scientific Papers 1, 209, *Phil. Mag.* **47** 81–3 and 193–205
[5] Guild J 1956 *The Interference Systems of Crossed Diffraction Gratings* (London: Oxford University Press)
[6] Walker C A Historical review of moiré interferometry *Exp. Mech.* **34** 281–99

CHAPTER 2

STRAIN MEASUREMENTS AT THE LIMIT—THE MOIRÉ MICROSCOPE

Chapter 2.1

Nonlinear analysis of interferometric moiré fringes

T W Shield
University of Minnesota, USA

The standard linear method for determining strain (or displacement) fields from interferometric moiré fringes suffers from two major problems:

(1) large strains are not measured properly; and
(2) Rigid body rotations appear as fictitious shear strains.

While the first of these may not be of concern in many applications, the second results in a significant complication of the experimental method. Because of the second item, the specimen grating must be exactly rotationally aligned with the interferometer to avoid calculation of fictitious shear strains from the fringe patterns. And this only assures that initially there will be zero shear strain due to rigid body rotation. If the loading results in rigid rotation of the specimen, this rotation will appear as a fictitious shear strain using linear analysis. Nonlinear analysis of the fringe patterns overcomes both of these problems. In fact, using nonlinear analysis makes it unnecessary to know the angle between the specimen grating and the interferometer at all. Instead all alignment is done between parts of the experimental apparatus, which does not need to be repeated for each new specimen. Using a nonlinear strain measure is simply a matter of performing a few additional calculations once the fringe number gradients have been determined. With modern PCs this does not significantly increase the data reduction effort.

This section is meant to be a practical guide to calculating the Almansi strain (defined later) from moiré fringe patterns as described in Shield and Kim [1]. This method has been used since then in Shield and Kim [2, 3] for experiments using a moiré microscope, which was also described in Shield and Kim [1] but use of this strain measure is not restricted to microscopic applications.

For a general deformation of a solid, let $x = (x, y, z)$ be the rectangular Cartesian position in the deformed configuration of a material point that was

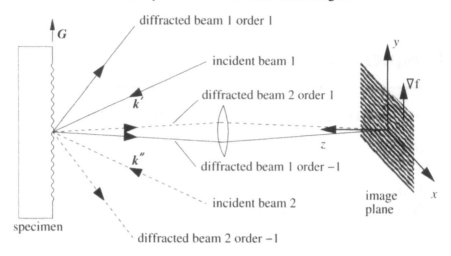

Figure 2.1.1. Typical interferometric moiré configuration. The plane of the drawing is the y–z plane.

originally at location $X = (X, Y, Z)$ in the reference (unloaded) configuration. The deformation gradient tensor, $\mathbf{F}$, has components

$$F_{iA} = \begin{bmatrix} \dfrac{\partial x}{\partial X} & \dfrac{\partial x}{\partial Y} & \dfrac{\partial x}{\partial Z} \\[2mm] \dfrac{\partial y}{\partial X} & \dfrac{\partial y}{\partial Y} & \dfrac{\partial y}{\partial Z} \\[2mm] \dfrac{\partial z}{\partial X} & \dfrac{\partial z}{\partial Y} & \dfrac{\partial z}{\partial Z} \end{bmatrix}. \tag{2.1.1}$$

The Almansi strain measure, $\mathbf{E}$, is then given by [4, pp 125–7] to be

$$\mathbf{E} = \tfrac{1}{2}[\mathbf{I} - (\mathbf{F} \cdot \mathbf{F}^{\mathrm{T}})^{-1}]. \tag{2.1.2}$$

An advantage of using this measure is that if the deformation is a rigid rotation, that is $\mathbf{F} = \mathbf{R}$, where $\mathbf{R}$ is a rotation matrix with $\mathbf{R}^{-1} = \mathbf{R}^{\mathrm{T}}$, then (2.1.2) gives $\mathbf{E} = 0$. Additionally, the Almansi strain is a current configuration measure, which requires less knowledge about the reference configuration. For this reason, only the specimen grating spacing will be needed in the analysis later.

Consider the interferometric moiré configuration shown in Figure 2.1.1. This figure shows a pair of laser beams diffracting from a grating on the specimen surface. To measure all three surface strain components, two pairs of laser beams are needed, one in the y–z plane, as shown, and one in the x–z plane, which diffract from a pair of crossed gratings on the specimen surface. In general the two pairs of beams need only lie in different planes, but a 90° angle between the planes is optimum. This figure also shows diffracted beams, with diffraction orders +1 and −1, imaged with a lens system (shown as a single lens for simplicity) on to an image plane. The image plane can either be photographic film or an electronic sensing device such as a CCD video camera.

The key part of this figure is the coordinate system. The coordinate system to use is the coordinate system of the image plane (the x–y plane). This is the natural coordinate system to use to analyse the fringe patterns. Locations of the fringes are measured in the image plane, either directly from a photograph or by pixel positions when using digital image processing. From now on it will be assumed that digital image processing is used but there is no significant difference when working from film. Because locations in a digital image are measured in pixels, it will be assumed that the optical system magnification is taken into account and that the pixel locations are converted directly into specimen units. Determining these scale constants is typically the first step in calibrating an interferometric moiré instrument.

Once the coordinate system of the instrument has been fixed, the incident laser beam wavevectors must be determined in this coordinate system. These vectors are labelled as k' and k'' in figure 2.1.1. They have the same direction as the laser beams and magnitude $2\pi/\lambda$ where λ is the wavelength of the laser beams. These two beams, in the absence of a specimen, would interfere and produce a uniform fringe pattern at the location of the specimen surface. This fringe pattern can be thought of as a virtual reference grating. The vector that describes the virtual reference grating is called the reciprocal virtual reference grating vector, g, and is given by

$$g = \frac{k'' - k'}{2\pi}.$$

(2.1.3)

For the special case of symmetric incident beams as shown in figure 2.1.1,

$$g = \frac{2\sin(\theta)}{\lambda}j$$

(2.1.4)

where θ is the common angle between the k vectors and the z axis. If the incident beam vectors are known accurately, then g can be directly calculated from (2.1.3). However, in practice, it is typically easier and more accurate to calculate g from a fringe pattern obtained from a known specimen grating. This process will be described later. The components of g in the x and y directions will be denoted as g_x and g_y. The magnitude of g is $|g| = 1/d^{vr}$, where $d^{vr} = \lambda/2\sin(\theta)$ is the virtual reference grating spacing.

The calculation of the nonlinear strain involves the effective specimen grating spacing, which is specified by three quantities: D^{spec}, the actual specimen grating spacing; N, the diffraction order imaged from beam 1; and M, the diffraction order imaged from beam 2. The effective specimen grating spacing, D^{es}, is given by

$$D^{es} = \frac{D^{spec}}{|N - M|}.$$

(2.1.5)

For the example in figure 2.1.1, the diffracted beam orders collected are $+1$ and -1, which gives $D^{\text{es}} = 0.5D^{\text{spec}}$, half of the actual grating spacing. If higher-order diffracted beams are imaged (these are not shown in figure 2.1.1), then a smaller effective specimen grating spacing is obtained with a corresponding increase in measurement sensitivity. The orientation of the specimen grating and its spacing can be combined to form the reciprocal specimen grating vector, G, which has magnitude $1/D^{\text{es}}$ and points normal to the ridges of the grating (as shown in figure 2.1.1) so that the shape of the grating is given by $\sin(2\pi G \cdot X)$, where X is a referential specimen position vector. The reciprocal specimen grating vector only needs to be known to calibrate the interferometer, it does not need to be determined for each specimen (although the specimen grating spacing, D^{spec}, must be known for each specimen).

The final pieces of data needed are information on the fringe pattern itself. Let $f(x, y)$ be the fringe number function, i.e. if the fringe pattern was a sinusoid with constant amplitude, A, then the intensity of the fringe pattern image would be given by

$$I(x, y) = A\cos(2\pi f(x, y)). \qquad (2.1.6)$$

Thus, the fringe pattern intensity is a maximum at values of $f(x, y) = n$, where $n = 0, \pm 1, \pm 2, \ldots$. Typically the locations of $f(x, y) = n$ are determined by an image processing technique, which will not be discussed here. It should be noted that it is more accurate to determine the locations of $I(x, y) = 0$, than the locations of the maximum (or minimum) intensity, see [1] for a discussion of phase shifting and fringe tracing. In either case, locations of specific values of $f(x, y)$ are determined and then an interpolation method is used to generate $f(x, y)$ at all points (x, y) in the field of view. It will be assumed that the fringe number function is suitably determined to allow all calculations to be performed.

The gradient of the fringe number function, $\nabla f(x, y)$, has components $\partial f/\partial x$ and $\partial f/\partial y$, in the x and y directions, respectively. These can be calculated with a finite difference formula and they are the same quantities that are needed for calculating linear strains. The coordinates used are those of the image plane (in specimen units) and, as such, represent current configuration specimen coordinates. Thus, these are Eulerian (or spatial) components of the fringe number function gradient. Linear strain calculations assume that these components are close to the Lagrangian (referential) components of the gradient, which is not always a reasonable approximation. The fringe number gradient points normal to the fringes as shown in figure 2.1.1.

The basic relationship between the reciprocal grating vectors and the fringe number function is

$$\nabla f = g - G \cdot F^{-1} \qquad (2.1.7)$$

where, F^{-1}, is the inverse of the deformation gradient. Equation (2.1.7) is useful for calibration of the instrument. If the specimen is undeformed, then $F^{-1} = I$, the identity matrix. Thus, we have

$$\nabla f = g - G \qquad (2.1.8)$$

when the specimen is undeformed. If a calibration specimen is carefully oriented relative to the image plane coordinate system so that G is known and then a fringe number function $f(x, y)$ is determined from the resulting fringe pattern (in practice all that is needed is the fringe spacing), then the virtual reference grating spacing and orientation can be calculated using (2.1.8).

There are two typical calibration cases. In the first, the virtual reference grating spacing and effective specimen grating spacing are fixed at the same value. In this case, careful alignment of g and G will produce a null fringe pattern ($\nabla f = 0$). Calibration using a null fringe pattern is not very accurate because it is hard to tell exactly when the null occurs. All that is known is that the fringe spacing is much greater than the field of view. Instead, rotational fringe multiplication can be used to produce a fringe pattern useable for calibration. By mounting the calibration specimen on an accurate rotation stage, the direction of G can be specified and once the fringe number gradient is determined the two equations in (2.1.8) can be solved for the magnitude and direction of g. Note that because there is an ambiguity in the sign of the fringe number gradient (which is discussed further later), there are two solutions to (2.1.8), one with the virtual grating spacing greater than (or equal to) the effective specimen grating spacing and one with it smaller. Mechanical measurements should be good enough to determine which is the relevant solution.

In the second typical calibration case, fringe multiplication is used to produce a fringe pattern with all the vectors in (2.1.8) aligned in the same direction (the y direction for the situation in figure 2.1.1). This is the case to use when the interferometer has the ability to adjust the direction of the incident beams and, hence, vary the virtual reference grating spacing, d^{vr}. The magnitude of ∇f is the reciprocal of the fringe spacing, d^{f}, and (2.1.8) becomes

$$d^{\mathrm{vr}} = \frac{1}{\frac{1}{\pm d^{\mathrm{f}}} + \frac{1}{D^{\mathrm{es}}}}. \tag{2.1.9}$$

There is a choice of the sign of the fringe spacing because the fringe number gradient can be in the same or opposite direction as g and G. Thus, whether d^{vr} is larger (minus sign) or smaller (plus sign) than D^{es} must be known in order to pick the proper sign on d^{f}. Again, this is usually easily determined from a rough measurement of the direction of the incident beams.

In practice, the fringe number gradient from a given fringe pattern is only known up to a sign. For example, from the fringe pattern shown in figure 2.1.1, ∇f is either in the positive y direction as shown or in the negative y direction. If enough information is known about the strain field, then the fringe gradient sign can be assumed. However, this does not allow a completely independent determination of the strains and, in some cases, there may not be enough information about the strain field to allow such an assumption to be made. If the interferometer has the capability to vary the direction of the illumination beams and, thus, vary g, then the fringe gradient signs can be determined absolutely using the following procedure: The virtual reference grating spacing, g, is varied until a null ($\nabla f = 0$) fringe pattern is obtained over part of the field of view. This virtual reference grating vector, denoted g_{null}, using (2.1.7) is approximately

$$g_{\text{null}} \approx G \cdot F^{-1}. \tag{2.1.10}$$

The interferometer is then adjusted to give a fringe pattern suitable for determining the fringe number gradient. The sign of this fringe number gradient (direction of increasing $f(x, y)$) is adjusted to match the sign of

$$g - g_{\text{null}} \tag{2.1.11}$$

in the part of the field of view where the null fringe pattern was originally obtained. This technique allows absolute determination of the strains without making any *a priori* assumptions. If the strains are small and a large amount of fringe multiplication is always used, then the approximation that $g_{\text{null}} \approx G$ can be used. This approximation cannot be used for interferometers that have g fixed and equal to G.

For calibration of the moiré microscope described in [1], a Fourier transform is used to calculate the average fringe spacing over a region of the fringe pattern, this value is then used in (2.1.9) for calibration. If the instrument has the ability to vary θ in equation (2.1.4), then the collinearity of the vectors ∇f, g and G can be tested by changing this angle and observing the fringe pattern. The direction of ∇f will only remain unchanged while θ is varied if the directions of g and G are also in the same direction. If the fringes are also aligned with image plane axes, this test assures that all three vectors are properly aligned with the image plane coordinate system. This process is necessary because solutions to (2.1.8) are possible that have ∇f in a coordinate direction while neither g nor G is.

Up to this point only a single set of illumination beams have been discussed. To measure all three in-plane strain components (E_{xx}, E_{xy} and E_{yy}) on the specimen surface, two sets of illumination beams are required to illuminate a pair of crossed gratings on the specimen surface. These two pairs of illumination beams are described by two reciprocal virtual reference grating vectors, $g^{(1)}$ and $g^{(2)}$. There is assumed to be a pair of perpendicular gratings with the same spacing, D^{spec}, deposited on the specimen surface and that the same diffraction orders are collected for both pairs of beams. If the specimen grating does not meet these requirements, the nonlinear strain measure can still be calculated; however, a more general formula must be used. See [1] for the these equations.

Each virtual reference grating will produce a fringe pattern from its corresponding specimen grating. These fringe patterns have fringe number functions that will be denoted, $f^{(1)}(x, y)$ and $f^{(2)}(x, y)$, respectively. The in-plane components of the Almansi strain, $\mathbf{E}$, are then given by

$$E_{xx} = \frac{1}{2}\left[1 - (D^{\text{es}})^2\left(\left(g_x^{(1)} - \frac{\partial f^{(1)}}{\partial x}\right)\left(g_x^{(1)} - \frac{\partial f^{(1)}}{\partial x}\right)\right.\right.$$

$$+ \left(g_x^{(2)} - \frac{\partial f^{(2)}}{\partial x} \right) \left(g_x^{(2)} - \frac{\partial f^{(2)}}{\partial x} \right) \right] \tag{2.1.12a}$$

$$E_{xy} = - \frac{(D^{es})^2}{2} \left(\left(g_x^{(1)} - \frac{\partial f^{(1)}}{\partial x} \right) \left(g_y^{(1)} - \frac{\partial f^{(1)}}{\partial y} \right) \right.$$

$$+ \left. \left(g_x^{(2)} - \frac{\partial f^{(2)}}{\partial x} \right) \left(g_y^{(2)} - \frac{\partial f^{(2)}}{\partial y} \right) \right) \tag{2.1.12b}$$

$$E_{yy} = \frac{1}{2} \left[1 - (D^{es})^2 \left(\left(g_y^{(1)} - \frac{\partial f^{(1)}}{\partial y} \right) \left(g_y^{(1)} - \frac{\partial f^{(1)}}{\partial y} \right) \right. \right.$$

$$+ \left. \left. \left(g_y^{(2)} - \frac{\partial f^{(2)}}{\partial y} \right) \left(g_y^{(2)} - \frac{\partial f^{(2)}}{\partial y} \right) \right) \right]. \tag{2.1.12c}$$

The Almansi strain measure is a nonlinear measure in the current configuration. If the measured strains are small (less than approximately 10%), then these strain components can be treated identically to the usual linear strain measure. However, a rigid body rotation of the specimen will not result in a calculated strain when (2.1.12) is used, as previously explained.

If the illumination beams are restricted to lie in the x–z and y–z planes and they produce virtual reference gratings with the same spacing (i.e. the illumination is aligned with the image plane coordinate system and is the same in both directions), then the reciprocal grating vectors reduce to

$$\mathbf{g}^{(1)} = \left(\frac{1}{d^{vr}}, 0 \right) \qquad \text{and} \, \mathbf{g}^{(2)} = \left(0, \frac{1}{d^{vr}} \right) \tag{2.1.13}$$

and (2.1.12) can be rewritten as

$$E_{xx} = \frac{1}{2} \left[1 - (D^{es})^2 \left(\left(\frac{1}{d^{vr}} - \frac{\partial f^{(1)}}{\partial x} \right) \left(\frac{1}{d^{vr}} - \frac{\partial f^{(1)}}{\partial x} \right) \right. \right.$$

$$+ \left. \left. \left(\frac{\partial f^{(2)}}{\partial x} \right) \left(\frac{\partial f^{(2)}}{\partial x} \right) \right) \right] \tag{2.1.14a}$$

$$E_{xy} = \frac{(D^{es})^2}{2} \left(\left(\frac{1}{d^{vr}} - \frac{\partial f^{(1)}}{\partial x} \right) \left(\frac{\partial f^{(1)}}{\partial y} \right) \right.$$

$$+ \left. \left(\frac{\partial f^{(2)}}{\partial x} \right) \left(\frac{1}{d^{vr}} - \frac{\partial f^{(2)}}{\partial y} \right) \right) \tag{2.1.14b}$$

$$E_{yy} = \frac{1}{2} \left[1 - (D^{es})^2 \left(\left(\frac{\partial f^{(1)}}{\partial y} \right) \left(\frac{\partial f^{(1)}}{\partial y} \right) \right. \right.$$

$$+ \left. \left. \left(\frac{1}{d^{vr}} - \frac{\partial f^{(2)}}{\partial y} \right) \left(\frac{1}{d^{vr}} - \frac{\partial f^{(2)}}{\partial y} \right) \right) \right]. \tag{2.1.14c}$$

This is a typical configuration for a non-adjustable interferometer instrument. Furthermore, if the reference gratings also have the same spacing as the effective specimen grating, i.e. $d^{vr} = D^{es} \equiv D$, then

$$E_{xx} = D\frac{\partial f^{(1)}}{\partial x} - \frac{D^2}{2}\left[\left(\frac{\partial f^{(1)}}{\partial x}\right)^2 + \left(\frac{\partial f^{(2)}}{\partial x}\right)^2\right] \tag{2.1.15a}$$

$$E_{xy} = \frac{D}{2}\left(\frac{\partial f^{(1)}}{\partial y} + \frac{\partial f^{(2)}}{\partial x}\right) - \frac{D^2}{2}\left[\frac{\partial f^{(1)}}{\partial x}\frac{\partial f^{(1)}}{\partial y} + \frac{\partial f^{(2)}}{\partial x}\frac{\partial f^{(2)}}{\partial y}\right] \tag{2.1.15b}$$

$$E_{yy} = D\frac{\partial f^{(2)}}{\partial y} - \frac{D^2}{2}\left[\left(\frac{\partial f^{(1)}}{\partial y}\right)^2 + \left(\frac{\partial f^{(2)}}{\partial y}\right)^2\right]. \tag{2.1.15c}$$

The first term in each of (2.1.15) is the usual linear term and the remaining terms are the nonlinear correction. Equations (2.1.15) make it clear that once the major work of calculating the fringe number gradients is done, it is very little extra work to calculate this nonlinear strain measure. Table 2.1.1 presents a summary of the assumptions for each set of equations.

2.1.1   Example fringe data reduction

To further illustrate the calculations outlined earlier, data reduction from an example set of fringe patterns will be presented. These data are from a tensile loading of a single crystal of a CuAlNi shape-memory alloy. The region imaged contains an austenite-martensite interface, which is the boundary between regions of untransformed and transformed material respectively. The tensile loading axis is vertical in the figures below and it is in the [12, 5, 0] crystallographic direction.

The experimental procedure is as follows:

Step 1. The specimen is prepared with a pair of orthogonal photolithographic gratings with spacings of 5.0 μm.

Step 2. The specimen is mounted in grips in an Instron testing machine and loaded.

Step 3. Four fringe patterns are obtained using a previously calibrated 'moiré microscope', which is described in Shield (1996). One null and one data image for two almost perpendicular directions of the virtual reference grating vector are obtained. The virtual reference grating vectors for the null images are chosen to give a null region in the lower left-hand portion of the images shown in figures 2.1.2–2.1.5. In these figures the virtual reference gratings are specified by the pair of values (d^{vr}, θ^{vr}), where $(g_x, g_y) = (\frac{1}{d^{vr}}\cos(\theta^{vr}), \frac{1}{d^{vr}}\sin(\theta^{vr}))$. For the data images in figures 2.1.2 and 2.1.4, g was chosen to give a fringe pattern suitable for the data reduction process.

Step 4. The images are binarized, retouched and smoothed. Note that the images shown in figures 2.1.2–2.1.5 have been further processed for publication.

Table 2.1.1.

	Equations (2.1.12)	Equations (2.1.14)	Equations (2.1.15)
Specimen gratings (D^{spec})	Orthogonal with same spacing	Orthogonal with same spacing	Orthogonal with same spacing
Virtual reference gratings ($g^{(1)}$ and $g^{(2)}$)	Any	Aligned with axis directions, equal spacings: $\lvert g^{(1)} \rvert = \lvert g^{(2)} \rvert = 1/d^{\mathrm{vr}}$	Aligned with axis directions, $d^{\mathrm{vr}} = D^{\mathrm{es}}$ for both reference vectors
Diffraction orders imaged (N and M)	Same for both (1) and (2) fringe patterns	Same for both (1) and (2) fringe patterns	Same for both (1) and (2) fringe patterns
Fringe multiplication allowed	Both spacing and rotational	Spacing only	None

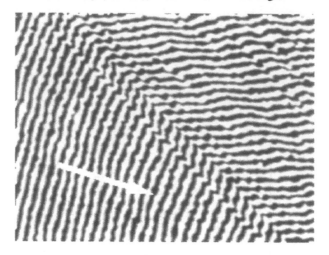

Figure 2.1.2. Data fringe pattern (2.1.1) for a reciprocal reference grating vector of $g^{(1)} = (2.3090\ \mu\text{m}, -2.001°)$.

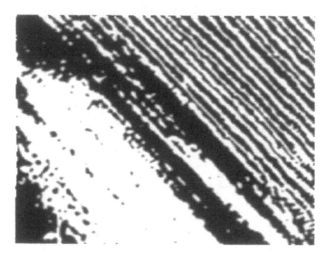

Figure 2.1.3. Null fringe pattern (2.1.1) for a reciprocal reference grating vector of $g^{(1)}_{\text{null}} = (2.4712\ \text{mm}, -0.701°)$.

Step 5. The direction of increasing fringe number function is determined for both pairs of images using (2.1.11). We find that

$$\nabla f \approx g^{(1)} - g^{(1)}_{\text{null}} = (0.2819\text{E} - 01, -0.1017\text{E} - 01)$$

and

$$\nabla f \approx g^{(2)} - g^{(2)}_{\text{null}} = (-0.2399\text{E} - 01, 0.2528\text{E} - 01)$$

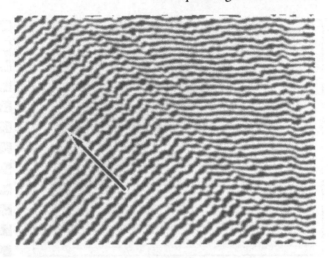

Figure 2.1.4. Data fringe pattern (2.1.2) for a reciprocal reference grating vector of $g^{(2)} = (2.3645 \text{ mm}, 92.500°)$.

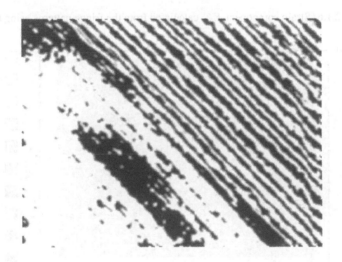

Figure 2.1.5. Null fringe pattern (2.1.2) for a reciprocal reference grating vector of $g^{(2)}_{\text{null}} = (2.5171 \text{ mm}, 89.200°)$.

for the two fringe patterns. These directions are indicated in figures 2.1.2 and 2.1.4 in the region of the fringe pattern where the null was obtained.

Step 6. The edges of the fringes are traced on the binarized images (not shown) to obtain the locations of $I(x, y) = 0$ in (2.1.6). These edges are assigned fringe numbers of 0, 0.5, 1, 1.5,... increasing in the direction determined in the previous step for each fringe pattern.

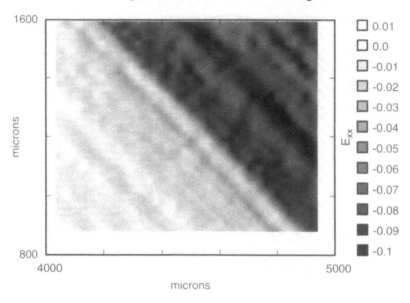

Figure 2.1.6. Strain component E_{xx} calculated from the fringe patterns using (2.1.12).

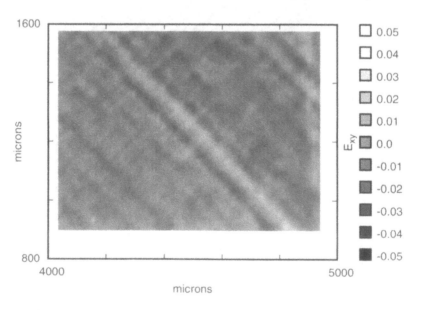

Figure 2.1.7. Strain component E_{xy} calculated from the fringe patterns using (2.1.12).

Step 7. These points are converted to data on an evenly spaced grid and finite difference formulae are used to determine $\nabla f^{(1)}$ and $\nabla f^{(2)}$. Equation (2.1.12) is used with $D^{es} = 2.5\ \mu$m (first-order diffracted beams were used) to give the

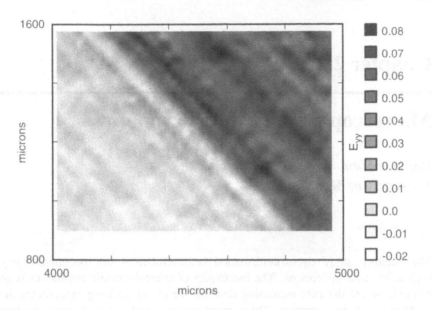

Figure 2.1.8. Strain component E_{yy} calculated from the fringe patterns using (2.1.12).

three in-plane Almansi strain components over the whole field of view. These are shown in figures 2.1.6, 2.1.7 and 2.1.8.

References

[1] Shield T W and Kim K-S 1991 Diffraction theory of optical interference moiré and a device for the production of variable virtual reference gratings: a moiré microscope *Exp. Mech.* **31** 126–34

[2] Shield T W and Kim K-S 1994 An experimental study of plastic deformation fields near a crack-tip in an iron-silicon single crystal *J. Mech. Phys. Solids* **42** 845–73

[3] Shield T W 1996 An experimental study of the plastic strain fields near a crack-tip in a copper single crystal during loading *Acta Metallurgica* **44** 1547–61

[4] Fung Y C 1977 *A First Course in Continuum Mechanics* 2nd edn (Englewood Cliffs, NJ: Prentice-Hall)

Chapter 2.2

Microscopic moiré interferometry

Bongtae Han
University of Maryland, USA

Many fields of study require deformation measurements of tiny specimens or tiny regions of larger specimens. The mechanics of microelectronic assemblies is an example where the ever-increasing demand for closer packing exacerbates the problems of thermal stresses. Other fields include crack-tip analyses in fracture mechanics, grain and intragranular deformations of metals and ceramics, fibre–matrix interactions in fibre-reinforced composites, interface problems etc.

The small size and the need for high spatial resolution require microscopic viewing of the specimen. Within a small field of view, the relative displacements are small even when the strains are not small. Accordingly, extremely high spatial resolution and displacement sensitivity are needed to map the tiny displacements in a microscopic field effectively. Microscopic moiré interferometry is an extension of moiré interferometry and it was developed to provide the desired spatial resolution and sensitivity required for micromechanics studies. The method employs an optical microscope for the required spatial resolution and enhances the sensitivity of moiré interferometry by an order of magnitude by utilizing an immersion interferometer and the optical/digital fringe multiplication (O/DFM) method.

2.2.1   Immersion interferometer

The immersion interferometer was developed to increase the basic sensitivity of moiré interferometry. Recall that the sensitivity of moiré interferometry increases with the frequency, f, of the virtual reference grating. Since

$$f = \frac{2}{\lambda} \sin \alpha \qquad (2.2.1)$$

the sensitivity increases with α (the angle of incidence of the illuminating beams) and it increases as the wavelength, λ, decreases. In air, the theoretical upper limit of sensitivity is approached as α approaches 90°. However, when $\alpha = 64°$, $\sin \alpha$ is already 90% of its maximum value and not much can be gained by greater angles. The basic sensitivity can be increased beyond the theoretical limit in air by creating the virtual reference grating inside a refractive medium, thus shortening the wavelength of light. Accordingly, the frequency of the virtual reference grating inside a refractive medium becomes

$$f_{\mathrm{m}} = \frac{2n \sin \alpha}{\lambda}. \qquad (2.2.2)$$

For a given light source of wavelength λgn vacuum, the frequency of the virtual reference grating in a refractive medium is increased by a factor of n. For example, the wavelength of green argon laser light is 514 nm in air but it is reduced to 338 nm in optical glass (BK7) of refractive index 1.52. Thus, the maximum frequency of the virtual reference grating is increased from 3890 lines/mm (in air) to 5910 lines/mm (in glass). Since sensitivity increases with f, the theoretical upper limit of sensitivity is also increased by the same factor, n.

2.2.1.1 Optical configuration

Various optical configurations that generate a virtual reference grating in a refractive medium can be devised [1]. Figure 2.2.1 is a schematic illustration of one configuration, where the shaded element represents a prism fabricated from optical glass. A coherent collimated beam enters through an inclined entrance plane. The lower portion of the beam impinges directly on the specimen grating while the upper portion impinges symmetrically after reflection from the mirrorized surface: they create a virtual reference grating on the specimen.

The gap between the interferometer and the specimen grating is filled with an immersion fluid. When the refractive index of the immersion fluid is not the same as that of the interferometer, angle α changes to α' as illustrated in the insert. However, this does not alter the frequency of the virtual reference grating, since $n' \sin \alpha' \equiv n \sin \alpha$ by Snell's law and equation (2.2.2) remains unaffected. The same result is obtained from the equations of diffraction and refraction without the conceptualization of a virtual reference grating. A reasonable match between the refractive indices of the immersion fluid and the immersion interferometer is desired but mismatch can be tolerated until it is so large that the critical angle of complete internal reflection occurs. The refractive index of the immersion fluid should be considered a limiting factor in equation (2.2.2), as well as the refractive index of the interferometer. In figure 2.2.1(a), the inclined entrance plane is provided by a separate prism bonded with optical cement to the main interferometer element.

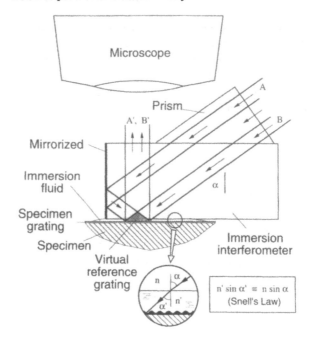

Figure 2.2.1. Optical paths in an immersion interferometer.

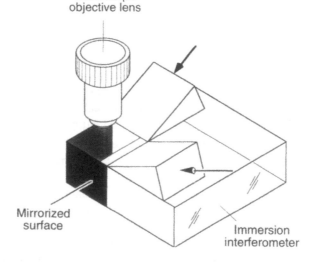

Figure 2.2.2. Four-beam immersion interferometer for U and V fields.

2.2.1.2 Four-beam immersion interferometer

The immersion interferometer concept was implemented in the form of a very compact four-beam moiré interferometer for microscopic viewing. It is illustrated

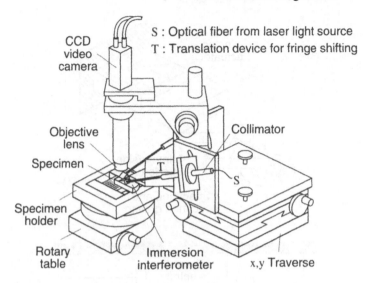

Figure 2.2.3. Mechanical and optical arrangement for microscopic moiré interferometry.

in figure 2.2.2, which is a four-beam version of the configuration. Two collimated beams enter the interferometer through the cemented prisms and they create the virtual reference gratings for the U and V fields. Since the interferometer requires accurately perpendicular surfaces of high optical quality, it was cut out of a corner cube reflector [1]. Corner cube reflectors are manufactured in high volume and, thus, they are inexpensive, which makes this an economical scheme.

The interferometer was designed for a virtual grating frequency of 4800 lines/mm. (122 000 lines/in). The parameters in equation (2.2.2) are $n = 1.52$, $\lambda = 514.5$ nm and $\alpha = 54.3°$. Angle α could be adjusted between $47°$ and $63°$ to provide a $\pm 10\%$ variation of frequency f_m; thus, carrier patterns could be used to subtract the uniform strain fields to extend the dynamic range of measurement to $\pm 10\%$ strain.

2.2.2 Mechanical configuration

The complete optical and mechanical arrangement is illustrated in figure 2.2.3. The specimen (with its specimen grating) is mounted in a loading fixture, which is mounted on a rotatory table for fine adjustments of angular orientation. The imaging or microscope system is comprised of a $10\times$ long-working-distance microscope lens and a CCD camera (without lens). Illumination is conducted by single-mode polarization-preserving optical fibres from an argon-ion laser to collimators that provide 5 mm diameter input beams to the interferometer. The collimators are attached to rotatable fixtures to vary the angle of incidence and, thus, accommodate large carrier patterns of extension when needed. The

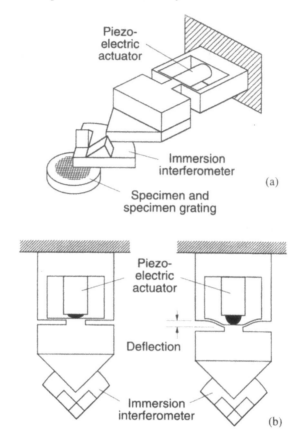

Figure 2.2.4. Piezoelectric translation device.

interferometer, illuminators and microscope are all connected to a heavy-duty *x,y* traverse to position the system over the desired portion of the specimen. Ancillary equipment includes a personal computer with a frame grabber board and TV monitors.

The microscope objective lens used here is infinity-corrected. The infinity correction means that it projects a well-corrected image regardless of the distance to the image. Accordingly, magnification is varied by changing the video camera position to change the lens-to-image distance and adjusting the lens for sharp focus. Since a CCD video camera was used, the normal specification of magnification is not meaningful. Instead, object size per TV frame is meaningful. Good results were obtained for object fields varying from 0.6 mm × 0.6 mm to 0.1 mm × 0.1 mm. This corresponds to object dimensions of 1.2 μm and 0.2 μm per pixel, respectively. The translation device is used for fringe shifting and multiplication [2].

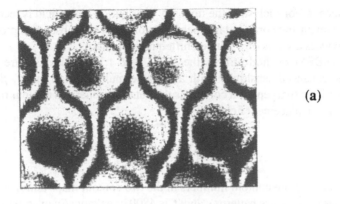

(a)

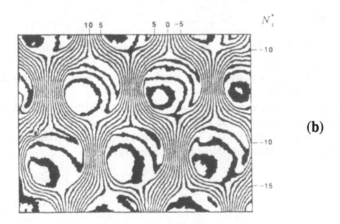

(b)

Figure 2.2.5. Example where (*a*) is a moiré pattern from the immersion interferometer with $f = 4800$ lines/mm, and (*b*) is the corresponding contour map produced by O/DFM for a multiplication factor of $\beta = 6$. The specimen is a unidirectional boron/aluminum metal-matrix composite. The deformation was induced by cooling the specimen from $122\,^{\circ}\text{C}$ to room temperature ($22\,^{\circ}\text{C}$).

2.2.3 Fringe shifting and O/DFM method

When the interferometer shown in figure 2.2.1 is translated horizontally by a fraction of the virtual reference grating pitch, g_m, the relative phase of rays A' and B' from each point in the field changes by the same fraction of 2π. Thus, the fringe positions are shifted by any translation of the interferometer relative to the specimen and a new intensity is present at every point on the specimen. In the configuration shown in figure 2.2.4, the piezoelectric actuator is computer controlled to move the interferometer by β equal steps of g_m/β each, where β is an even integer. For each step, the intensity distribution in the field is recorded by the CCD camera and compiled in a personal computer for subsequent processing, which produces a series of β fringe-shifted moiré patterns.

The O/DFM method utilizes the series of β fringe-shifted moiré patterns to provide enhanced sensitivity with a variable multiplication factor, β [3]. A robust algorithm sharpens and combines the data and generates contour maps that represent the displacement fields by

$$U = \frac{N_x^*}{\beta f} \qquad V = \frac{N_y^*}{\beta f} \tag{2.2.3}$$

where N_x^* and N_y^* are contour numbers in the two contour maps (similar to fringe orders in the basic moiré patterns), and f is 4800 lines/mm. An example is shown in figure 2.2.5. A detailed review is provided in [4].

References

[1] Han B and Post D 1992 Immersion interferometer for microscopic moiré interferometry *Exp. Mech.* **32** 38–41
[2] Han B 1992 Higher sensitivity moiré interferometry for micromechanics studies *Opt. Eng.* **31** 1517–26
[3] Han B 1993 Interferometric methods with enhanced sensitivity by optical/digital fringe multiplication *Appl. Opt.* **32** 4713–18
[4] Post D, Han B and Ifju P 1994 *High Sensitivity Moiré: Experimental Analysis for Mechanics and Materials* (New York: Springer)

Chapter 2.3

Localized thermal strains in electronic interconnections by microscopic moiré interferometry

Bongtae Han
University of Maryland, USA

As the components and structures involved in high-end electronic packaging are made smaller, the thermal gradient increases and the strain concentrations at electrical interconnections become more serious. When the deformations within those tiny regions are studied, the field of interest is small. Within such a small region, the relative displacements will be small even when the strains are large. Accordingly, extremely high spatial resolution and displacement sensitivity are needed to map the tiny displacements in a microscopic field effectively. A technique called microscopic moiré interferometry was used in this work, whereby the basic sensitivity of measurement was increased by a factor of two by means of an immersion interferometer. Then, fringe shifting and a robust algorithm (optical/digital fringe multiplication (O/DFM)) were used for a further increase by a factor of β [1, 2]. This chapter illustrates several applications of microscopic moiré interferometry to thermal deformation analyses of various electronic packaging interconnections [3, 4]. Normal strains and shear strains were determined effectively.

2.3.1 Specimen preparation

Microscopic moiré interferometry requires a plane of interest to be revealed and ground flat before specimen grating replication. Usually, a single package is trimmed from a fully populated assembly by using a low-speed high-precision diamond saw. The trimmed package is held in a precision vice as illustrated in figure 2.3.1. The position of the package is adjusted under a low magnification

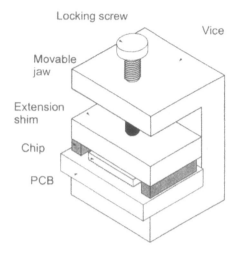

Figure 2.3.1. Schematic illustration of a package inserted in a precision vice for grinding.

stereomicroscope until a desired cross section is aligned with the side of the vice. Then, the specimen is ground by setting the vice onto a variable speed wet grinding wheel with a fine grit grinding paper (typically 600–1200 grit) until a desired microstructures appeared on the cross section.

2.3.2   Thin small outline package

The thin small outline package (TSOP) has been widely used in products that require very thin overall package dimensions. The TSOP modules consist of an active silicon chip and leaded frame assembly, encapsulated in a thin layer of plastic. The modules are then surface mounted on a printed circuit board (PCB), providing interconnections through the lead frame, emanating from its sides, by means of a soldering process. TSOP devices have a nominal height of 1.2 mm with a package to board clearance of about 0.05 mm. The device investigated here was a 14.4 mm × 5.6 mm TSOP module mounted on a FR4 PCB.

The module assembly was subjected to a uniform thermal loading of $\Delta T = -60\,^\circ\mathrm{C}$ and the local deformation of the interconnection was detailed by microscopic moiré interferometry [3]. Deformation of the module and the PCB was constrained mutually through the lead frame. The mutual constraints resulted in a system of forces exerted on the lead frame, as depicted schematically in the insert of figure 2.3.2. The bending moment produced significant stress and strain concentrations at the interface between the Alloy 42 lead and the solder fillet. The microscopic V-field pattern is shown in figure 2.3.2(a), where a fringe multiplication factor of $\beta = 6$ was used. The corresponding contour interval was 35 nm/fringe contour. The pattern revealed a highly localized strain concentration

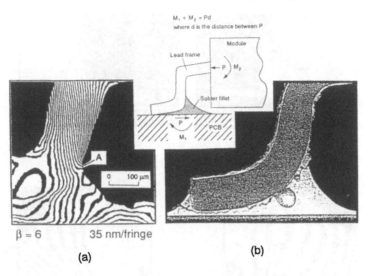

Figure 2.3.2. (*a*) Microscopic *V* displacement field at a typical joint of a TSOP device. (*b*) Cross section of a similar joint after thermal fatigue failure.

at the heel of the solder fillet (region A in figure 2.3.2(*a*)). The tensile strain, ε_y, in region A was 0.41%. This location matched that of the fatigue crack initiation point during the accelerated thermal cycle (ATC) test. Figure 2.3.2(*b*) depicts the cross section of a typical failed joint after the ATC test. The fatigue crack initiated in region A and it propagated along the interface between the lead frame and the solder joint. Inasmuch as modifying the lead frame was impractical, reliability of the interconnection was improved by reducing the strain concentration. This reduction was achieved by properly changing the shape of the solder fillet by controlling the amount of solder paste at each connection. Strain concentrations can be measured effectively with these techniques, even though the region of interest is extremely small.

2.3.3 Leadless chip carrier

Another common surface mount package is a leadless chip carrier (LCC) package. The LCC has metalized contacts at its periphery which are soldered to metalized contacts on the PCB. The package investigated here is a 84 I/O LCC (30 mm × 30 mm) mounted on an Epoxy/Kevlar PCB. The Epoxy/Kevlar PCB has a coefficient of thermal expansion (CTE) of 10 ppm °C^{-1}, which is lower than that of the regular Epoxy/Glass PCB ($\approx$ 17 ppm °C^{-1}). The Kevlar PCB was used to accommodate the large size of the high input/output (I/O) LCC module.

Microscopic moiré interferometry was utilized to investigate the shear strains in the solder joints. Figure 2.3.3 depicts the cross sections of solder joints and the

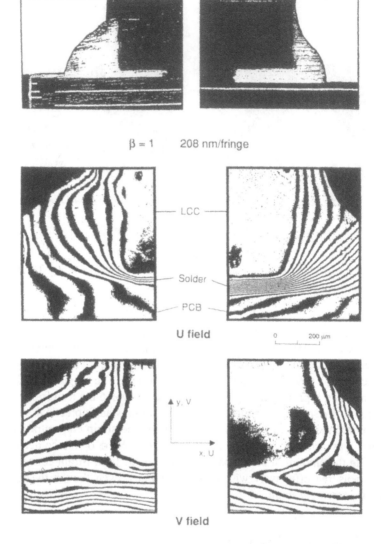

Figure 2.3.3. Cross sections of solder joints of an LCC package assembly. The fringe patterns show microscopic U and V displacement fields induced by thermal loading of $\Delta T = -60\,°\text{C}$.

corresponding U and V microscopic moiré fringe patterns at the left and right-hand side of the module, induced by a thermal loading of $\Delta T = -60\,°\text{C}$. For this experimental analysis, the module was tilted by $0.11°$ with respect to the PCB and, consequently, the thickness of solder was 40 μm for the left-hand side and 100 μm for the right-hand side. In general, the average shear strain is inversely proportional to the thickness of the solder. However, the shear strain determined from the fringe patterns revealed the contrary. The closely spaced fringes in the U field show that $\partial U/\partial y$ is substantially larger in the thicker joint: the shear strain at the thicker joint (2.0%) was greater than that at the thinner one (1.0%). This resulted from the global deformation of the assembly, wherein the neutral axis of bending shifted toward the thinner joint [3].

The relative number of cycles to failure of the joints can be estimated by employing the modified Coffin–Manson relationship [5], whereby the fatigue life is inversely proportional to the square of the shear strain. Thus,

$$\frac{N_{thick}}{N_{thin}} = \left(\frac{\gamma_{thin}}{\gamma_{thick}} \right)^2 = 0.25 \qquad (2.3.1)$$

where N is the number of cycles to failure and γ is the magnitude of shear strain. The fatigue life of the thicker joint is only 25% of that of the thinner joint. Furthermore, the thicker joint has a inherent narrower fillet (figure 2.3.2) because the amount of the solder dispensed to each solder pad was constant. The fatigue life of the thicker joint would be even worse in reality when the low cycle fatigue crack propagation is considered. The tests quantify the great importance of control of the solder thickness.

2.3.4 Effect of underfill encapsulation on flip chip solder bump

The flip chip technology has emerged as an important future chip technology to meet ever-increasing demand of a high I/O requirement. In conventional packages, such as wire bonding and tape automated bonding, the chip is positioned on its back surface and small diameter wires are used to connect the bonding pads located around the perimeter of the chip surface to the chip carrier. In chip carriers using flip chip connections, this conventional orientation of the chip is reversed. The chip is placed face downward and the connection between the chip and the chip carrier is achieved by a solder bump. Because of this orientation, the flip chip package has numerous advantages over the conventional packages.

The dominant deformation of solder joints in flip chip packages is shear strain produced by the thermal expansion mismatch between the chip and the carrier. The magnitude of the shear strain on each solder joint is proportional to its distance from the neutral point (DNP). Consequently, the size of the device and the CTE of the carriers must be limited to protect solder joints from premature failure. Recently, the flip chip technology has been extended for an organic substrate with an aid of underfill encapsulation technique, where the gaps between the solder joints are filled with an epoxy encapsulant. The effect of encapsulation on strains in flip chip joints was investigated by microscopic moiré interferometry [4].

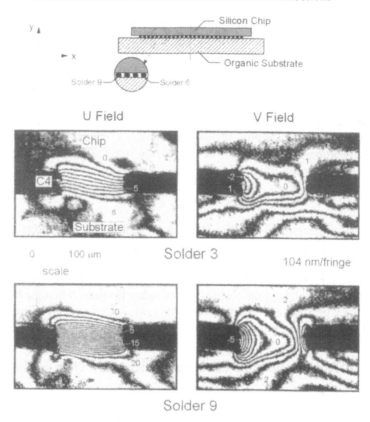

Figure 2.3.4. Microscopic U and V displacement fields of two representative solder bumps of the package without underfill. The contour interval is 104 nm per fringe. The inserts show micrographs of the region of interest.

The specimen is a flip chip package with a 9.5 mm square silicon chip attached to a 1.55 mm thick PCB. Two specimens were prepared: one with encapsulation and the other with no encapsulation. There were 12 solder bumps in the half of the package. Figure 2.3.4 shows the U and V field patterns of two representative solder bumps (solder 3 and 9) of the package without underfill, where a fringe multiplication of two was used and the corresponding contour interval was 104 nm/contour. The numbers shown in the patterns denote the fringe order. The large relative horizontal displacement between the top and bottom of the solder bumps is evident from the horizontal U-field fringes in the solder bump. The U and V displacements of a solder bump (solder 12) of the package with underfill are shown in figure 2.3.5, where the region of the solder bump is marked by a full line. A fringe multiplication of four was used and the corresponding contour interval was 52 nm/contour.

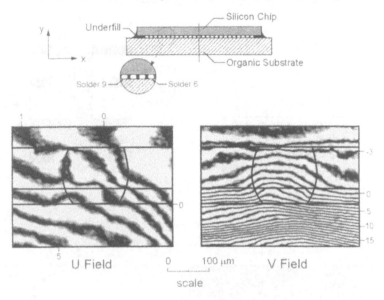

Figure 2.3.5. Microscopic U and V displacement fields of solder 12 of the package with underfill after cancelling rigid-body rotation. The contour interval is 52 nm per fringe.

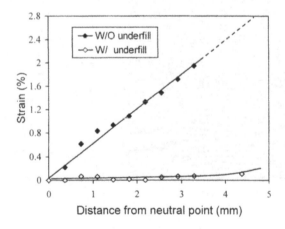

Figure 2.3.6. Average shear strain (γ_{xy}^{ave}) distributions.

Average shear strains of solder bumps were evaluated from the displacement fields. The results are plotted in figure 2.3.6. The result is striking. For the package without underfill, the average shear strain (γ_{xy}^{ave}) increases linearly as the DNP increases. The magnitude of γ_{xy}^{ave} of the solder 9 is about 2%, which is greater than that of the corresponding solder bump of the package with underfill by an order of magnitude. The effect of underfill is remarkable. The underfill

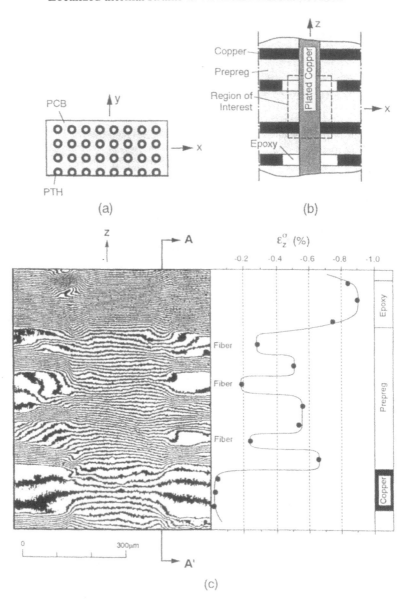

Figure 2.3.7. (*a*) Top and (*b*) cross-sectional view of the PCB specimen. (*c*) z displacement field and ε_z along AA', induced by thermal loading of $\Delta T = -80\,°\mathrm{C}$.

reduced the average shear strain significantly and its magnitude remained nearly constant, independent of the DNP.

2.3.5 Plated through hole

A multi-layer PCB is a very complex laminate containing layers of metal conductors separated by sheets of insulating material. Holes through the thickness of the laminate are copper plated for conductivity. Failures of the copper in plated through holes (PTH) are a major cause of failures of multi-layer PCB. The failures are caused by the thermomechanical stresses induced by the large CTE mismatch between the plated copper and PCB materials in the thickness direction, i.e. perpendicular to the plane of the PCB.

Microscopic moiré interferometry was employed to determine the deformation of plated copper during thermal cycling. The top view of the specimen is illustrated in figure 2.3.7(a), where the xz plane of PCB was ground until the largest area of the copper was exposed in the plane. The steady-state thermal deformation measurement was conducted for the xz cross section. The region of interest investigated by the microscopic moiré is shown by dashes in figure 2.3.7(b), where the plated copper adheres to three different materials of the PCB: epoxy, glass/epoxy laminate and copper. The epoxy is present where electrical connection to the PTH is not desired. Figure 2.3.7(c) represents the z direction displacement field for the region, induced by ΔT of $-80\,°C$. A contour interval of 52 nm per fringe contour was used.

The normal strain, ε_z, was determined from the pattern along the PTH wall (along line AA') and the result is plotted in figure 2.3.7(c). The maximum strain occurred along the copper/epoxy interface, which was indicated by the most closely spaced fringes. The strain ε_z over the interface was nearly uniform and its magnitude was -0.9%. Large fluctuations in the ε_z distribution were observed along the interface between the plated copper and glass/epoxy laminate. The strain in the plated copper at the interface with the glass fibre was smaller than at the interface with the epoxy-rich area. There was no evidence of copper delamination, which would be recognized as a discontinuity of fringe orders at the interfaces.

References

[1] Han B 1992 Higher sensitivity moiré interferometry for micromechanics studies *Opt. Eng.* **31** 1517–26

[2] Post D, Han B and Ifju P 1994 *High Sensitivity Moiré: Experimental Analysis for Mechanics and Materials* (New York: Springer)

[3] Han B and Guo Y 1995 Thermal deformation analysis of various electronic packaging products by moiré and microscopic moiré interferometry *J. Electron. Packaging, Trans. ASME* **117** 185–91

[4] Cho S and Han B 2002 Effect of underfill on flip-chip solder bumps: an experimental study by microscopic moiré interferometry *Int. J. Microcircuits Electron. Packaging* **24** 217–39

[5] Norris K C and Landzberg A H 1969 Reliability of controlled collapse interconnections *IBM J. Res. Dev.* **13** 1969

Chapter 2.4

Titanium in elastic tension: micromechanical deformation

Bongtae Han
University of Maryland, USA

2.4.1 Introduction

In a polycrystalline metal, individual grains are not subjected to a uniaxial stress system even when the specimen is subjected to a uniform tension [1]. These experiments investigated the deformations on the tensile surface of a polycrystalline titanium specimen in bending. Heterogeneous normal strains within grains were documented, together with a heterogeneous distribution of shear strains near grain boundaries. Heterogeneous distributions were observed at 57% of the yield point strain. The results corroborate earlier observations [2, 3] and reveal the behaviour in greater detail. Microscopic moiré interferometry was employed for the study. Its high sensitivity and spatial resolution provided a faithful account of the in-plane displacement field.

2.4.2 Specimen and loading fixture

The specimen material was a large grain beta alloy titanium. Its geometry and dimensions are illustrated in figure 2.4.1(a). Its yield point stress (0.2% offset) was 1450 MPa (210 000 psi) and its elastic modulus was 97 GPa (14×10^6 psi). Numerous grain sizes were present in the specimen, up to a maximum of 400 μm. The specimen was loaded as illustrated in figure 2.4.1(b), whereby a bending moment (plus a small compressive force) was applied, resulting in tension on the outside face of the specimen. The annealed copper gasket was designed to deform plastically to distribute contact forces. The forces, P, were applied by a displacement controlled loading fixture.

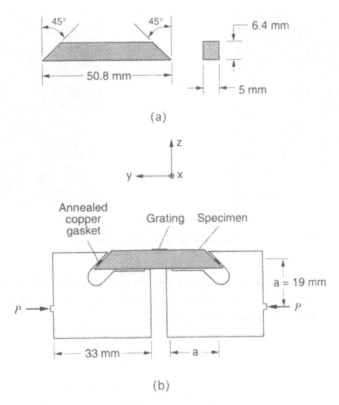

Figure 2.4.1. (*a*) Specimen geometry and (*b*) loading fixture.

2.4.3 Experimental procedure and fringe patterns

The top surface of the specimen was polished and an epoxy specimen grating was applied in a small central region. For the small region of the specimen, the thickness of the specimen grating was about 2 μm, which was slightly smaller than the spatial resolution of the imaging system employed in the micro moiré apparatus. The replicated specimen grating was examined by micro moiré interferometry to ensure its initial uniformity before a load was applied. Good null fields were obtained for both fields. Figure 2.4.2(*a*) illustrates the *V*-field pattern before loading, where carrier patterns of rotation were applied. Uniformly spaced straight fringes indicated that there were no initial irregularities.

The specimen was loaded until its front surface reached approximately half the yield point strain, ε_y^{yp}. Then, a typical region containing numerous grains of various sizes and shapes was selected. Although the specimen surface could not be seen through the opaque aluminum coating, the grain boundaries became visible when the microscope was slightly defocused: this effect is the result of large out-of-plane displacements that occurred along the grain boundaries. An

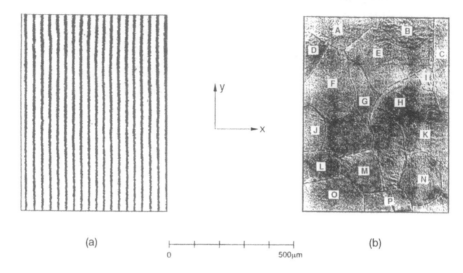

(a)   (b)

|————+———+———+———+———|
0   500μm

Figure 2.4.2. (*a*) *V* field pattern with carrier fringes of rotation before the loading: $f = 4800$ lines/mm, $\beta = 1$. (*b*) Microscopic image of the field of view, where the dashes represent loci of the grain boundaries.

image of the region investigated here was obtained with auxiliary lighting after the experiment was completed. It is shown in figure 2.4.2(*b*), where the dashes represent loci of the grain boundaries.

The *V*- and *U*-field fringe patterns are shown in figures 2.4.3(*a*) and (*b*). The average tensile strain (ε_y^{ave}) and the average transverse strain (ε_x^{ave}) was 0.85% and −0.30%, respectively. The corresponding average tensile stress was 825 MPa (119 000 psi), which is 57% of the yield point stress. In order to reveal heterogeneous strains associated with the individual grains, carrier fringes of extension (or compression) were applied to subtract most of the average strains, with the results shown in figures 2.4.3(*c*) and (*d*). The applied carrier fringes were equivalent to uniform strains of −0.76% for the *V* field and +0.28% for the *U* field, respectively. The number of fringes in figures 2.4.3(*c*) and (*d*) were, then, increased by the O/DFM method. Figures 2.4.3(*e*) and (*f*) depict the corresponding *V* and *U* displacements, where a fringe multiplication factor of $\beta = 6$ was used and the corresponding contour interval was 35 nm/contour. Fringe orders in the multiplied patterns represent N^*. The *V*- and *U*-field patterns (figures 2.4.3(*a*) and (*b*)) were inspected carefully to assign correct fringe orders to the multiplied patterns.

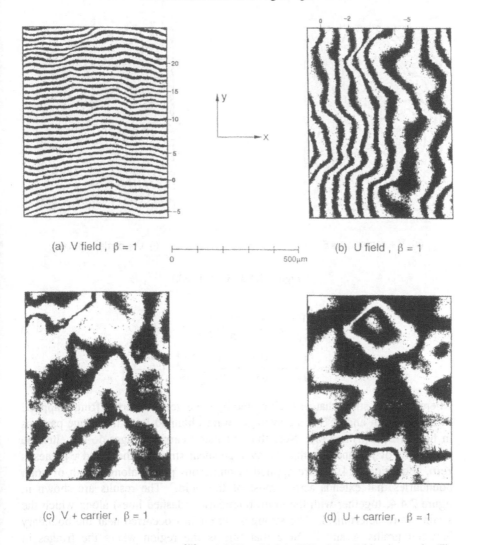

(a) V field , β = 1

(b) U field , β = 1

(c) V + carrier , β = 1

(d) U + carrier , β = 1

Figure 2.4.3. Fringe patterns at $\varepsilon_y^{\mathrm{ave}} = 0.85\%$, which is 57% of the yield point strain. The contour intervals are 208 and 35 nm/contour for $\beta = 1$ and 6, respectively.

2.4.4 Anomalous strains along the grain boundaries

The anomalous strains along the grain boundaries were extracted from the multiplied patterns by the relationships for engineering strains:

$$\varepsilon_x^{\mathrm{a}} = \frac{\partial U^{\mathrm{a}}}{\partial x} = \frac{1}{\beta f}\left[\frac{\partial N_x^*}{\partial x}\right] + \varepsilon_x^{\mathrm{c}} - \varepsilon_x^{\mathrm{ave}}$$

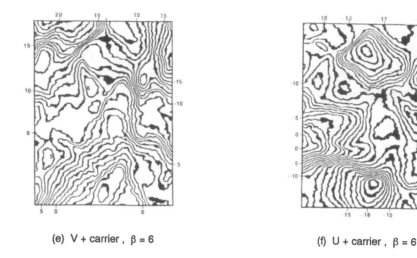

(e) V + carrier , β = 6   (f) U + carrier , β = 6

Figure 2.4.3. (Continued.)

$$\varepsilon_y^a = \frac{\partial V^a}{\partial y} = \frac{1}{\beta f}\left[\frac{\partial N_y^*}{\partial y}\right] + \varepsilon_y^c - \varepsilon_y^{ave} \tag{2.4.1}$$

$$\gamma_{xy} = \frac{\partial U^a}{\partial y} + \frac{\partial V^a}{\partial x} = \frac{1}{\beta f}\left[\frac{\partial N_x^*}{\partial y} + \frac{\partial N_y^*}{\partial x}\right]$$

where ε^c is the strain that was subtracted by the extension carrier fringes applied to each pattern and ε_y^{ave} is the average strain obtained from the fringe patterns in figures 2.4.3(a) and (b). Note that extension carrier fringes do not alter the magnitude of γ_{xy} and also that the average shear strain on the x and y planes is zero. Equation (2.4.1) were applied at numerous points along a path of grain boundaries that extends across most of the field. The results are shown in figure 2.4.4, together with the grain boundaries (dashed lines) along which the strains were determined. The strongest anomalies occurred near the boundary between grains A and E. Note that this is the region where the fringes in figures 2.4.3(e) and (f) are most closely spaced. At this location, the normal strains exceeded the yield point strain but the γ_{xy} shear strains remained very small.

To evaluate the anomalous shear strains associated with individual grains, the total shear strains in the tangential direction along the grain boundaries were calculated by the two-dimensional strain transformation equation

$$\gamma_{x'y'} = -(\varepsilon_x - \varepsilon_y)\sin 2\theta + \gamma_{xy}\cos 2\theta. \tag{2.4.2}$$

The prime in equation (2.4.2) indicates a local coordinate system where the x' axis is parallel to the tangential direction of the grain boundary; θ is the angle between the positive x axis and the grain boundary. The uniform part of the shear

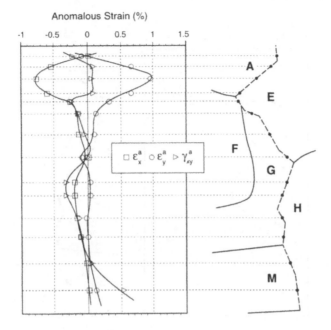

Figure 2.4.4. Anomalous strain distributions along the grain boundaries.

strains in the local coordinate systems was also calculated by

$$\gamma_{x'y'}^{\text{ave}} = -(\varepsilon_x^{\text{ave}} - \varepsilon_y^{\text{ave}}) \sin 2\theta. \tag{2.4.3}$$

The strain values obtained from equation (2.4.3) are the expected shear strains in the local coordinate systems when a uniform tensile member experiences the uniform tensile and transverse (Poisson) strains of $\varepsilon_y^{\text{ave}}$ and $\varepsilon_x^{\text{ave}}$, respectively. Then, the anomalous shear strains along the grain boundaries were determined by

$$\gamma_{x'y'}^{\text{a}} = \gamma_{x'y'} - \gamma_{x'y'}^{\text{ave}} \tag{2.4.4}$$

where superscript 'a' denotes the anomalous part. Figure 2.4.5 depicts the distribution of the anomalous shear strains along the grain boundaries (dashed lines).

2.4.5 Discussion

The heterogeneous strains were observed at an early stage of the loading. Near the grain boundaries, the degree of heterogeneity was generally greater for the transverse strain than for the tensile strain. This indicated that the local variation of Poisson's ratio was greater than that of Young's modulus.

The anomalous strains were observed within the global elastic limit. At 57% of the yield point strain, the strongest anomalies occurred near the boundary

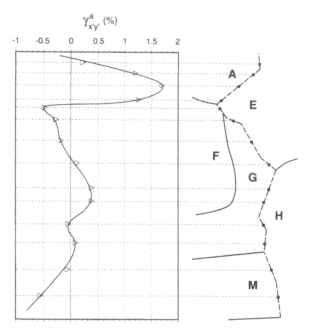

Figure 2.4.5.   Distribution of anomalous shear strains acting tangent to the grain boundaries.

between grains A and E. The grain boundary was almost straight and it was inclined by $40°$ with respect to the positive x axis. The magnitudes of the total normal strains at this location were very large: they were $\varepsilon_y = 1.81\%$, $\varepsilon_x = \mathrm{t}1.06\%$, respectively, which are substantially higher than yield point values. Although the shear strain γ_{xy} was very small (only 0.06%), the large biaxial normal strains produced a very large shear strain of $\gamma_{x'y'} = 2.84\%$ in the direction of the grain boundary. The anomalous part of the shear strain $\gamma^a_{x'y'}$ at this location was 1.70%, which was obtained from equation (2.4.4). The total shear strain $\gamma_{x'y'}$ increased as the load was increased [4]. However, the anomalous part of the shear strain $\gamma^a_{x'y'}$ did not increase as much as the shear strain $\gamma_{x'y'}$.

There was no evidence of grain slip, which would be recognized as a discontinuity of fringe orders at the grain boundary. Furthermore, the shear lag could not account for the shear within the grains since the grating thickness was about 2 μm. The large shear strain was centred at grain boundaries but strong tangential shear strains occurred in the grains, too, near the grain boundaries.

References

[1] Dieter G 1976 *Mechanical Metallurgy* 2nd edn (New York: McGraw-Hill) pp 190–208

[2] Mckelvie J, Mackenzie P M, McDonach A and Walker C A 1993 Strain distribution measurements in a coarse-grained titanium alloy *Exp. Mech.* **33** 320–5

[3] Post D, Han B and Ifju P 1994 *High Sensitivity Moiré: Experimental Analysis for Mechanics and Materials* (Berlin: Springer) pp 370–3

[4] Han B 1996 Micromechanical deformation analysis of beta alloy titanium in elastic and elastic/plastic tension *Exp. Mech.* **36** 120–6

Chapter 2.5

Micromechanical thermal deformation of unidirectional boron/aluminum composite

Bongtae Han
University of Maryland, USA

When a metal–matrix composite is subjected to a change of temperature, large thermal stresses develop as a result of the mismatch of coefficients of thermal expansion (CTE) between the fibre and matrix materials. In these experiments, thermal deformations of a unidirectional boron/aluminum metal–matrix composite were investigated for a free surface perpendicular to the fibres [1]. Isothermal loading was chosen to investigate the steady-state thermal deformation. The elevated temperatures were sufficiently low to ensure the linear elastic behaviour of the fibre and matrix materials. The fringe patterns clearly delineate the heterogeneous deformation associated with the individual fibres and surrounding matrix.

Microscopic moiré interferometry was employed for the study, whereby the virtual reference grating frequency was $f = 4800$ lines/mm and optical/digital fringe multiplication (O/DFM) provided a multiplication factor of $\beta = 6$. The boron fibres were approximately 200 μm in diameter. The high sensitivity and spatial resolution provided by the method made it possible to elucidate the micromechanical behaviour of the specimen at the fibre level.

In analytical or numerical models of metal–matrix composites, the fibre array is defined by a regularly repeating unit cell. However, various irregular arrangements of fibres are observed in actual specimens. A region that exhibited a hexagonal array of fibres and another region with a square array were investigated. Thermal deformations in these regions are reported in this chapter.

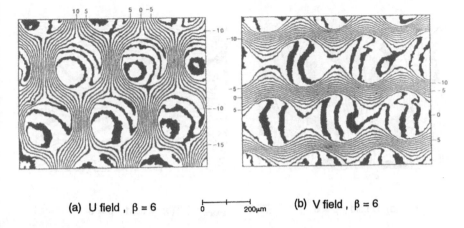

(a) U field , β = 6   0 ⊢——┼——⊣ 200μm   (b) V field , β = 6

Figure 2.5.1. Fringe patterns in the region of a hexagonal array of fibres: (*a*) *U* and (*b*) *V* patterns for $\beta = 6$, where the contour interval is 35 nm/contour.

2.5.1 Experimental procedure and fringe patterns

The specimen was a unidirectional boron/aluminum metal–matrix composite, specifically, boron fibres in an aluminum matrix. A small block was cut and a surface perpendicular to the fibres was ground and polished. The specimen was subjected to an isothermal loading [2]. In the experiment, the specimen grating was replicated in a small central region from an ultra low expansion (ULE) mould at an elevated temperature of 122 °C. The microscopic moiré apparatus was tuned with the same ULE mould to produce a null field condition. Then the deformed specimen was installed and observed at room temperature to document the deformations induced by an isothermal loading of $\Delta T = -100\,°C$. The specimen grating was located in the same plane as the ULE mould. This was achieved by adjusting the position of the specimen holder to regain the focus across the entire field after replacing the mould with the specimen.

After tuning the system, a region with a hexagonal array of fibres was selected. Although the specimen surface could not be seen through the opaque aluminum coating, the fibre boundaries became apparent when the microscope was slightly defocused: this effect is the result of large out-of-plane deformation that occurred along the fibre boundaries. The *U* and *V* patterns are shown in figures 2.5.1(*a*) and (*b*), where a fringe multiplication factor of $\beta = 6$ was used. The corresponding contour interval was 35 nm/contour. The numbers marked on the patterns are fringe contour numbers. The zero contours were selected arbitrarily in each case, since we are not interested in rigid body displacements. The aluminum exhibits closely spaced contours, consistent with its higher CTE.

After recording the patterns shown in figure 2.5.1, the micro moiré system was moved to view a region with a square array of fibres. The *U* and *V* patterns

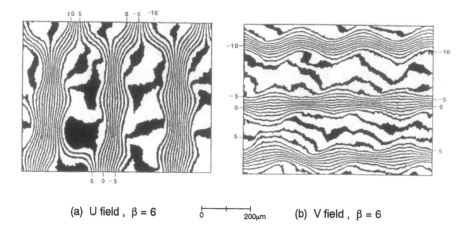

(a) U field , β = 6   0 ├────┼────┤ 200μm   (b) V field , β = 6

Figure 2.5.2. Fringe patterns in the region of a square array of fibres: (*a*) *U* and (*b*) *V* patterns for $\beta = 6$, where the contour interval is 35 nm/contour.

for this region are depicted in figures 2.5.2(*a*) and (*b*), where the contour interval is 35 nm/contour. The *U* and *V* patterns are similar because of the equivalent distribution of fibres in the *x* and *y* directions. In the *V* field, the contours are sparse near the centre lines of horizontal rows of fibres, since the free thermal contraction of the aluminum in these areas was constrained by the stiffer fibres. Similarly, vertical rows are sparse in the *U* field.

2.5.2   Analysis and results

The displacement fields obtained by microscopic moiré interferometry represent total displacements, which include the free thermal contraction part and the stress-induced part of displacements. The total strains can be extracted from the multiplied patterns by the relationship for engineering strains

$$\varepsilon^t_x = \frac{\partial U}{\partial x} = \frac{1}{\beta f}\left[\frac{\partial N^*_x}{\partial x}\right]$$
$$\varepsilon^t_y = \frac{\partial V}{\partial y} = \frac{1}{\beta f}\left[\frac{\partial N^*_y}{\partial y}\right] \tag{2.5.1}$$

where ε^t is the total strain and N^* is the fringe contour number. In order to evaluate the stress-induced part of strains, the free thermal contraction part was subtracted by

$$\varepsilon^\sigma = \varepsilon^t - \varepsilon^\alpha = \varepsilon^t - \alpha\Delta T \tag{2.5.2}$$

where ε^σ is the stress-induced part of strain, ε^α is the free thermal contraction part and α is the coefficient of thermal expansion of each material.

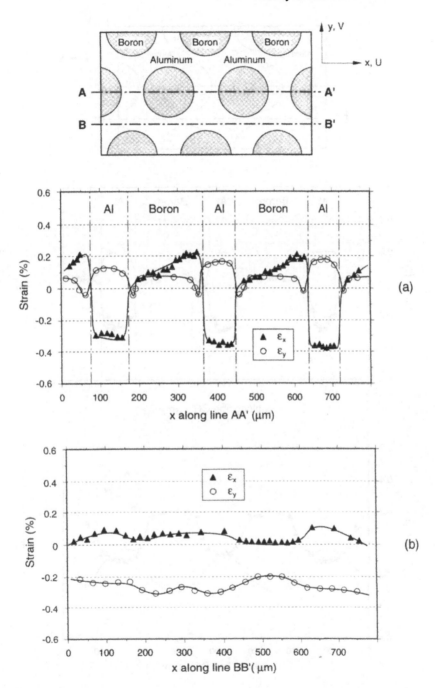

Figure 2.5.3. Stress-induced strain distributions in the region of a hexagonal array of fibres: (*a*) along AA′; and (*b*) along BB′.

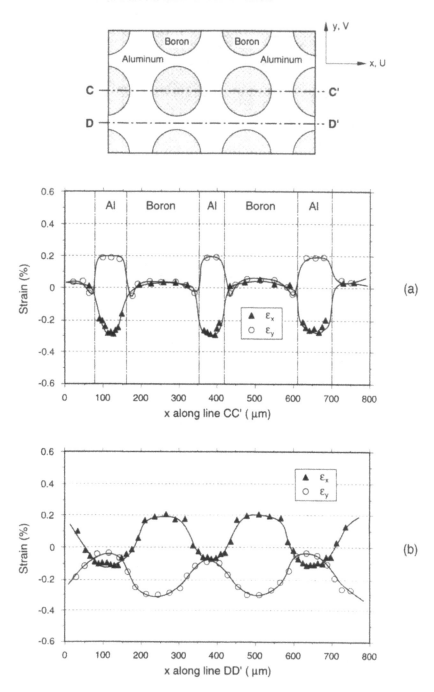

Figure 2.5.4. Stress-induced strain distributions in the region of a square array of fibres: (*a*) along CC′; and (*b*) along DD′.

The stress-induced strains, ε^σ, along the centre line through fibres and along the line between fibres, were calculated from the multiplied patterns by equations (2.5.1) and (2.5.2), where α is 8.3 and 23.6 ppm/°C for boron and aluminum, respectively. The results are given in figure 2.5.3 for the hexagonal array and figure 2.5.4 for the square array. As a result of the constraint by the stiff fibres, the strains in the aluminum matrix, near the interface and in the direction parallel to the fibre–matrix interface (ε_y^σ in figures 2.5.3(a) and 2.5.4(a), and ε_x^σ in figures 2.5.3(b) and 2.5.4(b)) were tensile. The largest tensile strain of the matrix occurred in the ligament between the square array fibres (ε_x^σ in figure 2.5.4(b)), where the height of the ligament was smallest. The maximum value of tensile strain was 0.20%. For the same reason, the strains in the fibres were compressive near the fibre–matrix interfaces (ε_y^σ in figures 2.5.3(a) and 2.5.4(a)). The maximum compressive strain was much smaller in magnitude, −0.05%, because of the larger Young's modulus of boron fibres.

However, the deformation of the aluminum matrix in the ligaments between fibres behaved differently in the direction perpendicular to the fibre–matrix interface. Large compressive strains were developed (ε_x^σ in figures 2.5.3(a) and 2.5.4(a) and ε_y^σ in figures 2.5.3(b) and 2.5.4(b)). The magnitude of the maximum compressive strain was almost twice as large as that of the maximum tensile strain. These large compressive strains are counterintuitive, inasmuch as one would expect compression in the boron and tension in the aluminum. The compression is caused by a free surface effect. Detailed explanations on the free surface effect can be found in [3].

2.5.3 Discussion

The micromechanical deformations obtained from the experiment were produced by the combined effects of (2.5.1) an interior effect wherein the fibre and matrix are mutually constrained due to the CTE mismatch; and (2.5.2) a free surface effect wherein the out-of-plane deformation contributes to the free surface strains of fibre and matrix. The results indicated that the free surface effect prevailed in the direction perpendicular to the fibre–matrix interface. A similar phenomenon was observed in a related experimental study on a bimaterial joint [4].

The patterns of figures 2.5.1 and 2.5.2 show deviations from the strict symmetry and unit cell repeatability that would be expected with ideal specimens. Accordingly, the results provide the deformation of real materials encountered in practice. They account for the variations of fibre spacing and fibre arrangements (hexagonal, square and non-uniform arrangements), variations of loading conditions and variations of material properties. For example, the eccentricity of fringe contours within the fibres in figure 2.5.1(a) is thought to result from different fibre arrangements surrounding the region viewed in the figure. This would be an example of variations of loading conditions.

References

[1] Han B 1996 Micromechanical thermal deformation analysis of unidirectional boron/aluminum metal-matrix composite *Opt. Lasers Eng.* **24** 455–66
[2] Post D and Wood J 1993 Determination of thermal strains by moiré interferometry *Exp. Mech.* **29** 18–20
[3] Guo Y, Post D and Han B 1993 Thick composites in compression: an experimental study of micromechanical behaviour and smeared engineering properties *J. Composite Mater.* **26** 1930–44
[4] Post D, Wood J D, Han B, Parks V J and Gerstle F P 1994 Thermal stresses in a bimaterial joint: an experimental analysis *J. Appl. Mech.* **61** 192–8

CHAPTER 3

FRACTURE MECHANICS

Chapter 3.1.1

Assessment of the shape of crack-tip plastic zones as a function of applied load

C A Walker and Peter M MacKenzie
Strathclyde University, Glasgow, UK

3.1.1.1   Introduction

The plastic yield zones around the end of a crack have been the subject of intense investigation, particularly since the definition of the HRR stress and strain fields some 20 years ago [1, 2]. The whole treatment of the fracture process really depends on the extent of these zones, in relation to the crack length, and the size of the uncracked ligament but the detailed description of the zones has received only passing discussion, largely as a result of the experimental problems involved in obtaining data which will allow an accurate evaluation, particularly in view of the three-dimensional nature of the problem. In some test situations this need give little cause for anxiety, since experimental conditions may be controlled to give a defined yield zone. In general, however, such control cannot be exercised and knowledge of details of the yield locus may well become important.

However, if we return to the general case in which, under a mixed loading regime, a crack-tip plastic field will develop, the definitive way of determining the yield locus would be through a knowledge of the strain fields and the use of a criterion which defines yield in terms of strain. For example, the von Mises criterion defines yield as

$$\sigma_{vM} = (\sigma_1^2 + \sigma_2^2 - \sigma_1\sigma_2)^{1/2} \qquad \text{(plane stress)}$$

such that when $\sigma_{vM} \geq \sigma_0$ (where σ_0 is the uniaxial yield stress), then the state of the three-dimensional stress is such that the material is in a post-yield condition.

In a situation of post-yield plane-stress, then it is possible to derive the stresses from the surface strain components. In this study, the surface strains have been measured using moiré interferometry [3] and the von Mises yield

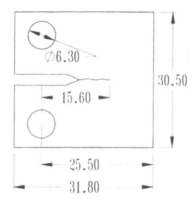

Figure 3.1.1.1. Specimen dimensions. The material is aluminium alloy RR58.

criterion assessed at a matrix of points around the crack-tip to detect values of $\sigma_{vM} \geq \sigma_0$. The yield locus then separates points where $\sigma_{vM} \geq \sigma_0$ from the region further from the crack-tip (in general), where $\sigma_{vM} < \sigma_0$. This mechaniztic definition is the one which is of interest in the calculation of energy-related fracture parameters: it may not produce the same result as would, for example, a surface polish-and-etch treatment to identify dislocation cores. In addition, it should be recalled that the plane-stress condition relates *not* to the specimen in general but to the region of the yield locus. For example, the detailed state of stress around the crack-tip is certainly not purely plane-stress and is probably undefined. It is a monument to the power of established fracture characterization methods, such as the J-integral, that a detailed knowledge of the three-dimensional state of stress is not a prerequisite for a satisfactory analysis.

3.1.1.2 Experimental details

The specimen geometry and properties are shown in figure 3.1.1.1. Prior to application of the grating, the initially machined notch was grown in fatigue, under conditions prescribed in ASTM E-813:81 [4]. A relief grating was cast on to the specimen surface [3].

The specimen was also fitted with a clip-gauge for monitoring the load-point displacement, so that the fracture parameters could be calculated from geometry, load and load-point displacement, independent of the measurements of J from the crack-tip strain fields which was the main object of these tests [5].

The specimen was loaded stagewise: at each load level, the moiré interferometric system [3] was used to record three interferograms, corresponding to in-plane displacement fields at $0°$, $45°$ and $90°$ to the plane of the crack, so that the strains could be analysed at any point in the field (figure 3.1.1.2). As the crack began to grow, the plastic crack-tip zones developed and the crack extended.

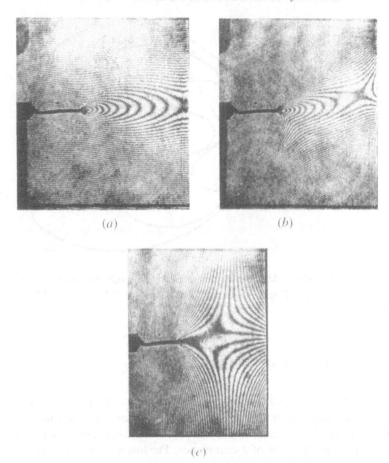

Figure 3.1.1.2. (*a*) 0°, (*b*) 45°, (*c*) 90° displacement fields in this case at an applied load of 2.6 kN $J = 7.2$ kN m^{-1}. The fiducial lines, used for scaling the interferograms, may be seen at the extreme edges of the picture above and below the crack.

3.1.1.3 Measurement of Von Mises yield locus

In the first instance, a rectangular array of lines at a pitch of 0.5 mm was defined at right angles to the crack direction. Along each line, measurements of σ_{vM} were made and plotted as a function of distance along the line: values of σ_{vM} above σ_0, the uniaxial yield stress, define points within the yield locus. It will be seen that the reason for recording displacement fields in three directions (0°, 45° and 90°, figure 3.1.1.2) is to allow an accurate evaluation of the shear strains [6].

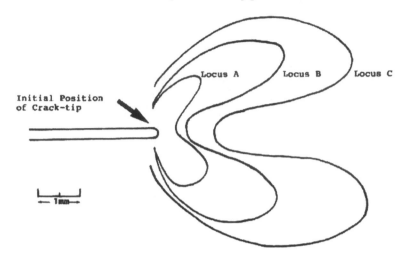

Figure 3.1.1.3. Von Mises yield loci for a range of applied loads and consequent J-integral values. Loci A, B and C were obtained for J values of 7.2, 10.9 and 17.2 kN m^{-1} respectively.

3.1.1.4   Discussion of results

The final von Mises yield loci are shown in figure 3.1.1.3 as a function of increasing J value. Over this regime of the development of plastic zones, the load was essentially constant. It is, perhaps, instructive, to compare the extent of these zones with the region of J dominance. The lower limit of J dominance has been estimated to be 3δ [7] (where δ = crack-tip opening displacement), while the upper limit is about $r_p/4$, where r_p is the approximate dimension of the plastic zone. The region of the HRR fields can be seen to be only a small zone within the overall plastic zone—i.e. a region with a dimension of about 1 mm diameter for the highest figure of applied J-integral.

3.1.1.5   Conclusions

Displacement fields around a crack-tip have been measured and were analysed to define the boundary between the yield zone and the elastic zone for the specimen surface. These yield loci have been measured as a function of increasing levels of the J-integral and compared with the region of J-dominance. The shape of the zones, for this geometry, has been shown to be a deep double lobe that was related in an approximately linear fashion to the imposed level of the J-integral and to the crack-tip opening displacement.

References

[1] Hutchinson J W 1968 Singular behaviour at end of tensile crack in hardening material *J. Mech. Phys. Solids* **16** 13–31

[2] Rice J R and Rosengren G F 1968 Plane strain deformation near crack-tip in power-law hardening material *J. Mech. Phys. Solids* **16** 1–12

[3] McKelvie J, Walker C A and MacKenzie P M 1987 A workaday moiré interferometer for use in stress analysis *Proc. Soc. Photo-Instrum. Eng.* **184** 464–74

[4] 1982 *Standard Test Method for J_{IcP}, a Measure of Fracture Toughness (ASTM-318: 81)* (Philadelphia, PA: American Society for Testing and Materials)

[5] MacKenzie P M *et al* 1989 Direct experimental measurement of the J-integral in engineering materials *Res. Mech.* **28** 361–82

[6] Theocaris P S 1969 *Moiré Fringes in Strain Analysis* (Oxford: Pergamon)

[7] Hutchinson J W 1983 Fundamentals of the phenomenological theory of non-linear fracture mechanics *J. Appl. Mech.* **50** 1042–57

Chapter 3.1.2

Deformation around fatigue cracks from moiré fringe measurement

Gianni Nicoletto
Universitá di Parma, Italy

3.1.2.1 Introduction

Considerable progress in fatigue design methodologies has been achieved by adopting the concepts of fracture mechanics. Central to this approach is a knowledge of the crack-tip deformation and its controlling field parameters, which permit one to avoid the detailed modelling of the often complex micro-structural mechanisms. The stress intensity factor range, ΔK, has been extensively used to correlate fatigue crack growth data although it may not be an accurate measure of the effective crack driving force in many circumstances. Crack-tip phenomena associated with plasticity-induced crack closure, residual stresses, corrosive environment and microstructure have motivated the development of effective ΔK definitions, ΔK_{eff}, to correlate fatigue crack growth behaviour.

The availability of powerful computational tools is increasingly motivating the global simulation of the fatigue crack propagation phenomenon from basic solid mechanics principles. As the history-dependent elastic–plastic material response, changing boundary conditions and residual inelastic deformations in the wake of the advancing crack are involved in the modelling, assumptions related to domain discretization and material constitutive laws will inevitably affect the predicted trends. Therefore, the parallel development and use of experimental techniques from mechanics to study crack-tip deformation should provide valuable insight into fundamental theoretical descriptions and support computational fracture modelling. As localized strain gradients and non-linear material response characterize crack-tip deformation, this task is a challenge for any experimental technique.

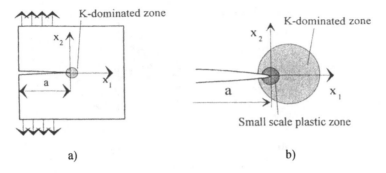

Figure 3.1.2.1. Basic fracture mechanics models: (*a*) linear elastic fracture mechanics; and (*b*) small-scale-yielding condition.

In this contribution, localized deformation about fatigue cracks, as obtained with moiré interferometry, is presented and discussed. Based on this experimental evidence, the applicability of crack models to actual deformation in fatigue is investigated using computer-generated moiré fringe patterns for a stationary crack in elastic and elastic–plastic solids. Discrepancies between experimental and theoretical patterns are used to point out several contributing factors to the complexity of the material response.

3.1.2.2 Basic crack-tip models

The present understanding of the local deformation at a fatigue crack-tip has largely evolved from the theoretical results for a stationary crack in an infinite plate of an elastic and elasto-perfectly-plastic materials under monotonic and cyclic loading. The main aspects are briefly reviewed as they provide a useful background to the subsequent experimental observations.

3.1.2.2.1 Stationary crack under monotonic loading

For the symmetrically loaded two-dimensional cracked body in figure 3.1.2.1(*a*), the elastic stress field is given by a series expansion about the crack-tip and its leading term is characterized by an inverse-square-root singularity [1]. The corresponding mode I displacement field equations in the polar coordinate system (r, θ) centred the crack-tip are:

$$\begin{vmatrix} u_1 \\ u_2 \end{vmatrix} = \frac{K}{G} \left(\frac{r}{2\pi}\right)^{1/2} \begin{vmatrix} \hat{u}_1 & (\theta, v) \\ \hat{u}_2 & (\theta, v) \end{vmatrix} - \frac{Tr}{2E(1+v)} \begin{vmatrix} -\cos\theta \\ v\sin\theta \end{vmatrix} \qquad (3.1.2.1)$$

where the first terms are associated with the singular part of the near-tip stress field (i.e. the K_1-field) and the second terms are associated with the first non-singular term T of the stress series expansion [2]. In equation (3.1.2.1), E and G are the

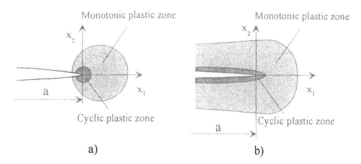

Figure 3.1.2.2. Stationary crack: (*a*) unloading of a stationary crack in an elastic–plastic material; and (*b*) fatigue crack.

Young's modulus of elasticity and shear modulus of elasticity, respectively, v is the Poisson's ratio and $\hat{u}_i(\theta, v)$ are non-dimensional functions, which depend on the stress state and are found in fracture mechanics textbooks [3]. The stress and displacement fields are uniquely controlled by the mode I stress intensity factor K which, conversely, is related to the overall geometry and loading of the cracked body.

In a structural material, the stress singularity of the elastic approach suggests the development of a plastic zone at the crack-tip. Localized yielding results in a stress redistribution and steeper strain gradients. However, if the plastic zone is contained well within the region of dominance of the singular elastic field, then the stress and deformation surrounding it are expected to be largely unaffected. It is often referred to as the small-scale-yielding (SSY) condition, schematically shown in figure 3.1.2.1(*b*), and it allows one to use the stress intensity factor to characterize the crack behaviour even in elastic–plastic materials. SSY conditions are important because they very often apply to sub-critical fatigue crack propagation [4].

3.1.2.2.2   Stationary crack under cyclic loading

Fatigue is associated to cyclic loading conditions. In the case of a load cycle between zero and a tensile load (i.e. the stress ratio $R = 0$), unloading will automatically remove all stresses at a crack in an elastic material. The effect of unloading on the local deformation of a stationary crack in an elasto-perfectly-plastic (EPP) solid was investigated with a plastic superposition method [5]. The development of two plastic zones ahead of the crack-tip is schematically shown in figure 3.1.2.2(*a*) and the generation of a residual stress distribution were thus predicted. The larger plastic zone is called the monotonic plastic zone and the smaller one is termed the cyclic plastic zone because it is characterized by reversed plasticity. Their sizes are theoretically proportional to the K at maximum load, K_{max}, and to the stress intensity factor range $\Delta K (K_{max} - K_{min})$, respectively.

The cyclic plastic zone is experimentally difficult to observe and measure due to its small size.

Cyclic loading conditions may promote sub-critical (i.e. stable) propagation of a fatigue crack. Therefore, according to the scheme in figure 3.1.2.2(*b*), two plastic zones and a wake of plastically deformed material characterize a fatigue crack. Residual stresses are expected in the plastic zones ahead of the crack-tip. The mechanism of anticipated crack closure due to the presence of the wake of permanently deformed material has been invoked to define an effective ΔK which successfully correlated crack growth at different stress ratios correlated crack growth at different stress ratios in thin sheets [6].

3.1.2.3   Experimental details

Moiré interferometry is well suited for localized full-field displacement and strain analysis about cracks in structural materials since it responds geometrically and with high sensitivity to material elastic–plastic deformation on an extended field of view. As both a tutorial treatment of moiré interferometry and its potential for measurements in fracture mechanics have been presented elsewhere in the monograph, here only the aspects peculiar to the present application are outlined. First of all, diffraction gratings (frequency $f = 2.4$ lines/μm) were used in all the experiments. Then, standard fracture specimens such as the Single Edge Notched Tensile (SENT) or the Compact Tension (CT) were used. The materials tested ranged from aluminium alloys to various construction steels [7, 8].

After the replication of the diffraction grating on the lateral surface of the specimen, through cracks were grown from starter notches using a fatigue-testing machine. At a prescribed crack length, the specimen was removed from the servo-hydraulic machine and inserted in a stiff, screw-driven loading rig mounted on an optical bench. The tensile load was then gradually increased up to its maximum level, while highly magnified fringe patters for two orthogonal directions were photographically recorded. Particular attention was devoted to the accurate focusing of the specimen grating to reduce uncertainties in the relative position of the fringes with respect to the crack-tip. Typical fringe patterns associated with the two orthogonal in-plane displacement components in a steel specimen are shown in figure 3.1.2.3. Displacements can be computed using the basic moiré equation

$$u_i = \frac{N_i}{f} \qquad i = 1, 2 \qquad\qquad (3.1.2.2)$$

where N_i is the fringe order and f is the reference grating frequency (i.e. $f = 2.4$ lines/μm). These patterns at fatigue cracks represent the experimental evidence to be discussed in the next section in the light of the previous theoretical framework and a parallel computational study.

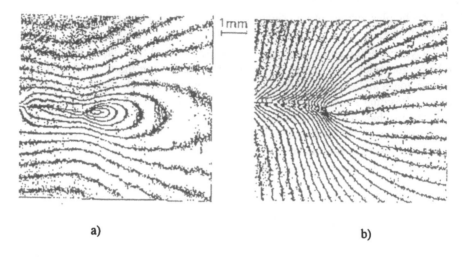

a)   b)

Figure 3.1.2.3. Moiré interferometric fringe patterns about a fatigue crack in steel: (*a*) x_i direction and (*b*) x_2 direction.

3.1.2.4   Fatigue crack-tip deformation

The complex nature of the previous fringes suggests a step-wise approach to their understanding. Therefore, the effect of various factors, namely (i) localized yielding, (ii) far-field stresses, (iii) cyclic plasticity and (iv) crack-tip mixed-mode deformation, will be identified in the fringe patterns and separately discussed.

3.1.2.4.1   Local yielding

As a plastic zone inevitably develops at the crack-tip in a structural material, the first aspect is the determination of the extent of the plastic deformation and its signature on the moiré fringe pattern. Two different approaches have been adopted for generating synthetic moiré interferometric fringe patterns under the plane-stress assumption. The moiré fringe patterns pertaining to the asymptotic solution for a monotonically loaded stationary crack in an elastic–plastic solid were obtained with a nonlinear finite-element code. Because of the high sensitivity of moiré interferometry and the required resolution at the crack-tip for a small plastic zone size in SSY, the boundary layer approach to this type of problem was used [9]. In both approaches the global applied loading was given by the nominal K as computed using the specimen geometry factor, the actual applied load in the loading rig and the measured crack length.

The effect on the in-plane displacement components is readily assessed by comparing the moiré fringe patterns for a linear elastic material and those for the elastic–plastic material as shown in the lower and upper halves, respectively, of the composite N_2 pattern of figure 3.1.2.4(*a*). Two different graphical formats are

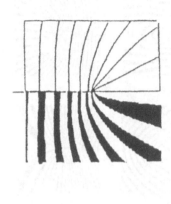
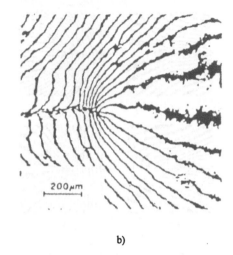

a)   b)

Figure 3.1.2.4. (*a*) Combined elastic (below) and elastic–plastic (above) N_2 fringe patterns. (*b*) N_2 fringe pattern about a fatigue crack-tip.

used to distinguish between the numerically-generated elastic and elastic–plastic fringes: in the first case, integer order fringes are depicted as broad black–white fringes; in the second case they are represented as thin black lines. Marked differences between the two patterns are observed, especially for the fringes ahead of the crack-tip where the plastic zone develops. It is found that they are not parabolic (i.e. displacements vary according to $r^{1/2}$ and strains according to $r^{-1/2}$) as predicted by linear elastic fracture mechanics but are appreciably linear and convergent to the crack-tip. It would imply a displacement variation proportional to r and an r^{-1} strain field. This trend is also discernible in the magnified experimental fringes of figure 3.1.2.4(*b*). From a point-wise strain analysis, the strain gradient ahead of the crack-tip would be easily determined and used for a correlation with asymptotic elastic–plastic field descriptions.

3.1.2.4.2 Non-singular stresses

The singular K field associated with the first terms of equation (3.1.2.1) is dominant only very close to the crack-tip. Inevitably, the subsequent terms of the series, especially the T stress in equation (3.1.2.1) [2], become increasingly important as one extends the field of investigation about the crack-tip or increase the applied load. The elastic T-stress is specimen-geometry-dependent [10] and similarly to the stress intensity factor, it is also crack-length-dependent. Interestingly, a degree of inherent crack-tip bi-axiality is present even in uniaxially loaded specimens and it affects the crack-tip deformation. The effect of both positive and negative values of the elastic T-stress on the elastic fringe patterns is shown in figure 3.1.2.5. From a qualitative standpoint, the fringe patterns of

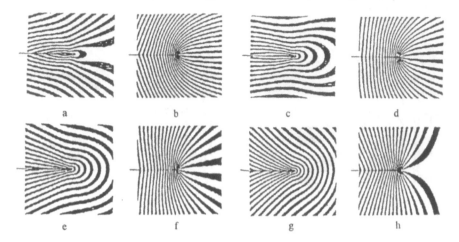

Figure 3.1.2.5. Elastic T-stress effect on crack-tip displacements: (*a*) N_1 and (*b*) N_2 fringes for high negative T; (*c*) N_1 and (*d*) N_2 fringes for low negative T; (*e*) N_1 and (*f*) $N_2 - N_2$ fringes for low negative T; and (*g*) N_i and (*h*) N_2 fringes for high positive T.

figure 3.1.2.3 and those of figure 3.1.2.5 would reveal the presence of negative T stress in the specimen. Indeed, the moiré pattern was obtained on an SEN specimen.

3.1.2.4.3 Cyclic plasticity

Further inspection of figure 3.1.2.4(*b*) shows that the fringes behind the crack-tip are initially perpendicular to the crack plane and suddenly kink at a distance. The computed fringe pattern of figure 3.1.2.4(*a*) does not show this peculiar behaviour. It is therefore attributed to different plastic histories near to and far from the crack plane due to the advancing crack. Invoking the scheme of figure 3.1.2.2(*b*), the kinks would define the limit of the cyclic plastic zone: actual measurements of the zone size were found to agree with theoretical and computational predictions [11]. Furthermore, if an overload was applied in a constant amplitude load sequence, the cyclic plastic zone is initially found to reduce from its pre-overload size, it then grows again to its undisturbed size. This trend matches the evolution of the crack growth rate and supports the importance of the cyclic plastic zone size as a predictive parameter.

3.1.2.4.4 Local mode-mixity

Another deviation of the fringe pattern of figure 3.1.2.4(*b*) from the theoretical fringes is worth mentioning: mixed-mode displacements may develop in the near-tip region even for a mode I fatigue crack loaded in purely uniaxial cyclic tension. They may be associated to local deflections in crack path due to various

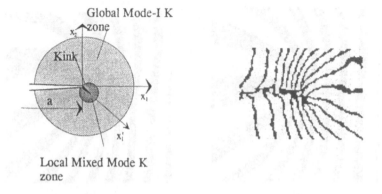

Figure 3.1.2.6. Kinked crack-tip: (*a*) scheme and definitions; and (*b*) magnified crack-tip N_2 fringes from figure 3.1.2.4(*b*).

mechanical, micro-structural and environmental factors. Deviation of the crack from the mode I growth plane leads to alterations in the effective stress intensity and in the crack growth rates.

Considering the scheme in figure 3.1.2.6, mode I crack deformation is expected on a macroscopic level and mixed-mode (i.e. mode I and mode II) deformation would occur at the tip of the kinked crack. Estimate of the changes in nominal K_I introduced by the deflection mechanism can be obtained considering local opening (mode I) and sliding (mode II) stress intensity factors, k_1 and k_2, respectively,

$$k_1 \approx \cos^3 \beta K_1$$
$$k_2 \approx \sin \beta \cos^2 \beta K_1$$

where β is the angle formed between the kink and the main crack portion. These equations predict that a tilt of 15 degrees induces a significant mode mixity at the crack-tip. Fringe patterns for two degrees of mixed-mode loading are shown in figure 3.1.2.7. They were generated by superposition of the asymptotic, elastic, mode I and mode II displacement fields. Comparison with near-tip moiré fringes of figure 3.1.2.4(*b*) confirms the presence of crack-tip mixed- mode deformation in the experiment.

3.1.2.5 Summary

Experimental and simulated fringe patterns about crack-tips have been correlated in this chapter with the aim of sorting out several factors contributing to fatigue crack-tip deformation. Moiré interferometry has demonstrated adequate measuring sensitivity and spatial resolution. Simulations considered only stationary cracks in elastic and elastic–plastic solids. Factors such as material

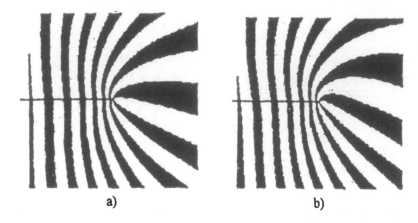

a)   **b)**

Figure 3.1.2.7. Mixed-mode fringe patterns for a linear elastic material: (*a*) N_2 fringes for $k_{II}/k_I = 0.2$ or $\beta = 11$ deg; and (*b*) N_2 fringes for $k_{II}/k_I = 0.4$ or $\beta = 22$ deg.

plasticity, specimen-geometry-dependent deformation and localized mixed-mode growth conditions have been pointed out.

Acknowledgment

Partial financial support of this research by the Ministero dell'Università e della Ricerca Scientifica (MURST) is gratefully acknowledged.

References

[1] Williams M L 1957 On the stress distribution at the base of a stationary crack *Trans. ASME J. Appl. Mech.* **24** 109–14
[2] Larsson S G and Carlsson A J 1973 Influence of non-singular stress terms and specimen geometry on small-scale yielding at crack-tips in elastic–plastic materials *J. Mech. Phys. Solids* **21** 263–77
[3] Kanninen M F and Popelar C H 1985 *Advanced Fracture Mechanics* (New York: Oxford University Press)
[4] Rice J R 1967 *Mechanics of Crack-Tip Deformation and Extension by Fatigue, Fatigue Crack Propagation (ASTM STP 415)* (Philadelphia, PA: American Society for Test and Materials) pp 247–311
[5] Rice J R 1968 Mathematical analysis of the mechanics of fracture *Fracture–An Advanced Treatise* vol II, ed H Liebowitz (New York: Academic Press) pp 191–311
[6] Sanford R J 2002 *Principles of Fracture Mechanics* (Englewood Cliffs, NJ: Prentice-Hall)
[7] Nicoletto G 1986 Experimental crack-tip displacement analysis under small-scale yielding conditions *Int. J. Fatigue* **8** 83–9

[8] Nicoletto G 1987 Fatigue crack-tip strains in 7075-T6 aluminium alloy *Fatigue Fracture Eng. Mater. Struct.* **10** 37–49

[9] Nicoletto G 1990 Biaxial stress effects on the elastic–plastic crack-tip displacement fields *Meccanica* **25** 99–106

[10] Kfouri A P 1986 Some evaluations of the elastic t-term using Eshelby's method *Int. J. Fract.* **30** 301–15

[11] Nicoletto G 1989 Platic zones about fatigue cracks in metals *Int. J. Fatigue* **11** 107–15

Chapter 3.2.1

Applications of moiré to cellulosic (paper and wood) and concrete materials

R E Rowlands
University of Wisconsin, USA

3.2.1.1   Introduction

The fact that wood and concrete are two of society's most common building materials and the widespread domestic/industrial applications of cardboard boxes and paper necessitate an ability to stress analyse components fabricated from these materials adequately.   Individual members of wood structures are often connected by bolts and these connections are frequently the weakest element in the structure.   While there is a desire to conserve materials, development of improved design procedures has been inhibited by insufficient knowledge of the stresses in the wood surrounding the bolt-loaded holes.   Theoretical stress analyses of realistic bolted joints having finite-size members are almost non-existent and reliable numerical analyses of such problems are aggravated by the complicated nonlinear contract conditions between bolt and wood. Such aspects of wood as its poor thermal conductivity reduce the attractiveness of using strain gauges. Optional methods for measuring displacements/strains in paper products are certainly no greater than with wood. The relatively high compliance of paper essentially precludes contacting the paper with or bonding to the paperboard any comparatively stiff transducer.   The brittle nature and low tensile strength of concrete motivates the need for a better understanding of this material's fracture processes.   The lack of alternative viable methods for strain/stress analysing cellulosic materials and the advantages such as its full-field nature for fracture analyses render moiré well suited for these applications [1–6].

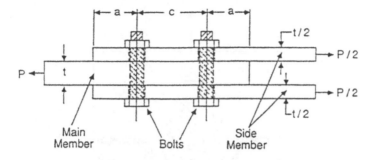

Figure 3.2.1.1. Cross-section of double-bolted mechanical fastener.

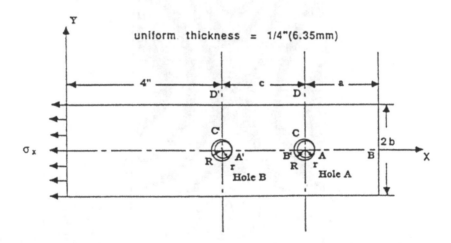

Figure 3.2.1.2. In-plane view of double-bolted mechanical fastener.

3.2.1.2   Analyses

3.2.1.2.1   Bolted wood connections

A cross section of a double-bolted joint whose middle member is twice as thick as the side members is shown in figure 3.2.1.1, while figure 3.2.1.2 contains a schematic diagram of the in-plane geometry associated with figure 3.2.1.1. In addition to measuring and predicting the onset of failure, single- and double-bolted wood joints of geometry of figures 3.2.1.1 and 3.2.1.2 have been stress analysed by finite element analysis and moiré under various loading conditions. Figure 3.2.1.3 is a moiré fringe pattern of the vertical displacements in the wood in the neighbourhood of the bolts of a double-bolted joint in Sitka Spruce [2]. All of the vertical load, $P = 445$ N (100 lbs), applied to the wood in figure 3.2.1.3 is carried by the top bolt-loaded hole.

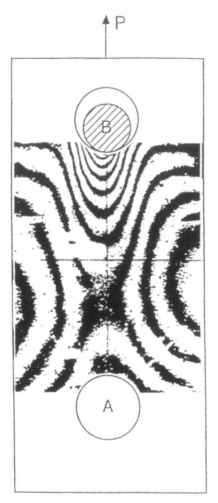

Figure 3.2.1.3. Vertical displacement field in the neighbourhood of the bolts of a double-bolted joint.

Figure 3.2.1.4 compares the vertical displacements in a single-bolted joint of Douglas Fir as obtained from finite element methods (FEM) (left-hand side) and moiré interferometry (right-hand side) [3]. The circular bolt here can be considered as fixed in space and the wood as being pulled ($P = 356$ N, 80 lbs) vertically upward around the stationary bolt.

Moiré images such as that of figure 3.2.1.3 were obtained by contact printing a 200 dot cm^{-1} array onto a very thin, smooth, epoxy layer which had been applied to the wood. Reflectivity was enhanced by depositing an aluminum film onto the epoxy prior to printing the moiré rulings/dots. Contact printing

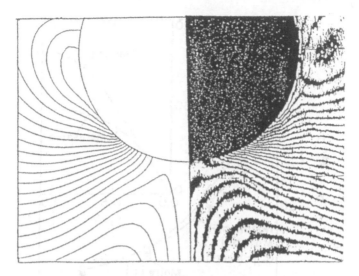

Figure 3.2.1.4. Vertical displacement fields near the bolt of a single-bolted joint—a comparison of FEM (left-hand side) and moiré (right-hand side).

used Kodak Photoresist (KPR). One-to-one replicas of the wood dot array were recorded photographically at various stages of loading. These replicas were subsequently analysed at a fringe multiplication of ten in an optical set-up containing a blazed diffraction ruling [7]. The fringe pattern of figure 3.2.1.4 was recorded using interferometric moiré in which the 1200 lines/mm ruling on the wood was interrogated with a 2400 lines/mm virtual ruling to produce a moiré resolution of 416 nm/fringe [8]. The surface of the Douglas Fir of figure 3.2.1.4 was again prepared with a thin, smooth epoxy layer onto which the 1200 lines/mm ruling was contact printed using KPR [3].

Figure 3.2.1.5 illustrates the effects of the nonlinear stress–strain response of the Sitka Spruce on the longitudinal stress distribution along the line (line AB of figure 3.2.1.2) connecting the point of initial bolt contact and the end of the wood. This figure compares the results obtained from moiré and FEM.

3.2.1.2.2 Fracture of linerboard

Linerboard is the paper material which forms the outside surfaces of the walls of cardboard boxes. The inelastic, orthotropic and compliant mechanical behaviour of paperboard complicates its fracture response [4]. Figure 3.2.1.6 shows the moiré fringes (interferometric moiré, effective line density of 600 lines/mm) for the vertical displacements in a vertically-loaded linerboard tensile strip (1.9 cm wide by 0.25 mm thick) containing 3 mm long, symmetrical edge cracks. The strong/stiff/machine direction of the paperboard specimen of figure 3.2.1.6 is also vertical. Figure 3.2.1.7 shows the moiré fringes associated with the residual

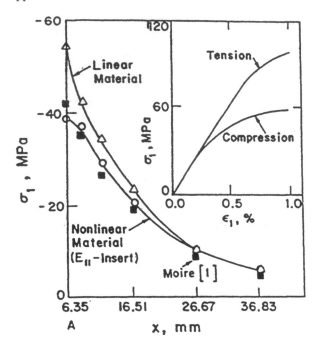

Figure 3.2.1.5. Effect of material nonlinearity of spruce on longitudinal stress along line bolted joint ($a = 7.6$ cm, $r/R = 0.936$, $P = 1.1$ kN).

deformations in such a notched linerboard tensile specimen which failed between its edge cracks after being loaded vertically for 2.1 min at 89 N.

Materials such as paperboard necessitate some specific moiré techniques, particularly in providing the specimen rulings. Motivated by an earlier experience [5], the notched tensile strips of figures 3.2.1.6 and 3.2.1.7 were first provided with an RTV silicone surface into which the 300 lines/mm phase ruling was moulded. The 300 lines/mm master used to create the specimen ruling was formed by exposing a holographic plate to the virtual grating produced by interfering two collimated light beams at the appropriate angle. Independent verifications were conducted to ensure that the RTV has minimal effect on the paperboard's material response.

3.2.1.2.3 Fracture processes in concrete

Relative to an effort to understand the fracture processes in concrete better, figure 3.2.1.8 is a schematic diagram of a three-point bending specimen of mortar or concrete containing a bottom, vertical edge crack [6]. Moiré interferometry (effective ruling of 2400 lines/mm) was used to study the region around the crack-tip of specimens of figure 3.2.1.8. Figure 3.2.1.9 contains a representative moiré

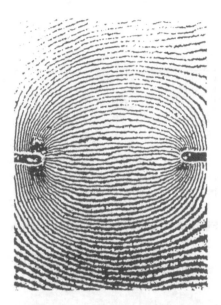

Figure 3.2.1.6. Moiré fringes (600 lines/cm) in vertically-loaded linerboard tensile AB of single-specimen.

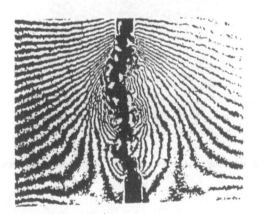

Figure 3.2.1.7. Moiré recorded residual vertical displacements in notched linearboard tensile specimen which failed after 2.1 min at 89 N

fringe photograph of the region surrounding and ahead of the crack in such a three-point concrete specimen.

Results from several such specimens of mortar and concrete reveal that a large FPZ (fracture process zone) occurs ahead of a crack-tip in concrete/mortar. The size of this FPZ can significantly affect the magnitude of the stress intensity factor, K_I. At relatively low loads where the damage zone ahead of the crack

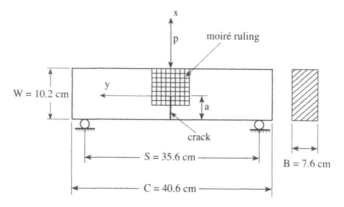

Figure 3.2.1.8. Three-point bend specimen

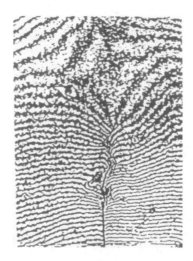

Figure 3.2.1.9. Moiré fringes in concrete three-point bend specimen ($a = 5.08$ cm, $P = 2.4$ kN).

is small, the classically predicted value of K_1 agrees with that from moiré interferometry. However, the former becomes less reliable as the size of the damage zone increases. Stress intensity factors were obtained from images such that of figure 3.2.1.9 by hybridizing the moiré data with FEM concepts described in [9].

3.2.1.3   Summary

Moiré interferometry is well suited for analysing engineering problems involving wood, paper and concrete/mortar. Individual aspects of these materials somewhat restrict the choice of suitable methods for measuring their displacement/strains, as well as necessitating some specific moiré techniques relative to applying the active rulings/dots. Among other features, the FEM results correlate well with the present moiré analyses of bolted joints and moiré interferometry quantifies the complex fracture behaviour in paperboard and concrete/mortar.

References

[1] Wilkinson T L and Rowlands R E 1981 Analysis of mechanical joints in wood *Exp. Mech.* **21** 408–14

[2] Rahman M U, Chiang Y J and Rowlands R E 1991 Stress and failure analysis of double-bolted joints in Douglas fir and Sitka spruce *Wood and Fibre Sci.* **23** 567–89

[3] Shih (Stone) J 1992 Experimental-numerical analysis of bolted joints in finite composites with and without inserts *PhD Thesis* University of Wisconsin

[4] Hyzer J B 1985 Photomechanical analysis of paperboard *MS Thesis* University of Wisconsin

[5] Rowlands R E, Gunderson D E and Beazley P K 1983 Moiré strain analysis of paper *TAPPI* **66** 81–4

[6] He S, Feng Z and Rowlands R E 1997 Fracture process zone analysis of concrete using moiré interferometry *Exp. Mech.* **37** 367–73

[7] Post D and MacLaughlin T F 1971 Strain analysis by moiré fringe multiplication *Exp. Mech.* **11** 408–13

[8] Post D, Han B and Ifju P 1994 *High Sensitivity Moiré* (Berlin: Springer)

[9] Feng Z, Sanford R J and Rowlands R E 1991 Determining stress intensity factors from smoothing finite-element representation of photomechanical data *Eng. Fracture Mech.* **40** 593–601

Chapter 3.2.2

Mixed-mode stress intensity factors in finite, edge-cracked orthotropic plates

C A Walker[1] *and Jamasri*[2]
[1] *University of Strathclyde, Glasgow, UK*
[2] *Gadjah Mada University, Indonesia*

3.2.2.1   Introduction

The linear elastic fracture mechanics (LEFM) approach has been successfully used to analyse isotropic materials, where stress intensity factors (SIFs) of a crack with length $2a$ in a finite plate for various geometries have been generalized as the stress intensity factor for an infinite plate by using a finite correction factor. This procedure has been adopted for composite materials by a number of investigators [1, 2]; alternatively a finite plate of composite materials may be treated as an infinite plate by taking such a factor equal to 1 [3–5].

However, the use of a finite correction factor adopted from isotropic materials is basically unsound, since the SIFs in composite materials are dependent on their directional properties, as investigated computationally by Parhizgar *et al* [6] and Lin [7]. A number of analytical/computational models of fracture in composites have been published [8, 9] but none of these has been subjected to experimental validation. In essence, the situation is that the investigations to date have shown that SIFs are influenced significantly by orthotropic material properties but, as yet, no general solution to the problem has been proposed. Much of the work that has been published related to material in which the orthotropic axes are aligned parallel and perpendicular to the loading direction. This is a simplification which will not, in general, occur and so a requirement for methods of determining the SIFs under generalized loading conditions may be discerned, which will also validate the analytical and computational models.

It is the aim of this present chapter to describe such a method for measuring, by experiment, the mixed-mode stress-intensity factors in orthotropic materials under generalized loading conditions. The basic methodology has already been shown to be a robust tool for the analysis of near-crack-tip strain distributions, both for the measurement of K_I and K_{II} in metals and for the measurement of the J-integral from a knowledge of the strain field around the contour of integration [10, 11]. Given a successful application to the generalized loading problem, the method may then be used as validation for the computational and analytical predictions and also to measure the stress intensity factors on real structural elements.

Since the use of a finite correction factor in composite materials is not inherently accurate, the experimental method should ideally look at near-crack-tip data. Hence, a tool that provides a way to record the whole field information around the crack-tip with sufficient accuracy is required. Smith and co-workers [12], for example, used high-sensitivity moiré interferometry to record whole-field displacements around the crack-tip of isotropic materials. The moiré patterns were then analysed in order to determine the mode I SIF. A similar approach has been used [10] to assess the mode I and mode II SIFs in a biaxially loaded centre-cracked aluminium panel.

The aim of the present work is to extend Smith's idea applied to composite materials. Since the displacement fields around the crack-tip of orthotropic materials are substantially different from those in isotropic materials, equations must, first of all, be derived for the mixed-mode stress intensity factors as a function of displacement data.

3.2.2.1.1 Notation

a_{ij}	Elastic constants
$\bar{a}_{ij}$	Components of the compliance matrix
E_{11}, E_{22}	Stiffness moduli
K_I	Mode I stress intensity factor
K_{II}	Mode II stress intensity factor
r, θ	Radial coordinates with crack-tip as origin
u, v	Displacements in x- and y-coordinate directions
α	Angle between fibre direction and the normal to the loading direction
v_{12}	Poisson's ratio

3.2.2.2 Theoretical analysis

3.2.2.2.1 Displacement fields around the crack-tip

The displacement fields around a crack-tip results from mode I and mode II loadings, shown in figure 3.2.2.1, will now be considered as follows [13].

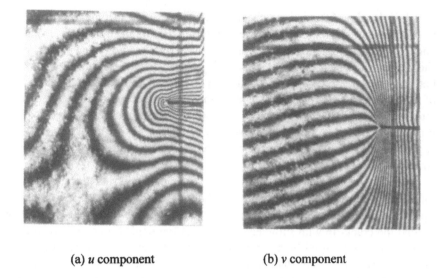

(a) *u* component   (b) *v* component

Figure 3.2.2.1. Moiré interferograms (scale 5:1) of deformation field with $\alpha = 90°$ under 3.2 kN far-field uniaxial loading. Each fringe represents 1.05 μm displacement

Mode I loading:

$$u = K_{\mathrm{I}} \left(\frac{2r}{\pi}\right)^{1/2} \mathrm{Re}\left[\frac{1}{s_1 - s_2}\{s_1 p_2 (\cos\theta + s_2 \sin\theta)^{1/2}.\right.$$
$$\left. - s_2 p_1 (\cos\theta + s_1 \sin\theta)^{1/2}\}\right] \tag{3.2.2.1}$$

$$v = K_{\mathrm{I}} \left(\frac{2r}{\pi}\right)^{1/2} \mathrm{Re}\left[\frac{1}{s_1 - s_2}\{s_1 q_2 (\cos\theta + s_2 \sin\theta)^{1/2}\right.$$
$$\left. - s_2 q_1 (\cos\theta + s_1 \sin\theta)^{1/2}\}\right] \tag{3.2.2.2}$$

Mode II loading:

$$u = K_{\mathrm{II}} \left(\frac{2r}{\pi}\right)^{1/2} \mathrm{Re}\left[\frac{1}{s_1 - s_2}\{p_2 (\cos\theta + s_2 \sin\theta)^{1/2}\right.$$
$$\left. - p_1 (\cos\theta + s_1 \sin\theta)^{1/2}\}\right] \tag{3.2.2.3}$$

$$v = K_{\mathrm{II}} \left(\frac{2r}{\pi}\right)^{1/2} \mathrm{Re}\left[\frac{1}{s_1 - s_2}\{q_2 (\cos\theta + s_2 \sin\theta)^{1/2}\right.$$
$$\left. - q_1 (\cos\theta + s_1 \sin\theta)^{1/2}\}\right] \tag{3.2.2.4}$$

where p_i and $q_i (i = 1, 2)$ are defined as

$$p_1 = \bar{a}_{11}s_1^2 + \bar{a}_{12} - \bar{a}_{16}s_1 \qquad p_2 = \bar{a}_{11}s_2^2 + \bar{a}_{12} - \bar{a}_{16}s_2$$

$$q_1 = \frac{\bar{a}_{12}s_1^2 + \bar{a}_{22} - \bar{a}_{26}s_1}{s_1} \qquad q_2 = \frac{\bar{a}_{12}s_2^2 + \bar{a}_{22} - \bar{a}_{26}s_2}{s_2} \qquad (3.2.2.5)$$

and the roots of the characteristic equation (s_i) may be obtained from the following expression:

$$s_i^4 - \left(\frac{2\bar{a}_{16}}{\bar{a}_{11}}\right)s_i^3 + \left[\frac{2\bar{a}_{12} + \bar{a}_{66}}{\bar{a}_{11}}\right]s_i^2 - \left(\frac{2\bar{a}_{26}}{\bar{a}_{11}}\right)s_i + \left(\frac{\bar{a}_{22}}{\bar{a}_{11}}\right) = 0. \qquad (3.2.2.6)$$

Lekhnitskii [14] has shown that equation (3.2.2.6) could have roots that are either complex or purely imaginary and could not have real roots. For an orthotropic material, the roots are:

$$s_1 = \alpha_1 + i\beta_1 \qquad s_2 = \alpha_2 + i\beta_2 \qquad s_3 = \bar{s}_1 \qquad s_4 = \bar{s}_2 \qquad (3.2.2.7)$$

where the values of α_i, $\beta_i (i = 1, 2)$ have been calculated using a numerical implementation of Bairstow's method (for example, see the application of Bairstow's method described in [15] and also appendix 3.2.2.1): the values required for the present study are listed in table 3.2.2.1.

The components of the compliance matrix $(\bar{a}_{ij})$ are obtained from

$$\bar{a}_{11} = a_{11}\cos^4\alpha + (2a_{12} + a_{66})\sin^2\alpha\cos^2\alpha + a_{22}\sin^4\alpha$$
$$\bar{a}_{12} = a_{12}(\sin^4\alpha + \cos^4\alpha) + (a_{11} + a_{22} - a_{66})\sin^2\alpha\cos^2\alpha$$
$$\bar{a}_{22} = a_{11}\sin^4\alpha + (2a_{12} + a_{66})\sin^2\alpha\cos^2\alpha + a_{22}\cos^4\alpha \qquad (3.2.2.8)$$
$$\bar{a}_{66} = 2(2a_{11} + 2a_{22} - 4a_{12} - a_{66})\sin^2\alpha\cos^2\alpha + a_{66}(\sin^4\alpha + \cos^4\alpha)$$
$$\bar{a}_{16} = (2a_{11} - 2a_{12} - a_{66})\sin\alpha\cos^3\alpha - (2a_{22} - 2a_{12} - a_{66})\sin^3\alpha\cos\alpha$$
$$a_{26} = (2a_{11} - 2a_{12} - a_{66})\sin^3\alpha\cos\alpha - (2a_{22} - 2a_{12} - a_{66})\sin\alpha\cos^3\alpha$$

where

$$a_{11} = 1/E_{11} \qquad a_{22} = 1/E_{22} \qquad a_{12} = -v_{12}/E_{11} \qquad \text{and} \qquad a_{66} = 1/G_{12}.$$

The values of $\bar{a}_{ij}$ are listed in table 3.2.2.2.

3.2.2.2.2 Mixed-mode fracture analysis

The mixed-mode SIFs may be derived by combining the displacement fields resulting from mode I and mode II loadings as expressed in equations (3.2.2.1)–(3.2.2.4), appropriately, which can be written as follows:

$$u = \left(\frac{2r}{\pi}\right)^{1/2}(K_1 g_{1x} + K_{II} g_{2x}) \qquad (3.2.2.9)$$

Table 3.2.2.1. The roots of the biharmonic equation (s_i, $i = 1, 2$).

	$s_1 = \alpha_1 + i\beta_1$		$s_2 = \alpha_2 + i\beta_2$	
α (deg)	α_1	β_1	α_2	β_2
75	0.1026	1.1597	−0.2541	0.2349
90	0.0	1.1925	0.0	0.2199

Table 3.2.2.2. The components of the compliance matrix ($\bar{a}_{ij}$)(m^2 N^{-1}).

α (deg)	$\bar{a}_{11}$	$\bar{a}_{12}$	$\bar{a}_{22}$	$\bar{a}_{16}$	$\bar{a}_{26}$	$\bar{a}_{66}$
75	0.9874E–4	−0.4690E–4	0.1603E–4	−0.1496E–4	−0.3279E–4	0.1447E–3
90	0.1026E–3	−0.2116E–5	0.7052E–5	0.0	0.0	0.1550E–3

$$v = \left(\frac{2r}{\pi}\right)^{1/2} (K_I g_{1y} + K_{II} g_{2y}) \qquad (3.2.2.10)$$

where

$$g_{1x} = \mathrm{Re}\left[\frac{1}{s_1 - s_2}\{s_1 p_2(\cos\theta + s_2\sin\theta)^{1/2} - s_2 p_1(\cos\theta + s_1\sin\theta)^{1/2}\}\right]$$

$$g_{2x} = \mathrm{Re}\left[\frac{1}{s_1 - s_2}\{p_2(\cos\theta + s_2\sin\theta)^{1/2} - p_1(\cos\theta + s_1\sin\theta)^{1/2}\}\right]$$

$$\qquad\qquad (3.2.2.11)$$

$$g_{1y} = \mathrm{Re}\left[\frac{1}{s_1 - s_2}\{s_1 q_2(\cos\theta + s_2\sin\theta)^{1/2} - s_2 q_1(\cos\theta + s_1\sin\theta)^{1/2}\}\right]$$

$$g_{2y} = \mathrm{Re}\left[\frac{1}{s_1 - s_2}\{q_2(\cos\theta + s_2\sin\theta)^{1/2} - q_1(\cos\theta + s_1\sin\theta)^{1/2}\}\right]$$

By rearranging equations (3.2.2.9) and (3.2.2.10), the following equations may be obtained:

$$K_I = \left(\frac{vg_{2x} - ug_{2y}}{g_{2x}g_{1y} - g_{1x}g_{2y}}\right)\lim_{r\to 0}\left(\frac{\pi}{2r}\right)^{1/2} \qquad (3.2.2.12)$$

$$K_{II} = \left(\frac{ug_{1y} - vg_{1x}}{g_{2x}g_{1y} - g_{1x}g_{2y}}\right)\lim_{r\to 0}\left(\frac{\pi}{2r}\right)^{1/2}. \qquad (3.2.2.13)$$

Equations (3.2.2.12) and (3.2.2.13) may be simplified as follows:

$$K_I = (C_1 u + C_2 v)\lim_{r\to 0}\left(\frac{\pi}{2r}\right)^{1/2} \qquad (3.2.2.14)$$

Table 3.2.2.3. The coefficients C_1, C_2, C_3 and C_4 (N/m^2).

α (deg)	θ (deg)	C_1	C_2	C_3	C_4
75	0	0.4036E+5	0.0	0.0	−0.4036E + 5
	180	0.4856E+4	0.2502E+5	0.8204E+5	0.4856E+4
90	0	0.4036E+5	0.0	0.0	−0.4036E + 5
	180	0.0	0.2633E+5	0.6903E+5	0.0

$$K_{\text{II}} = (C_3 u + C_4 v) \lim_{r \to 0} \left(\frac{\pi}{2r} \right)^{1/2}. \qquad (3.2.2.15)$$

These equations show that both K_{I} and K_{II} are dependent upon u and v; therefore, the u and v data have to be acquired simultaneously at the same distance from the crack-tip. The values of the coefficients C_1, C_2, C_3 and C_4 have been calculated as listed in table 3.2.2.3, whereas the u and v may be obtained from moiré patterns. K_{I} and K_{II} are, therefore, obtainable from the experimental data. In fact, it has been found convenient to set u or v equal to zero and to gather the data from one set of interferograms—using $\theta = 0°$ or $\theta = \pm180°$, i.e. using the data from directly in line with the crack or along the crack face. In taking the limit as $r \to 0$, it was felt preferable to carry out a best-fit nonlinear extrapolation, since the measurements were make to within 0.2 mm of the crack-tip and there was no evidence of near-crack-tip damage [12].

3.2.2.3 Experiments

3.2.2.3.1 Specimen preparation

The material used was a CFRP (carbon fibre reinforced plastic) composite system with an eight-ply unidirectional laminate; after moulding, the cured laminate was 1 mm thick. The specimens were double-edge-cracked plates with fibres orientated at 75° and 90° with respect to the x-axis. Each of the test specimens had a width of 25 mm and a length of approximately 236 mm. A pair of aluminium tabs 38 mm in length were mounted on each end of the specimen, the gauge length of the specimen being approximately 160 mm. Artificial cracks were introduced using an 0.18 mm thick fret-saw blade; the length of the cut was 3.8 mm. Each of these crack-tips was subsequently sharpened using a razor blade.

Finally, a 475 line/mm orthogonal grating, which consisted of an epoxy resin (STYCAST parts A and B in which the proportion by weight was 100 of parts A and 28 of parts B), was cast on to the region around the crack-tips. To avoid the effect of the grating thickness on the specimen behaviour, the grating was made as thin as possible (between 0.05 and 0.1 mm) and the resin, which had seeped

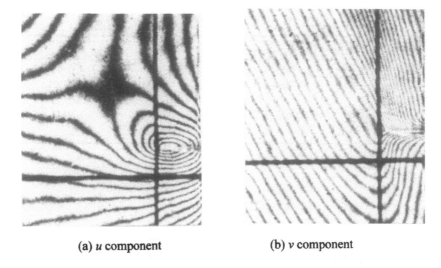

(a) u component   (b) v component

Figure 3.2.2.2. Moiré interferograms (scale 5:1) of deformation field with $\alpha = 75°$ under 1.76 kN far-field loading. Each fringe represents 1.05 μm displacement

between the crack faces, was removed by recutting the cracks [16], by careful use of the same fret-saw blade. Since the epoxy resin is cured into a tough matrix, the grating was not damaged by this process, even close to the crack faces (see, for example, figures 3.2.2.2 and 3.2.2.3, where the fringes meet the crack faces).

3.2.2.3.2 Specimen testing

A moiré interferometer which has been described elsewhere [17] was used to record the whole-field surface displacements around the crack-tip. The interferometer was designed to allow ease of use and alignment. Since the optics are insensitive to vibration, the use of the interferometer is akin to the use of a camera; once aligned, the fringe patterns are stable.

The moiré patterns resulting from the moiré interferometer were then recorded on film (Tri-X pan processed to 400 ASA) with a 35 mm camera with an 80–210 mm zoom. Zero-load fringe patterns were recorded; however, the accuracy of the grating replication was such that there was no requirement to subtract zero-load fringes from the on-load interferograms.

Due to the effect of the unsymmetrical material properties, the specimens may exhibit end rotation, especially for those specimens for which the principal axes are not coincident with the load direction. To avoid such an effect of end rotation, self-aligning grips were employed instead of fixed grips, as has been suggested by Pagano and Halpin [18].

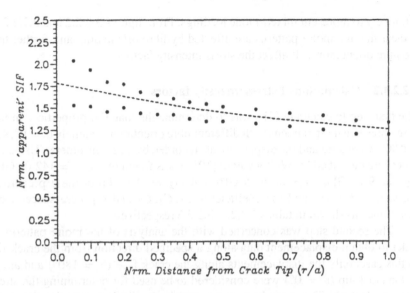

Figure 3.2.2.3. Mode I normalized apparent SIF curve as a function of distance from the crack-tip ahead of the crack for $\alpha = 90°$ (fibre orientation parallel to the direction of loading).

3.2.2.4 Analysis of the results

3.2.2.4.1 Moiré patterns

Selected moiré patterns resulting from the experiment are shown in figures 3.2.2.2 and 3.2.2.3, in which each figure consists of moiré interferograms representing the u and v components. It may be seen in figure 3.2.2.2 that for the fibres oriented at 90° (i.e. the principal axes of the material used coincide with the coordinate system) the moiré patterns around the crack-tip are symmetric with respect to the coordinate system. This indicates that only pure mode I loading is given by this type of specimen. A reference grating with a pitch of 20 mm was superimposed on the moiré grating. These reference lines appear as a black cross in the interferograms. Since the 475 line/mm grating lines cannot be seen visually, the reference lines were used to establish the orientation of the grating during the casting process.

The moiré patterns around the crack-tip were no longer symmetric with respect to the coordinate system when the principal axes of the material used were not coincident with the coordinate system (as shown in figure 3.2.2.3). This may be explained by the effect of the unsymmetrical material properties. Since the materials used consist of fibres and matrix in which fibres are much stronger compared to the matrix, the properties of the composite are dominated by the orientation of the fibres. The unsymmetrical moiré patterns may be seen intuitively to represent mixed-mode loadings. From figures 3.2.2.2 and 3.2.2.3, it is clear that the moiré patterns are affected by fibre orientation, and further that the fibre orientation will affect the stress intensity factors.

3.2.2.4.2   Calculation of stress intensity factors

The first step to calculate SIFs was to determine the material properties. In this case, three tensile specimens with different fibre orientations, namely $\alpha = 0°, 45°$ and $90°$, were used and the displacements recorded by an extensometer. Using the procedure described by Whitney *et al* [19], it was found that $E_{11} = 142.00$ GPa, $E_{22} = 9.75$ GPa, $G_{12} = 6.45$ GPa and $v_{12} = 0.3$. From these properties, the values of s_i, $\bar{a}_{ij}$ and the coefficients C_1, C_2, C_3 and C_4 were successively determined as shown in tables 3.2.2.1–3.2.2.3 respectively.

The second step was concerned with the analysis of the moiré patterns in order to obtain displacement data along a specified direction from the crack-tip. In this case, only the displacements along the crack face ($\theta = 180°$) and ahead of the crack-tip ($\theta = 0°$) were considered to be used for determining the stress intensity factors. The procedure used to obtain the displacement data has been described in detail by Nicoletto *et al* [12] and by Smith [20].

By substituting the values given in tables 3.2.2.1–3.2.2.3 into equations (3.2.2.14) and (3.2.2.15) and taking into account the displacement data, the 'apparent' SIF graphs as a function of distance from the crack-tip may be obtained. By extrapolating these graphs back to $r = 0$, the stress intensity factors of the crack-tip may be determined. The results of the 'apparent' stress intensity factor measurements, which were normalized by $\sigma_y^\infty \sqrt{a}$, are shown in figures 3.2.2.4–3.2.2.6.

3.2.2.5   Discussion and conclusions

Moiré patterns resulting from the experiments of the double-edge cracked specimens of unidirectional CFRP composites have been shown in figures 3.2.2.2 and 3.2.2.3 and these moiré patterns have been analysed in order to determine mode I and mode II SIFs. The essence of the method is that the moiré patterns represent contours of displacement and that the displacement data as a function of distance from the crack-tip must be obtained prior to the calculation of 'apparent' SIF data (see equations (3.2.2.14) and (3.2.2.15)). It would be anticipated that the overall accuracy of the results should depend on the accuracy achieved in acquiring those displacement data.

The major problem encountered in measuring such displacements was the determination of the point coinciding with the crack-tip. Any error in identifying this point caused an extraneous error in determining the SIFs. In addition, the accuracy of the results depended upon the accuracy in determining the centre of each fringe corresponding to the distance from that fringe to the crack-tip, r. This applied especially to the fringes close to the crack-tip, where the density

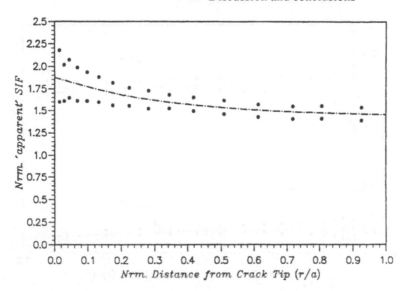

Figure 3.2.2.4. Mode I normalized apparent SIF curve as a function of distance from the crack-tip along the crack face for $\alpha = 90°$ (fibre orientation parallel to the direction of loading).

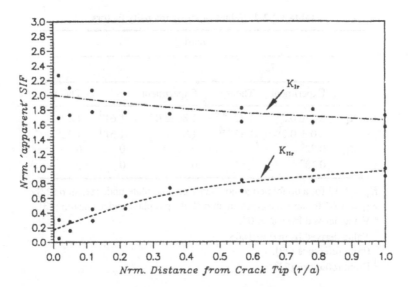

Figure 3.2.2.5. Normalized apparent SIF curve as a function of distance from the crack-tip ahead of the crack for $\alpha = 75°$.

of fringes is extremely high. To minimize this error, the measurements at each point were repeated to ascertain the experimental error from this source. A

Mixed-mode stress intensity factors

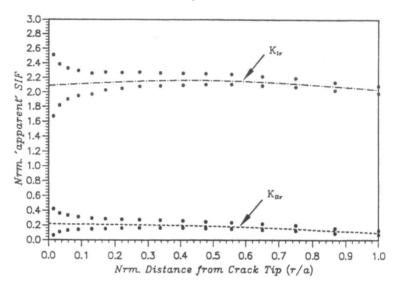

Figure 3.2.2.6. Normalized apparent SIF curve as a function of distance from the crack-tip ahead of the crack for $\alpha = 75°$.

Table 3.2.2.4. Values of stress intensity factors.

	α (deg)				
	75		90		
	Experiment	Theory	Experiment		Theory
K_I	2.2 ± 0.2^a	$1.12*^{a,c}$	1.86 ± 0.2	1.61^d	1.12^c
	2.0 ± 0.2^b	$1.12^{b,d}$	1.8	1.61^d	1.12^c
K_{II}	0.22^a		0	0	0
	0.18^b		0	0	0

$K_I = 1.12$ for a double-edge-notched isotropic plate under tension [22].
$K_{II} = 1.61$ for an edge crack in an orthotropic plate under tension [23].
[a] Value derived from $\theta = 0°$.
[b] Value derived from $\theta = 180°$.
[c] From reference [22].
[d] From reference [23].

Monte Carlo simulation of a similar complex data analysis has indicated that the overall error should be about $\pm 8\%$ in the measurement of the SIF [21]. The mean values of the normalized 'apparent' SIF data points have been plotted together with their probable errors. Actual values of the stress intensity factors are shown in table 3.2.2.4, where, for comparison, values of the SIFs derived from an analytical solution are also shown, in addition to the SIF for an isotropic material. It is evident that significant differences do exist between the SIFs measured in isotropic and orthotropic materials, since, as far as the isotropic data are concerned, it is a factor of two different (for $\alpha = 75°$), while for the $\alpha = 90°$ case, the difference is some 60%. The use of isotropic corrections, therefore, may seem to be a source of error.

It will be seen that the experimental and theoretical orthotropic evaluations are in broad agreement, although the actual theoretical model used did not duplicate the experimental measurements. It is, therefore, possible to proceed with a degree of confidence to apply this method for experimental validation.

Overall, then, the mixed-mode SIFs have been successfully determined experimentally using local crack-tip data. The method here is an alternative method to avoid the use of a finite correction factor adopted from isotropic materials.

Acknowledgments

The second author is indebted to the Indonesian Government which has provided financial support through the Second University Development Project (World Bank XVII and World Bank XXI). Thanks are due to Mr J Sivewright in carrying out the experimental work.

Appendix: Theory of the numerical approach for solving the characteristic of the biharmonic equation

The biharmonic equation may be simplified to become a characteristic equation that consists of a fourth-degree polynomial equation, as shown in equation (3.2.2.6). This equation has to be solved further in order to obtain its roots, which are required for analysing the fracture behaviour in composite materials. Since the polynomial equation is fully populated, it is extremely difficult to solve analytically; therefore, a numerical approach may be used as an alternative solution.

There are several methods available for solving a polynomial equation. However, Bairstow's method was chosen for the present study since the method is quite powerful to solve any degree of polynomial equation, whether it contains imaginary roots or not. Basically, Bairstow's method solves any polynomial equation by gradually reducing its degree with two-degree polynomials. The user may then easily solve those two-degree polynomials using the usual method.

Let a polynomial equation be considered as follows:

$$P_m(x) = x^m + a_1 x^{m-1} + a_2 x^{m-2} + \cdots + a_{m-1} x + a_m \qquad (3.2.2.16)$$

where m is the degree of the polynomial equation and the a_m are the coefficients of the polynomial equation. If equation (3.2.2.16) is divided by a two-degree polynomial, $Q(x) = x^2 + px + q$, it becomes

$$P_m(x) = Q(x)P_{m-2}(x) + b_{m-1}(x) + b_m \qquad (3.2.2.17)$$

where b_{m-1} and b_m are algebraic functions and $P_{m-2}(x)$ is a new function after being reduced by the two-degree polynomial and may be expressed as follows:

$$P_{m-2}(x) = x^{m-2} + b_1 x^{m-1} + b_2 x^{m-2} + \cdots + b_{m-3}x + b_{m-2}. \qquad (3.2.2.18)$$

The goal is now to find the values of the parameter functions p, q so that $b_{m-1} = b_m = 0$. Approximations of the functions p, q which satisfy this condition are given by the following recurrent relations:

$$p_r = p_{r-1} + \Delta p_k \qquad q_r = q_{r-1} + \Delta q_k \qquad (3.2.2.19)$$

where

$$\Delta p_k = \frac{\nabla_{k,p}}{\Delta k} = \frac{b_{k,m-1}c_{k,m-2} - b_{k,m}c_{k,m-3}}{c_{k,m-2}^2 - d_{k,m-1}c_{k,m-3}}$$

$$\Delta q_k = \frac{\nabla_{k,q}}{\Delta k} = \frac{b_{k,m}c_{k,m-2} - b_{k,m-1}d_{k,m-1}}{c_k^2, m-2 - d_{k,m-1}c_{k,m-3}} \qquad \text{for } k = 1, 2, 3, \ldots, r$$

$$(3.2.2.20)$$

where r is the number of repetition given by the user in order to achieve the accuracy needed. Hence, the accuracy must be specified by the user in the forms of $(\Delta p_k - \Delta p_{k-1})$ and $(\Delta q_k - \Delta q_{k-1})$, noting that p_0 and q_0 are the initial estimations which are given by

$$p_0 = \frac{a_{m-1}}{a_{m-2}} \qquad q_0 = \frac{a_m}{a_{m-2}}. \qquad (3.2.2.21)$$

Coefficients in equations (3.2.2.20) are again related by recurrent relations as follows:

$$b_{k,m} = a_{k,m} - p_k b_{k,m-1} - q_k b_{k,m-2}$$

$$c_{k,m-1} = p_k c_{k,m-2} - q_k c_{k,m-3}$$

$$d_{k,m-1} = c_{k,m-1} - b_{k,m-1}. \qquad (3.2.2.22)$$

To start with equation (3.2.2.22), the following initial values may be used:

$$b_{k,0} = a_{k,0} = 1$$

$$b_{k,1} = a_{k,1} - p_k b_{k,0} \qquad (3.2.2.23a)$$

$$c_{k,0} = b_{k,0} = 1$$

$$c_{k,1} = b_{k,1} - p_k c_{k,0}. \qquad (3.2.2.23b)$$

The numerical calculations dealing with the extraction of the roots of the biharmonic equation, which are based on Bairstow's method, were performed using the FORTRAN 77 language [9].

References

[1] Cruse T A 1973 Tensile strength of notched composites *J. Composite Mater.* **7** 219–29

[2] Sih G C, Hilton P D, Badaliance R, Shenberger P S and Villareal G 1973 Fracture mechanics of fibrous composites *ASTM STP* **521** 99–132

[3] Donaldson S L 1985 Fracture toughness testing of graphite/epoxy and graphite/PEEK composites *Composites* **16** 103–12

[4] Williams J G and Birch M W 1976 Mixed-mode fracture in anisotropic media *ASTM STP* **601** 125–37

[5] Hahn H T and Johannesson T 1983 A correlation between fracture energy and fracture morphology in mixed-mode fracture in composites *Mechanical Behaviour of Materials IV* ed Carlson J and Ohlson N G (London: Pergamon) pp 431–8

[6] Parhizgar S, Zachary L W and Sun C T 1982 Application of the principles of linear fracture mechanics to the composite materials *Int. J. Fracture* **20** 3–15

[7] Lin Y 1989 An energy release rate criterion for crack growth in unidirectional composite materials *Eng. Fracture Mech.* **32** 861–9

[8] Yam Y-T and Hong C-S 1991 Stress intensity factors in finite orthotropic plates with a crack under mixed-mode deformation *Int. J. Fracture* **47** 53–67

[9] Woo C-W and Wang Y-H 1993 Analysis of an internal crack in a finite anisotropic plate *Int. J. Fracture* **62** 203–18

[10] MacKenzie P McA, Walker C A and Giannetoni R 1993 A technique for experimental evaluation of mixed-mode stress intensity factors *Int. J. Fracture* **62** 139–48

[11] MacKenzie P McA *et al* 1989 Direct experimental measurement of the *J*-integral in engineering materials *Res. Mechanica* **28** 361–82

[12] Nicoletto G, Post D and Smith C W 1982 Moiré interferometry for high sensitivity measurements in fracture mechanics *Joint Conf. of the Society For Experimental Analysis and the Japanese Society of Mechanical Engineers on Experimental Mechanics* (Bethel, CT: SEM) pp 266–70

[13] Sih G C, Paris P C and Irwin G R 1963 On cracks in rectilinearly anisotropic bodies *Int. J. Fracture* **1** 189–203

[14] Lekhnitskii S G 1963 *Theory of Elasticity of an Anisotropic Body* ch 3 (San Francisco, CA: Holden-Day)

[15] Gerald A and Wheatly G 1984 *Applied Numerical Analysis* 3rd edn (Reading, MA: Addison-Wesley) pp 27–31

[16] Jamasri 1993 The fracture characterization of orthotropic composites using a hybrid experimental/numerical technique *PhD Thesis* University of Strathclyde, Glasgow

[17] Walker C A 1988 Moiré interferometry for strain analysis *Opt. Lasers Eng.* **8** 1–50

[18] Pagano N J and Halpin J C 1968 Influence of end constraint in the testing of anisotropic body *J. Composite Mater.* **2** 18–31

[19] Whitney J M, Daniel I M and Pipes R B 1982 *Experimental Mechancis of Fibre Reinforced Composite Materials* (Englewood Cliffs, NJ: Prentice-Hall)

[20] Smith C W 1984 Stress intensity factor distributions by optical means *Application of Fracture Mechanics to Materials and Structures* ed Sih, Sommers and Dahl (The Hague: Martinus Nijhof) pp 709–23

[21] MacKenzie P M 1994 Influence of errors in an interferogram analysis *Exp. Mech.* **34** 243–8

[22] Misitani H 1975 Tension of a strip with symmetric edge cracks or elliptical notches *Trans. Japan. Soc. Mech. Eng.* **49** 2518–26
[23] Kaya A C and Erdogan F 1980 Stress intensity factors and COD in an orthotropic strip *Int. J. Fracture* **16** 171–90

Chapter 3.2.3

A hybrid experimental/computational approach to the assessment of crack growth criteria in composite laminates

C A Walker and M Jam
University of Strathclyde, Glasgow, UK

Notation

$R(r, \phi)$	Value of the normalized stress ratio
ϕ	Radial distance from the end of the crack
r	Azimuthal angle around the crack-tip
σ_ϕ	Normal stress acting on the radial plane defined by the angle ϕ
$T\phi$	Tensile strength of the composite on the ϕ-plane
	$T\varphi = X_t \sin^2 \beta + Y_t \cos^2 \beta$
β	Angle from ϕ to the fibre axis
X_1	Axial strength of the composite
Y_1	Transverse strength of the composite
E_{11}	Longitudinal stiffness modulus (aligned with the fibres)
E_{22}	Transverse stiffness modulus(transverse to the fibre direction)

3.2.3.1 Introduction

The application of fracture mechanics in metals has concentrated largely upon the use of the stress intensity factor, which has been extended to include elastic–plastic effects, and upon the single-parameter characterization of more extensive elastic/plastic effects (the J-integral). The course of fracture in composite mechanics has, however, taken quite a different path. In the first instance, composites, particularly layered structures, are inherently more complex than metals. Furthermore, the stress state around a crack-tip is defined not only by the loading and geometry—as it is in metals—but also by the relationship of the

89

fibre direction to the loading axis. While in an ideal world, one would wish to analyse realistic laminated materials—i.e. multi-layer composites, with a lay-up sequence designed to give a desired set of properties. It is at this stage preferable, in terms of complexity, to examine the behaviour of individual plies which are unidirectional and, therefore, maximally orthotropic. Given an understanding of the behaviour of individual plies, the overall behaviour of a laminate sequence may be predicted by assembling a suitable model. In passing, one might note that at this stage of development, the propagation of cracks in the plane of orthotropy is not a major source of composite component failure—delamination at free edges and fastener holes being of much greater practical importance. Nevertheless, with more and more composite structures entering service usage, it is of real engineering importance, for structural integrity assurance, to be able to predict crack growth under such complex loading regimes as exist in the interior of the laminate.

This chapter discusses the problems of making an assessment of the forecasting capability of one particular crack growth criterion, the normalized stress ratio (NSR) [1], which has been proposed as a predictor of the direction of crack growth. The logical background to the NSR is discussed and a validation routine is outlined, by using finite element modelling combined with moiré interferometry, to show how the NSR may be plotted as a function of angle around a crack to predict the most likely direction of crack propagation.

3.2.3.2   Crack growth criteria—theoretical background

The literature on crack growth criteria is extensive [2–8]. A number of criteria have been shown to have validity under certain limited conditions. Since there is not yet an accepted universal criterion, for the purpose of this study, the normal stress ratio (NSR), first proposed by Buczek and Herakovich in 1963, has been selected, since it has been shown to be an accurate predictor of crack growth direction and incorporates consideration of both local applied stress and yield strength parameters. The model, as proposed, suggests that the direction of crack extension corresponds to the radial direction of the maximum value of the ratio $R(r, \phi)$ defined by

$$R(r, \phi) = \frac{\sigma_{\phi\phi}}{T_{\phi\phi}} \qquad \text{(see figure 3.2.3.1)} \qquad (3.2.3.1)$$

where $\sigma_{\phi\phi}$ corresponds to the normal stress acting on the radial plane defined by the angle ϕ at a distance r, from the crack-tip. $T_{\phi\phi}$ is the tensile strength on the ϕ-plane, defined by

$$T_{\phi\phi} = X_t \sin^2 \beta + Y_t \cos^2 \beta$$

where B is the direction from ϕ to the fibre axis.

The definition of $T_{\phi\phi}$ is such as to satisfy the following conditions:

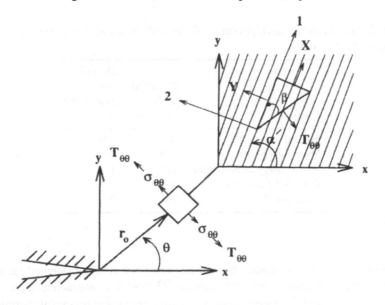

Figure 3.2.3.1. Coordinate system for analysis of fracture around a pre-cut crack, considering the fibre direction.

(1) for an isotropic material, $T_{\phi\phi}$ is independent of ϕ;
(2) for crack growth parallel to the material fibres, $T_{\phi\phi}$ must equal the transverse tensile strength Y_t; and
(3) for crack growth perpendicular to the fibres, $T_{\phi\phi}$ must equal the axial tensile strength X_t.

As an example of its application, table 3.2.3.1 shows a comparison of results for the direction of crack growth as determined by different criteria, discussed in [3]. The criteria were the strain energy density theory, the strain energy destiny ratio, the Tsai–Hill extended criterion, the Norris extended criterion and the NSR; and these were compared with experimental results for a centre-notched 30° off-axis coupon. These results show that the NSR correctly predicts the crack growth direction, when a comparison is made between the experiment and the finite element implementations of the several criteria.

3.2.3.3 Crack growth criterion in use—the experimental problem

The actual use of the criterion indicates that the ratio $R(r, \phi)$ should be plotted for fixed values of r, as a function of ϕ, i.e. the azimuthal angle around the crack-tip: the value of ϕ for which R is a maximum will then define the direction of crack growth when the load has gone to a level such that the fracture stress has

Table 3.2.3.1. Predicted crack growth directions and deviations from the experimentally determined value (108°).

Criterion	Predicted angle (degree)	Deviation from experiment (deg)
Normal stress ratio	98	−10
Strain energy density	155	47
Strain energy density ratio	109	1
Extended Tsai–Hill	105	−3
Extended Norris	106	0

been reached at the crack-tip. As will be seen from results presented later, the dependence of R upon r follows the usual $r^{0.5}$ rule; in consequence, a plot of R versus ϕ at values of r greater than, say 1 mm, will show no useful modulation which may be used as a prediction. In fact, as it turns out, a value for r less than 100 μm is really required, even in a highly orthotropic material. For an experimental assessment of the NSR to be made, the stress needs to be measured at gauge level, therefore, of a few micrometres on a radius around the crack-tip. For such precise spatial discrimination, moiré interferometry is the system of choice. As will be evident from the work of McKelvie [9], however, this level of spatial resolution is at or beyond the limit of existing moiré interferometry, particularly as the stress field is varying rapidly close to the crack-tip. At this point in time, it is not practical to design a moiré system with the dynamic range to quantify both the near- and far-field deformation levels. These considerations lead to the following proposal for a hybrid experimental/computational method for measuring R.

3.2.3.4   Outline assessment method

Plainly, one may model the crack-tip using the finite element (FE) method. However, for reliance to be placed upon its predictions, an experimental validation step is really required. It is a simple matter to generate displacement contours from FE models (see figures 3.2.3.2–3.2.3.4) likewise, an equivalent set of data may be generated experimentally using moiré interferometry (figure 3.2.3.4). The experimental deformation field may be compared with the FE predicted deformation field point by point or, more accurately, by overlaying the contour lines both in the far field, and in the near field, close to the crack-tip (down to <1 mm). The FE modelling process may be adjusted until an adequate fit is obtained. If the mesh around the crack-tip is sufficiently refined, the FE model

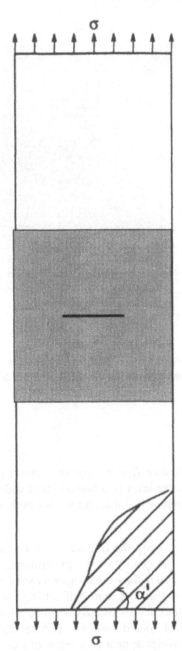

Figure 3.2.3.2. Specimen schematic diagram showing the centre crack, the area covered by the grating and the relationship of the ply-angle to the loading axis.

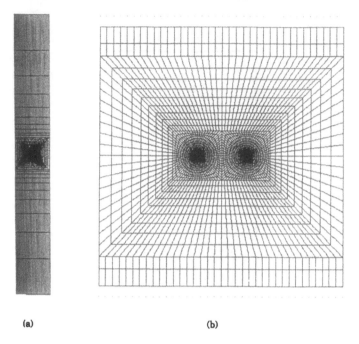

(a)   (b)

Figure 3.2.3.3. Finite-element mesh. Since the specimen is not symmetrical, a whole model is analysed: (*a*) complete mesh and (*b*) detail of region around crack.

may then be used to calculate R-value contours closer to the crack-tip, until a sufficiently useful predictive level is achieved. It is with the implementation of this outline assessment routine that this chapter is chiefly concerned.

For the assessment to carry the full weight of conviction, one should, first of all, be satisfied as to the quality of the experimental data. Before use, the specimen grating was viewed in the interferometer to detect any zonal aberrations: over the whole field there was less than half a fringe of departure from a zero field (aside from the obvious flaws—usually small bubbles in the epoxy moulding material). Since the optical system itself was checked against an NPL reference grating, the fringe fields which appear as a result of loading may be regarded as absolute contours of deformation, and the accuracy measurement depends upon the location of the fringe centres as a function of gauge length [10]. The use of the fringes in the way outlined here is much more powerful than simply comparing strain values from point to point, since the comparison was being made over whole areas at a time and also used data from many points on each fringe.

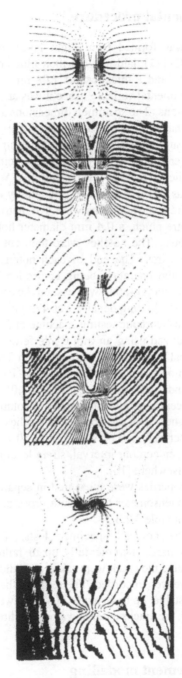

Figure 3.2.3.4. Deformation fields, in the directions X (top), Y (centre) and $\theta = 45°$ (bottom). Computed by finite element and measured experimentally by moiré interferometry. Fringe interval 1.05 μm.

3.2.3.5　Experimental procedure

The basic material was formed from eight layers of carbon/epoxy prepreg (Fibredeux T300/914C) [11], laid up with their fibres parallel and cured to give a fibre volume of 60% and a thickness of 1.08 mm: the fibre length was not considered as an experimental variable, since it was an order of magnitude larger than the crack dimensions. The specimen dimensions are shown in table 3.2.3.2. The specimens used had a length/width ratio of 12:1 to avoid end effects, in accordance with the suggestions by Pagano and Halpin [12]; in addition, the aluminium end tabs were glued to the ends of the specimens, to avoid their being crushed in the hydraulic jaws of the testing machine. Within the specimen in the testing machine, it was possible to assess the quality of the loading regime by ensuring that the moiré interferograms were symmetrical about the crack. For the cutting of the centre notch, a 0.4 mm diameter hole was first drilled in the centre of the specimens. The notches were then cut using 0.005 in diameter diamond-impregnated wire. The end of each notch was carefully sharpened using a scalpel, with the operation being controlled by viewing it through a microscope. A thin (0.1 mm) grating was cast on the centre region containing the crack (figure 3.2.3.2); the grating density was 475 line/mm [10]. The specimens were loaded in a servo-hydraulic testing machine and the deformation contours under load were visualized using an moiré-interferometric technique described previously [10]. Typical interferograms are shown in figures 3.2.34(a)–(c). These show deformation contours at an interval of 1.05×10^{-3} mm, i.e. each point on the contour line has undergone a displacement of 1.05×10^{-3} mm with respect to each point on adjacent contours; the spatial resolution achievable was about 50×10^{-3} for the highest level of strain (this implies that strain measurements could be made at a discrimination between points 50×10^{-3} mm or more apart). The relationship between contour interval, strain level and spatial resolution has been fully discussed elsewhere [10].

The specimen properties were measured in separate tests, using either the moiré system or an extensometer to measure gross deformation. The material properties are shown in table 3.2.3.3.

During the centre notched specimen tests, a slow cross-head speed (0.1 mm min^{-1}) was used, incrementally up to failure. Photographs of the deformation fields around the crack were taken periodically and the load at crack initiation and on failure were recorded. The result of these tests showed the crack growth was slow and stable, up to an angle of 45° between direction and loading. As was to be expected, the cracks extended along the fibre direction to the nearest free edge.

3.2.3.6　Finite element modelling

Due to the anistropic behaviour of the material, symmetry of deformation does not exist above and below the crack as it does in homogeneous materials; a

Table 3.2.3.2. Specimen configuration and dimensions.

Specimen no.	Laminate orientation	Width (mm)	Gauge length (mm)	Thickness (mm)
1	$(0)_{4S}$	12.7	152.4	1.06
2	$(90)_{4S}$	25.5	152.4	1.06
3	$(0)_{4S}$	12.8	160	1.06
4	$(45)_{4S}$	25.2	160	1.06
5	$(\pm 45)_{4S}$	25.35	126	1.1
6	$(\pm 45)_{4S}$	25.6	134	1.1
7	(α)	25.5	160	1.06

Table 3.2.3.3. Mechanical properties obtained by experiment.

Longitudinal modulus E_{11} (N mm^{-2})	135×10^3
Transverse modulus, E_{22} (N mm^{-2})	8.8×10^3
Shear modulus, G (N mm^{-2})	6.94×10^3
Poisson ratio, ν_{12}	0.31
Ultimate unidirectional tensile strength, X (N mm^{-2})	1.67×10^3
Ultimate transverse tensile strength, Y (N mm^{-2})	26
Ultimate longitudinal compressive strength, X_1 (N mm^{-2})	1350
Ultimate transverse compressive strength, Y_1 (N mm^{-2})	230
Ultimate tensile strength of $45°$ off-axis, U_{45} (N mm^{-2})	618
Ultimate in-plane shear strength, S (N mm^{-2})	96

complete specimen model was required: the full model, with crack-tip detail is shown in figure 3.2.3.3 with 4284 elements in all. Around the crack-tip an array of 72 singular 'quarter-point' elements were used to model the square root singularity [13]. The remaining nodes were spaced on each line so that the size of the elements grew as the square of the distance from the crack-tip, with the midside nodes at the midside of the elements. This gave a suitable gradiation of elements near the crack-tip. It will be noted that the elements in the far field, away from the crack-tip have an unusually large aspect ratio, in order to reduce the number of elements in that region. It is well known that such elements can lead to numerical instability if they are of irregular shape. This effect was assessed, both

Table 3.2.3.4. A comparison between the normalized stress intensity factor obtained by different finite element methods and the closed-form solution for the isotropic case.

Method	Normalized stress intensity factor K_1	Deviation from closed-form solution (%)
Closed-form solution	1.816	—
Stress method ($\theta = 0$)	1.85	−1.87
Displacement method ($\theta = 0$)	1.84	−1.32
J-integral method	1.813	+0.16

by computation, and by comparison with experiment, and was found negligible. The model was solved using the abacus package [14].

3.2.3.7   Validation of the finite element model

3.2.3.7.1   Stress intensity factors from the closed form solution (isotropic material assumption)

The stress intensity factor at the crack-tip was calculated, from the FE model, and from an available closed form solution [15], on the assumption, for the purpose of comparison, that the material was steel, homogeneous and isotropic. The results are shown in table 3.2.3.4. It will be seen that there is close agreement between the several computations: on this basis, one may place a certain confidence in the basic model.

3.2.3.7.2   Comparison with experimental data

If we now turn to the real material, figures 3.2.3.4(a)–(c) show a comparison between computed and measured deformation fields for a range of ply angles α. It will be seen that the general form of the deformation surface is closely reproduced by the computation and that the FE model reproduces the fringe distribution both in the far field and close to the crack. On this basis also, one may proceed with faith in the model's ability to handle the orthotropic case. The basic problem may be seen from figure 3.2.3.4(a): in this moiré interferogram, although the overall deformation field may be visualized. There are actually few fringes close to the crack-tip, so that evaluating any function at a radius r from the crack-tip is actually quite inaccurate unless the magnitude of r is several millimetres. This restriction does not apply, in principle, to a validated FE model with an adequate density of near crack-tip elements, since the strain field has been characterized with much finer spatial resolution.

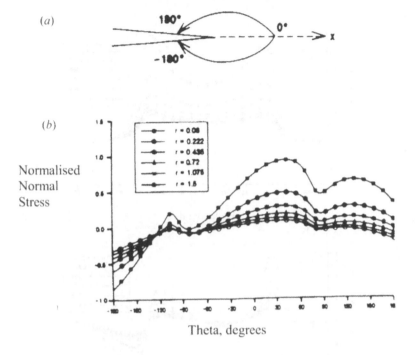

Figure 3.2.3.5. (*a*) Coordinate system for NSR at the crack-tip. (*b*) Normal stress as a function of angle and distance from the crack-tip for a ply-angle of 80°.

3.2.3.8   Discussion of the computational results

As the singular stress field at the crack-tip still obeys a $r^{1/2}$ dependence, it is evident that the stress levels increase rapidly close to the crack-tip. Also, from equation (1), it will be seen that the level of the NSR will rise close to the crack-tip. An inspection of equation (3.2.3.1) will show that the sensitivity of the NSR, i.e. its ability to predict the crack growth direction, it is also strongly dependent upon r. In fact, at any one value of Φ, the NSR value shows an r^{05} dependence. Figure 3.2.3.5 shows the dependence of the stress components (normal stress, transverse stress and shear stress) as a function of angle around the crack-tip, for different distances from the crack-tip. One interesting and important aspect of the NSR is that its actual prediction of the crack growth direction does not vary with the value of r—only the accuracy with which the prediction is made. This is an attractive feature of the NSR, since not all crack growth criteria show this behaviour. It will be seen, then, that the effect of the crack is only evident at small distances away from the crack (say, less than 0.4 mm). For a strong dependence upon, therefore, the value of r chosen for the evaluation of the NSR value was 0.08 mm. The value of the NSR for a range of fibre angles is shown in figure 3.2.3.6. The actual ply angles are shown marked on the ordinate of

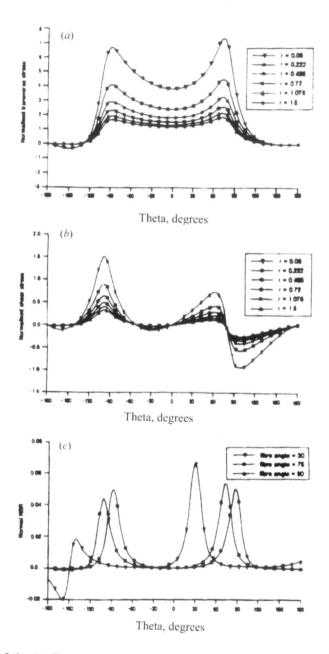

Figure 3.2.3.6. (*a*) Transverse stress as a function of angle θ and distance from the crack-tip for a ply-angle of 80°. (*b*) Normalized shear stress as a function of angle θ and distance from the crack-tip for a ply-angle of 80°. (*c*) NSR as a function of θ for different ply-angles at $r = 0.08$ mm from the crack-tip.

figure 3.2.3.6. It will be seen that, for this instance, the NSR is a good predictor of crack growth angle. The point to be made, though, is not whether or not the NSR is a valid criterion but that, in this study, its predictions are validated predictions. In more complex situations, where the NSR's lack of sensitivity to shear stress may become important, the NSR may be a less useful parameter. Although not all of the points are shown, the data for figure 3.2.3.6 were calculated at 2.5° intervals, so that the curve would be generated with precision.

3.2.3.9 Conclusion

A general methodology for evaluating crack growth criteria in orthotropic composite materials has been developed. The implementation involves the validation of a FE model, both by comparison with a pre-existing analytical solution for isotropic material and with experimentally determined deformation fields derived from moiré interferograms. It has been shown that the normal stress ratio may be evaluated in this manner and that the NSR is a useful and accurate predictor of crack growth direction in this specific orthotropic, unidirectional carbon/epoxy composite.

These data may now be used to model multi-layer plies and to predict crack growth in such complex systems. The basic routine may also be used to evaluate other crack growth criteria.

References

[1] Herakovich C T, Gregory M A and Beuth J L 1985 Crack growth direction in unidirectional off-axis graphite-epoxy *Euromech 182, Mechanical Characterizations of Load Bearing Fibre Composite Laminates (Brussels)* (New York: Elsevier)
[2] Ye L 1989 An energy release rate criterion for crack growth in unidirectional composite materials *Eng. Fracture Mech.* **32** 861–9
[3] Buckzck M B and Herakovich C T 1983 *Direction of crack growth in fibrous composites Mechanics of Composite Materials (AMD-58)* (New Jersey: American Society of Mechanical Engineers) pp 79–82
[4] Gregory M A and Herakovich C T 1986 Predicting crack growth direction in unidirectional composites *J. Composite Mater.* **20** 67–85
[5] Zang S Y 1989 Extending Tsai-Hill and Norris criterion to predict cracking direction in orthotropic materials *Int. J. Fracture* **40** R101–4
[6] Gdoutos E E and Zaharopolos D A 1989 Mixed mode crack growth in anistropic media *Eng. Fracture Mech.* **34** 337–46
[7] Hahn H T 1983 A mixed mode fracture criterion for composite materials *Composite Sci. Technol. Rev.* **1** 26–9
[8] Ellyin F and Elkadi H 1990 Prediction of crack growth direction in unidirectional composite laminae *Eng. Fracture Mech.* **36** 27–37
[9] McKelvie J 1986 On the limits of moiré interferometry *Proc. 1986 SEM Spring Conf. (New Orleans)* (Bethel, CT: SEM)

[10] Walker C A 1988 A review of moiré interferometry *Opt. Lasers Eng.* **8** 213–62

[11] 1984 *Ciba-Geigy Bonded Systems Data sheet FTP 49f (London)*

[12] Pagano N J and Halpin J C 1968 Influence of end constraint in the testing of anistropic bodies *J. Composite Mater.* **2** 18–31

[13] Hussain M A, Lorensen W E and Pfegel G 1976 Quarter-point quadratic isoparametric elements as a singular element for crack problems *TM-X-3428* (Santa Ana, CA: MSC Software) p 419

[14] Hibbit, Karlsson and Sorensen 1989 'ABAQUS Users' Manual, Version 4-7, Pawtucket, R1, USA

[15] Brown W F and Srawley J E 1966 Plain strain toughness testing of high strength materials *ASTM STP* **410** 10–18

Chapter 3.3.1

Experimental and computational assessment of the T^* fracture parameter during creep relaxation

C A Walker and Peter M MacKenzie
University of Strathclyde, Glasgow, UK

The limitations of experimental measurement and advantages of a hybrid approach to the problem

3.3.1.1   Introduction

While the use of the J-integral has become widespread in fracture mechanics, progress has been slow towards a new parameter that will be valid for conditions where J is, by definition, outside its area of relevance. New fracture parameters such as the T^*-integral have been developed [1] and investigated by means of computational models but the difficulties in carrying out valid experiments have meant that few studies have attempted to correlate theory, computation and experimental measurements [2, 3]. It is only by such means that confidence may be established and such new formalisms pass into routine use.

In a previous study the authors have shown how strain-field data from a crack growth experiment may be analysed to show that in reality J and T^* are indeed equal in situations of modest crack growth as would be expected [3]. The main conclusion from this work was a validated routine for assessing T^* from the strain field components around the crack-tip as visualized by moiré interferometry. The next phase in the investigation of T^* is to evaluate it in a situation where either extensive plasticity or crack growth occur or where time-dependent deformation takes place. This study is concerned with time-dependent deformation.

The particular situation is one of stress relaxation where the specimen is held on a fixed grip configuration at a load sufficient to cause creep at the test temperature (190 °C for the aluminium 7075 T6). In this experiment, the load steadily reduced as the specimen deformed. The crack in essence did not extend which meant that the data analysis was a simple matter compared to that for an extending crack.

3.3.1.2 T^* integral fracture parameter

T^* may be defined as a summation along the loading history of the incremental form ΔT^*, i.e.

$$\Delta T^* = \int_\Gamma [\Delta W \delta_{in} - (T_i + \Delta T_i)\Delta u_{i,l} - \Delta T_i - u_{i,l}] \, ds.$$

The path Γ should be taken as small as possible but still contain (and avoid) the process zone, where intense microcracking will take place and the strain-field parameters will not be well defined.

It can be shown that T^* may be calculated from the strain-field parameters alone, with a knowledge of the constitutive relations for the material. Moiré data may, therefore, be used to measure T^*.

3.3.1.3   Integration path

In the summation of ΔT^* along the loading history, the path Γ stays stationary in relation to the cracked tip, i.e. the Γ moves with the crack-tip.

This implies that data need to be recorded ahead of the crack-tip for use eventually as the crack propagates. This is the one feature of the grating method that is highly attractive, as the whole crack-tip field is recorded as a unity. Given that no crack growth was intended in this study, a fixed contour could be used (dimensions are given in figure 3.3.1.2).

3.3.1.4   Experimental procedure

The specimen material was a high strength aluminium alloy (7075 T6), in the shape of a compact tension fracture specimen (figure 3.3.1.3). The starter crack was fatigue precracked from a machined notch, although, in fact, due to the degree of creep deformation envisaged, this precaution was probably unnecessary. After precracking a 500 line/mm cross grating was cast on to the area around the crack-tip. An epoxy resin was used although at 190 °C, this was close to the upper operational limit for epoxy gratings. Perceptible degradation took place in the gratings over the period of the test (24 h).

A servo hydraulic test frame applied the initial load of 12.5 kN and, thereafter, the displacement was held constant. An oven with a low-expansion glass window of optical quality maintained the temperature of the test. A shroud to minimize convection currents surrounded the windows. With this system,

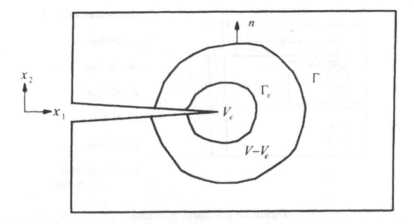

Figure 3.3.1.1. Integration contours for calculation of a path integral in a cracked body: n, outward normal; Γ, outer integration contour; Γ_ϵ, inner integration contour; V_ϵ, volume defined by Γ; $V - V_\epsilon$, volume defined by Γ_ϵ; and x_1, x_2, coordinate system.

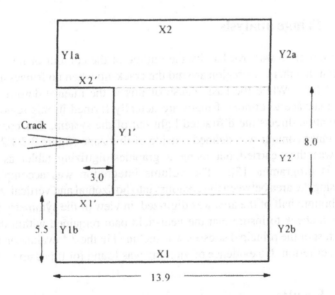

Figure 3.3.1.2. Path of integration (distances in mm).

it was found that high-quality interferograms could be recorded using a moiré interferometer that has been previously described [4]. This system is largely immune to ambient vibration: interferograms were recorded in three directions $(0°, 90°, 45°)$ so that a complete record of deformation was available for any point in the field for the times recorded ($t = 0, 3, 8, 15,, 23, 79, 283, 1318$ min).

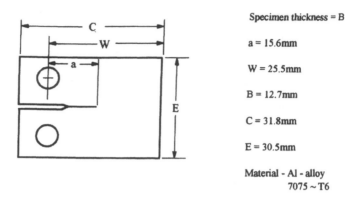

Specimen thickness = B

a = 15.6mm

W = 25.5mm

B = 12.7mm

C = 31.8mm

E = 30.5mm

Material - Al - alloy
7075 ~ T6

Figure 3.3.1.3. Specimen details.

(Sample interferograms are shown in figure 3.3.1.4). Over the period of the tests, the load relaxed to 5.3 kN, i.e. to less than 50% of its original level.

3.3.1.5 Fringe analysis

Each interferogram was overlain by the outline of the contour of integration. It will be seen that there is a region around the crack-tip when no fringes are visible (figure 3.3.1.4). While the basic cause of this is the intense deformation that causes the surface to dimple, fringes are actually formed in this region but the surface rotation directs the diffracted light out of the system. For convenience, then, the inner contour was defined outside of this zone (figure 3.3.1.2). Fringe analysis was then carried out using a graphics-digitizing tablet as input to an analysis programme [5]. The volume integration was accomplished by decomposing the area between the contours into horizontal and vertical stripes [6]. Only the bottom half of the area was digitized, in view of the symmetry about the crack. As a check to insure that the near-field path remained within the elastic regime, a test of the principal stresses was included in the T^* evaluation using the von Mises criterion. No evidence of yielding was found for times up to 23 min.

3.3.1.6 Results

The calculated values for $t = 0$–23 min are shown in table 3.3.1.1. Up to this point, the creep zone evolved close to the track tip. Beyond this time, the creep zone spread beyond the integration contour. It is possible to calculate T^* in this regime, if the creep rates are known as a function of time at each position and one is able to presume that the deformation is primarily creep. An interpolation of the creep rates from 79 to 1318 min showed unstable results. The fact remains that there are two simple situations—one where the creep zone is limited and the other

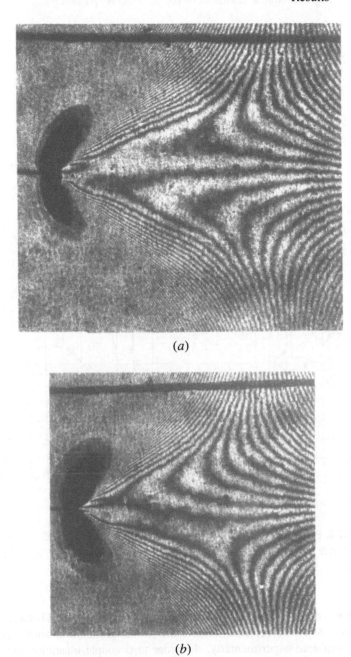

(*a*)

(*b*)

Figure 3.3.1.4. (*a*) Deformation contours around the crack-tip after 3 min (deformation contour interval—1.05 μm, linear magnification 8.5×). (*b*) Deformation contours around the crack-tip after 24 h (deformation contour interval—1.05 μm, linear magnification 8.5×).

Table 3.3.1.1.

Time (min)	0	3	8	15	23
Y_{1b}	5.90	1.90	−2.51	1.68	0.87
X_1	17.26	0.81	−0.10	0.14	−0.42
Y_2'	3.74	1.03	−0.41	0.64	0.05
Y_{2b}	9.53	−0.18	−0.83	1.48	3.30
ΔT_1	69.14	6.11	−7.39	7.40	7.52
$\Delta T2$	−28.31	1.77	−9.12	−19.87	−18.76
ΔT^*	97.45	4.34	1.73	27.28	26.29
T^*	97.45	101.79	103.52	130.80	157.09

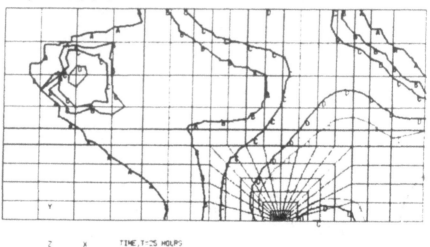

Figure 3.3.1.5. Predicted creep contours after 25 h: A, 0% creep strain; B, 0.005% creep strain; C, 0.01% creep strain; D, 0.05% creep strain; E, 0.1% creep strain; and F, 0.5% creep strain.

where creep is dominant throughout the specimen: in the transition between these two neither elastic not creep deformations are dominant and the strain field cannot be easily measured experimentally. In order to decouple plasticity and creep, one could unload the specimen periodically, removing the elastic component (which is time-dependent due to the reducing load and the spreading creep field). Alternatively, the moiré data could be used to validate an elastic–plastic finite-element model of the specimen. This option was adopted [7], and with the model, the spread of the creep zone (defined as a 0.1% creep strain) from a small area around the crack at 15 min to a zone extending to the uncracked border by 203 min was analysed and predicted (figure 3.3.1.5).

It should be noted that the results achievable with a combined approach are greater than with either of the techniques on its own, since the finite-element model gained its credibility from the comparison with the experimental data.

References

[1] Atluri S N and Nagaki M 1984 An incremental path independent integral in inelastic and dynamic fracture mechanics *Eng. Fracture Mech.* **20** 209–44

[2] Atluri S N *et al* 1996 *T*-Integral Analysis of Fracture Specimens, Proc. ESDA (Montpellier) (ASME 76)* (New Jersey: ASME) pp 89–94

[3] Walker C A and Mackenzie P M 1996 *ECF 11 (Poitiers)* (Warley: EMAS) pp 1957–64

[4] MacDonach *et al* 1983 *Exp. Techniques* **7** 20–4

[5] MacKenzie P M *et al* 1989 *Res. Mechanica* **28** 361–82

[6] Houssin S 1992 Experimental evaluation of the T* integral parameter under conditions of creep relaxation *MSc Thesis* University of Strathclyde, Glasgow

[7] Le Helloco M 1992 An evaluation of the T* integral under conditions of stress relaxation *MSc Thesis* University of Strathclyde, Glasgow

Chapter 3.3.2

Moiré interferometry and crack closure

University of Strathclyde, Glasgow, UK

3.3.2.1   Introduction

3.3.2.1.1   Crack closure

A characteristic of all cracks propagated by fatigue is that, to a greater or lesser degree, they display the phenomenon known as crack closure. Zones of localized plasticity are left behind in the wake of the propagating crack-tip. These 'stretched layers', as Elber [1] described them, act as props in compression supporting the surrounding bulk of elastic material. As the crack is loaded up, complete opening will be delayed by their presence: similarly, during the decreasing load phase, the surfaces come together early.

It is quite obvious that where there is delayed opening, the stress intensity conditions experienced by the material at the crack-tip may be quite different during a given applied stress range than would otherwise be the case. Since its history of stress intensity fluctuations determines the rate at which a given fatigue crack progresses, it follows that the presence of crack closure will have a substantial impact on the rate at which real cracks grow.

The closure effect is difficult to model. In a complex and nonlinear manner, it depends on factors such as crack size, fit between surfaces and the loading history which generated the 'stretched layers' in the first place.

3.3.2.1.2   Moiré interferometry

In laboratory investigations of crack closure, an important feature of cracked bodies, that their compliance changes markedly between the closed and open states, has been the measurement used by most observers of the phenomenon. The effect is typically identified by remote measurements of displacements using extensometers or nearer to the crack-tip using resistance strain gauges. It is the nature of these discrete measurements that they provide little with which we can assess crack-tip stress fields or to which we can cross reference contact effects. In essence, the information provided is incomplete and consequently difficult to interpret.

By contrast, moiré interferometry can be especially productive in enabling direct observation of much that interests us as we seek to gain practical information on closure behaviour.

The technique applied for the demonstration described here involved the use of the Strathclyde design of robust, compact, self-contained interferometer [2]. It is important to note some features of this approach. As a whole-field analysis tool, its principal characteristic was that it allowed the observations to be conducted in a conventional materials test laboratory set-up: specimens could be examined *in situ* on standard test machines without exceptional regard for vibration isolation of the optical layout. Displacement fields at 0°, 45° and 90° to the datum axis and 50 mm in diameter were viewed and recorded using conventional 35 mm SLR photography. One of the penalties for this convenience of use is that the set-up has a fixed sensitivity (1/950 mm per fringe at 0° and 90°, 1/1344 mm per fringe for the 45° field): the system depended on the use of thin, high-efficiency specimen gratings which were cast onto the specimen surface by way of a replication technique. The combination of interferometer and grating limited light losses sufficiently to allow exposure times down to 1/1000 s.

3.3.2.2 Example applications

3.3.2.2.1 Specimens

The aim here is to demonstrate some of the potential of the moiré approach in relation to crack closure. To this end, sample results taken from studies involving quite different materials are discussed, one being a high-yield strength titanium alloy (IMI 685) and the other, a fairly ductile structural steel (BS 4360-50D). Two specimen configurations were represented, CTS for the titanium, and three-point bend SEN for the steel. Mechanical properties for each material are given in table 3.3.2.1 and the specimen dimensions are shown in figures 3.3.2.1 and 3.3.2.2.

3.3.2.2.2 Controlled crack development

Different procedures were applied to the control of fatigue crack growth for the two materials. The titanium was fatigued following requirements of ASTM E-813 [3], or, in other words, it was fatigue pre-cracked as for a standard J-integral fracture specimen. The loading regime for the steel was more tightly controlled to give near-constant values of R and ΔK throughout over 15 mm of crack growth, the aim being to obtain consistent 'stretched layers' along the crack surfaces. This was achieved by a process of progressive load reduction to obtain a constant ΔK

Table 3.3.2.1. Mechanical properties of test materials

Material	0.1% Proof stress σ_y, (MN m^{-2})	Young's modulus E, (GN m^{-2})	Poisson's ratio ν
Titanium alloy IMI 685	985	120	0.32
Structural steel BS 4360:50D	440	208	0.28

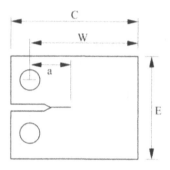

Specimen thickness = B

Specimen material	a	W	B	C	E
IMI 685	5.4	25.5	12.7	31.8	30.5

Figure 3.3.2.1.

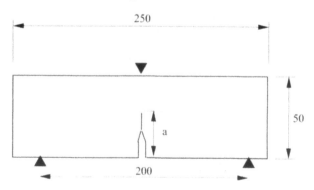

Specimen thickness = 15.43
Specimen : BS 4360 : 500 steel

Figure 3.3.2.2.

as the crack extended, the loads being derived from elastic–plastic calculations of K using a finite-width Dugdale model [4]. In the case of the steel specimen used in this work, a fatigue crack was extended 17.7 mm into the specimen from the starter notch by cycling at a ΔK of 30 MN m$^{-3/2}$ with an R ratio of only 0.01.

3.3.2.2.3 Grating limitations

As previously indicated, gratings, of epoxy casting resin, were applied using a replication process [2] in the areas of interest. Note that, for both materials, this was done *after* the fatigue cracking process. The cast grating is an integral part of the high-efficiency optics on which the flexibility of the Strathclyde system depends but one of its drawbacks in this particular application is that the grating material also experiences the effects of crack closure during the fatigue cracking process. A narrow zone of stretched grating material along the crack becomes damaged and locally debonded by being repeatedly collided against the opposite face during cyclic loading. Although this zone is small (generally < 1 mm), interferometric information within it is degraded to the point of being useless and its exact extent is difficult to define. Conversely, valuable data can be extracted from this region by applying a pristine grating for interrogation during monotonic loading, post-fatiguing and this is described later.

3.3.2.2.4 Crack closure effects revealed in moiré interferograms

Three important measures of the effects of closure can be extracted from the in-plane displacement fields. It should be noted that each method is quite independent of the others, using essentially separate zones of the displacement fields, as we will see. The usual important caveat applies: in common with virtually all strain measurement techniques, moiré interferometry does not allow us to observe directly the behaviour in the interior of the specimen and this must be inferred.

3.3.2.2.4.1 *Crack face separation profiles*

Direct approach

Even for those with only a little experience of analysing moiré displacement fields, the pattern produced by the component normal to the crack surface is easy to comprehend. (In the system used here, this is the 90° field.) A sequence of displacement contours, traced from the original interferograms for clarity, for increasing load is shown in figure 3.3.2.3, the specimens being one of the steel SENs.

The dominant fringe pattern is caused by gross rotation on either side of the crack. Pure rotation alone would produce vertically oriented fringe lines: normal strain results in horizontal fringes and, very clearly, there is an element of such in all the diagrams in figure 3.3.2.3, as the fringes are slightly inclined.

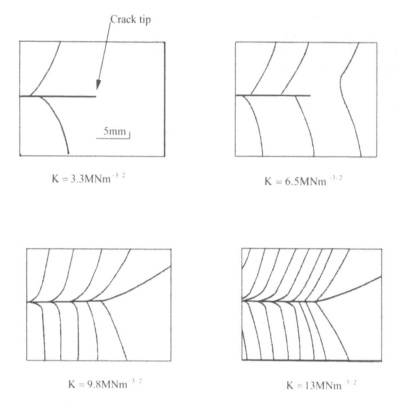

Figure 3.3.2.3. Direct interferograms showing crack-face release, steel SEN specimen.

Of special note, for our purpose here, is the strong horizontal distortion that occurs progressively at the crack border, indicating a significant local strain in this region. Now remembering that the field we are looking at is for the component of displacement normal to the crack surface, we can reasonably believe that this strain is due to the liberation of the local compressive contact between the faces of the crack. By numbering the fringes, we can see that the sense of the strain ties in with this.

It is interesting to see the development of crack face release as this sequence progresses. Note that at 13 MN m$^{-3/2}$, the distinctive kink in the rotational fringes is absent as we approach the crack-tip along the crack boundaries, indicating that there is still no normal strain and that the compressive contact is yet to be released.

Mismatch technique

A drawback to the 'direct approach' in which the raw contours of displacement are observed and recorded occurs when fringes are sparse. One of the few dependable methods of interpolation involves introducing a mismatch in either frequency or rotation between the reference grid and the specimen grid [5]. In either case, the effect is to introduce a controlled density of uniformly spaced fringes within the undeformed field. These act as 'carrier' fringes in that the initial uniform pattern is 'frequency modulated' by the phase information generated when the underlying fringe pattern is, for example, strained. With mismatch, fringe intensity maxima and minima can be readily identified in the subsequent analysis and the data corresponding to the undeformed stage are then subtracted. The use of mismatch does not increase the basic sensitivity or accuracy of the system but it does improve the spatial resolution and is, hence, useful at the lower end of the measurement range. Indeed, the accuracy is degraded since a differencing process is employed.

Frequency mismatch can be achieved quite simply by altering the angle of intersection of the interferometer beam pair in use by means of a micrometre adjuster on the mount of one of the output beam mirrors. *Rotational* mismatch is imparted by relative rotation of the optics, the entire interferometer in our case, in the plane of the specimen grid: again, a precision adjuster is built into the system for this purpose.

For the present application, the aim was to improve the spatial resolution of measurements of displacement perpendicular to the crack surface, at points along the crack. This was achieved by introducing rotational mismatch between the virtual grating produced by the intersecting beams from the interferometer and the physical grating bonded to the specimen. Again using the steel SEN specimen, a typical interferogram is shown in figure 3.3.2.4. The results of the analysis are given in figures 3.3.2.5 and 3.3.2.6 in two scales of increasing intensity. The accuracy in crack face separation measurements was estimated to be ± 1 μm and the positional accuracy of the measurements along the crack length was ± 0.1 mm. The graphs suggest that the crack faces begin to separate near to the mouth end at 5 MN m$^{-3/2}$. However, the characteristic elastic profile of an open crack, consisting of straight flanks and elliptical transitions to the tip, does not appear until about 20 MN m$^{-3/2}$. This implies the crack was opening gradually from about 5 MN m$^{-3/2}$ upwards.

(It is important to note that whether the 'direct' or the 'mismatch' approach is employed, the sharp definition of crack face release can only be observed, with the system described here, when the grid material has been fixed *after* fatigue cracking: if the grid had been fixed before cracking, the local elastic effects would have been swamped by crack border plasticity and out-of-plane 'dimpling' movements of the grid.)

3.3.2.2.4.2   *Crack-tip stress intensity evaluation (displacement method)*

Nicolletto *et al* [6] have shown that it is possible to use the relation between the stress intensity factor, K, and the displacements in the region $r \ll a$ to analyse fringe data for this purpose. The most convenient displacement data to use were those in the component parallel to the crack (i.e. in the 0° field in our system),

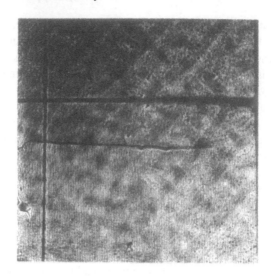

Figure 3.3.2.4. Mismatch interferogram steel specimen.

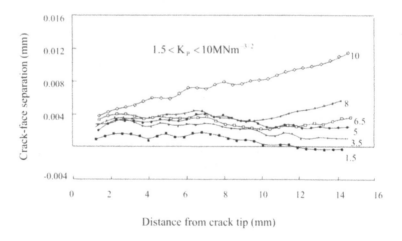

Figure 3.3.2.5. Crack-face opening displacements evaluated using mismatch specimen C50D-2.

taken at various points along a line radiating from the crack-tip at right angles to the crack plane. The stress intensity factor 'K_u' is given for plane strain by

$$K_u = \lim_{r \to 0} \frac{2\sqrt{\pi} E u_x}{(3 - 4v)(1 + \gamma)\sqrt{r}}.$$

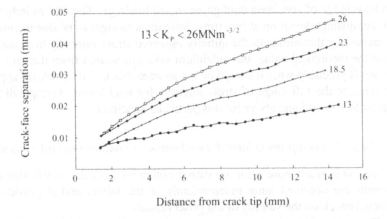

Figure 3.3.2.6. Crack-face opening displacements evaluated using mismatch specimen C50D-2.

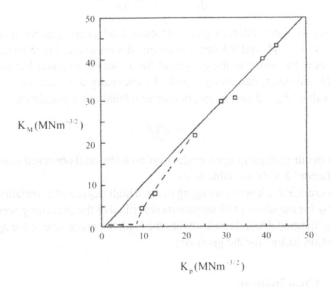

Figure 3.3.2.7. Results of moiré fringe analysis (Ti 685). Comparison of moiré evaluation K_M with theoretical stress intensity K_p.

The limiting value is determined in practice by plotting K as a function of $r^{1/2}$ and extrapolating to the intercept at $r = 0$ to obtain the observed value 'K_M'. The extrapolation must be made from a region that is unambiguously linear elastic.

Here, we employed one of the high-strength titanium CTS specimens. Figure 3.3.2.7 shows the correlation of the observed values with the theoretical stress intensity factor 'K_P' (calculated, on the assumption of an 'open' crack, from knowledge of specimen configuration and loading). The dashes indicate a small but distinct deviation at low stress intensities thought to be due to closure. It is interesting that although the initially observed stress intensity increase lags behind the theoretical value, the maximum value, measured from the zero-load baseline, is precisely what it should be for an open crack. The crack-tip appears to experience the full range of stress intensity for load levels corresponding to about 20 MN m$^{-3/2}$ and above for this particular specimen.

3.3.2.2.4.3   *Crack-tip stress intensity evaluation (J contour integral method)*

This process is much more laborious than the other methods described. However, its results are obtained quite independently of the others and it provides an additional check on the validity of our observations.

The J-integral [7] is given by

$$J = \int_\Gamma W \, \mathrm{d}y - T_i \frac{\partial u_i}{\partial x} \, \mathrm{d}s.$$

Using a set of three interferograms at each load point (giving displacement components at 0, 45 and 90 degrees to our datum axis), MacKenzie *et al* [8] described how the terms in the integrand for J can be obtained for any location in the field and integrated along a path, Γ, enclosing the crack-tip to obtain an observed value 'J_M'. J and K are, of course related quite simply by

$$K = \sqrt{(JE)}.$$

Hence, from our contour integral evaluation, an additional observed value of stress intensity factor 'K_J' is available to us.

In figure 3.3.2.8 it is encouraging to see a high degree of correlation between K_J and K_M for the series of observations described in the preceding section. Note that neither method depends on knowing the load, the crack size or the appropriate 'configuration factor' for the geometry.

3.3.2.3   Conclusions

Using moiré interferometry, it is possible to detect the effects of crack closure on stress intensity with substantial sensitivity.

Crack face contact and opening (at least, at the specimen surface) can be readily detected by direct examination of the fringes corresponding to the component of displacement normal to the crack surface. Accurate measurements of crack profiles can be made simply in most cases by reading the fringe order at the points of intersection of the fringes and the crack borders. At low loads, spatial resolution can be enhanced by using rotational mismatch. The grid should be fixed to the specimen *after* fatigue cracking for either approach.

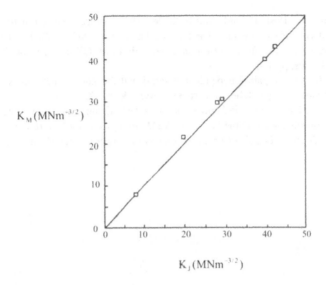

Figure 3.3.2.8. Results of moiré fringe analysis (Ti 685). Comparison of K_M derived from U_X with K_J from contour evaluation.

The stress intensity factor can be evaluated by analysis of the displacement fields in the vicinity of the crack. Both techniques used here, the *displacement* method and the *J-Integral* method, appear to be equally sensitive for essentially linear elastic cases, the former being much simpler to apply. There is initially a distinct lag between the stress intensities measured using both approaches and the theoretical value: this appears to be due to closure. At higher loads, the measured values are precisely in agreement with those predicted for a fully open crack.

References

[1] Elber W 1971 *The Significance of Fatigue Crack Closure (ASTM STP 486)* (Conshohocken, PA: American Society for Testing and Materials) pp 230–47
[2] McKelvie J, Walker C A and MacKenzie P M 1987 A workaday moiré interferometer: conceptual and design considerations, operation, applications, variations, limitations *Proc. SPIE, Int. Conf. on Photomechanics and Speckle Metrology (San Diego, CA)* vol 2 (Bellingham, WA: SPIE) 464–75
[3] E-813 1981 *Standard Test Method for J_Ic, a Measure of Fracture Toughness* (Conshohocken, PA: American Society for Testing and Materials)
[4] Gray T G F 1977 A closed form approach to the assessment of practical crack propagation problems *Fracture Mechanics in Engineering Practice* ed P Stanley (London: Applied Science) pp 123–47
[5] McKelvie J 1986 On the limits to the information available from a moiré fringe pattern *Proc. SEM Spring Conf. on Experimental Mechanics (New Orleans, LA)* (Hartford, CT: Society for Experimental Mechanics) pp 971–80

[6] Nicoletto G, Post D and Smith C W 1982 Moiré interferometry for high-sensitivity measurements in fracture mechanics *Proc. SESA, JSME Spring Conf. on Experimental Mechanics (Honolulu, HI)* (Hartford, CT: Society for Experimental Mechanics) pp 1–5

[7] Rice J R 1968 A path-independent integral and the approximate analysis of strain concentration by notches and cracks *J. Appl. Mech.* **35** 379–86

[8] MacKenzie P M, McKelvie J and Walker C A 1986 Evaluation of the J-Integral using a high-sensitivity moiré method *Proc. SEM Spring Conf. on Experimental Mechanics (New Orleans, LA)* (Hartford, CT: Society for Experimental Mechanics) pp 773–80

CHAPTER 4

ELECTRONIC PACKAGING

Chapter 4.1

Determination of effective coefficients of thermal expansion (CTE) of electronic packaging components

Yifan Guo
Motorola Semiconductor Products Sector, Tempe, USA

4.1.1 Introduction

Applications of moiré interferometry are introduced for the determination of effective CTEs (coefficients of thermal expansion). In electronic packaging, materials and components with very different properties are assembled at elevated temperatures and tested by temperature cycles. CTE mismatch in a packaging system can cause significant thermal strains and consequent fatigue failures in the interconnections during the assembly processes and the temperature cycle tests. It can also cause poor dimensional stability and difficulties in assembly processes [1–3]. It is crucial, therefore, to determine the CTE values in the electronic package designs and reliability predictions accurately. In order to determine the CTE values of electronic components and their variation from region to region, thermal deformations of samples have to be measured with high sensitivity and resolution. As a whole-field technique, moiré interferometry is an ideal method. Global and local CTE values can be determined from the whole-field thermal deformations provided by the moiré interferometry technique.

4.1.2 Definitions of effective CTE

In an isotropic and homogenous material, the CTE value is determined by

$$\alpha = \frac{\Delta L}{L} \frac{1}{\Delta T} \tag{4.1.1}$$

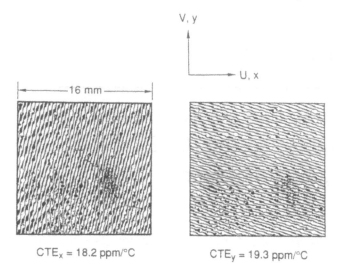

CTE$_x$ = 18.2 ppm/°C   CTE$_y$ = 19.3 ppm/°C

Figure 4.1.1. U and V displacement fields of a printed circuit board, induced by thermal loading of $\Delta T = -60\,°C$.

where α is the CTE value, ΔL is the thermal displacement, L the gauge length and ΔT the temperature change. The thermal displacement ΔL is measured by the change in fringe order, ΔN, between two points on the specimen, where

$$\Delta L = \frac{1}{f}\Delta N \qquad (4.1.2)$$

The reference grating frequency was $f = 2400$ lines/mm, whereby $1/f = 0.417\ \mu m$. Typically, ΔN is measured to within one-quarter of a fringe order, which means that ΔL is resolved to within about $0.1\ \mu m$.

Many inhomgeneous materials such as particle and fibre reinforced materials are used in electronic packaging. Very often, the thermal deformations of the constituents are unequal and the local deformations of the composite materials are not uniform. In these cases, the CTE is defined as the average thermal mechanical strain expressed in a normalized unit, ppm/°C, which is the same unit used for CTE in homogenous materials. In the following discussions, the term CTE is also used for characterizing the average thermal expansions of electronic components and elements that are inhomogeneous or composite materials. Figure 4.1.1 is an example, where the zigzag pattern is not optical noise but real data representing displacements in the glass fibres and epoxy matrix of a printed circuit board.

4.1.3 Procedures of CTE determination

4.1.3.1 High-temperature replication of specimen gratings

In this method, the specimen and the grating mould are heated to a uniform elevated temperature, the adhesive is applied and the grating is replicated on the specimen at the elevated temperature. The grating mould is removed after the adhesive is cured, then the specimen is cooled to room temperature and analysed in a moiré interferometry [4–6]. The specimen grating formed at the elevated temperature T_1 deforms with the specimen when the temperature decreases to T_2. The changes in the specimen grating represent the thermal deformation of the sample during the temperature change $\Delta T = T_2 - T_1$ and the CTE values are calculated by equation (4.1.1).

Different materials have been used to replicate the specimen gratings according to the different measurement requirements. High-temperature curing epoxies are the most common materials. However, when the samples are thin and soft, the mechanical constraint from the epoxy could affect the true material properties of the testing samples. In this case, the specimen gratings have to be made from very soft materials in order to reduce the influence of the grating stiffness. Very thin silicon rubber (RTV) gratings, which have a very low Young's modulus, can be used effectively on thin and soft materials.

4.1.3.2 Real-time measurements of the thermal expansion

In this method, a specimen grating is applied at room temperature and the specimen is placed in an oven that has a glass window. The temperature of the oven is controlled systematically and a thermocouple mounted on the sample is monitored during the measurements . The thermal deformations and the sample temperatures are recorded manually or automatically. Moiré fringe patterns are recorded at desired temperature levels, providing data for CTE determination. A curve of the thermal displacement *versus* temperature can be generated during the test. The CTE value as a function of the temperature can be calculated from the displacement–temperature curve.

4.1.4 Application I: evaluation of thin small outline packages

A relative assessment of solder joint reliability was made for thin small outline packages (TSOP), designed for the same electronic function, but supplied by different manufacturers. It was a quick and inexpensive guideline for the selection of the package with the most reliable solder joints when subjected to accelerated thermal cycle (ATC) tests. The relevant dimensions of the packages are summarized in table 4.1.1. In the package, an active chip is bonded to the lead frame's central platform and connected to the radiating beams of the frame with fine-diameter gold wires. The assembly is then overmoulded to form a final

Table 4.1.1. TSOP dimensions and CTE values determined from moiré experiments.

Packages	Package dimensions (mm)		Chip dimensions (mm)		Average CTE (ppm/°C)	
	length	width	length	width	α_x	α_y
Specimen I	18.61	8.21	14.1	5.59	6.9	9.3
Specimen II	18.36	8.08	5.59	3.78	12.2	13.8
Specimen III	18.4	8.07	14.73	5.79	7.2	9.0
Specimen IV	18.41	7.96	14.32	5.59	9.1	15.7

package. Two different lead frame materials were used in the packages: Alloy 42 for specimen I, III, IV and copper alloy for specimen II.

The CTE of the TSOP packages and a printed circuit board (PCB) fabricated for the packages were measured separately and the information was used to estimate the maximum horizontal relative displacements between each package and the PCB. In the experiment, a thin specimen grating (typically, 10–20 μm thickness) was replicated on the bottom surface of the packages at 82 °C by using a high-temperature curing epoxy (F320, Tri-Con Inc) and the deformation was documented at room temperature ($\Delta T = -60$ °C). Figure 4.1.1 shows the U and V displacement fields for the PCB. The CTEs, in the x and y directions were determined from the fringe patterns by using equations (4.1.1): they were $\alpha_x^{PCB} = 18.2$ ppm/°C and $\alpha_y^{PCB} = 19.3$ ppm/°C.

The packages listed in table 4.1.1 were tested under the same loading condition. The U and V displacement fields for the four packages are shown in figure 4.1.2. The effect of a silicon chip is evident as sparse fringes in the central regions of the fringe patterns. The average values of the CTEs in the x and y directions of each package were determined from the fringe patterns and the results are also summarized in table 4.1.1. It is important to note that, whereas the overall dimensions of the packages are nearly the same, the size of the chip of specimen II is significantly smaller than that of the other three, which results in the high average CTE of specimen II. It is also interesting to note that in spite of having nearly the same size of chip and the same lead frame material as in specimen I, III and IV, specimen IV has the highest CTE, especially in the y direction. This was ascribed to the higher CTE moulding compound used in specimen IV.

If the TSOP lies freely on the PCB, the relative horizontal displacement, U_{max}^{rel}, per unit temperature change (°C) between the package and PCB in the diagonal direction can be expressed as:

$$U_{max}^{rel}(\mu m \, °C^{-1}) = \sqrt{\left(\Delta CTE_x \frac{L}{2}\right)^2 + \left(\Delta CTE_y \frac{W}{2}\right)^2} \qquad (4.1.3)$$

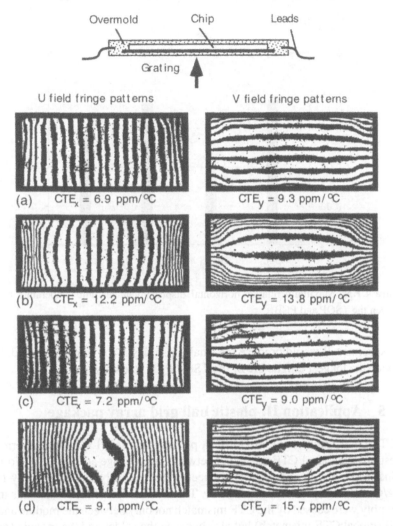

Figure 4.1.2. U and V displacement fields of TSOP specimens, induced by thermal loading of $\Delta T = 60\,^\circ\text{C}$: (*a*) specimen I, (*b*) specimen II, (*c*) specimen III and (*d*) specimen IV.

where L and W are the length and width of the module in metres, respectively, and ΔCTE is in ppm/°C. The magnitude of $U_{\text{max}}^{\text{rel}}$ was calculated from the moiré results and compared in figure 4.1.3. Specimen II has the minimum value of $U_{\text{max}}^{\text{rel}}$, as expected from the largest CTE. The magnitude of the actual relative horizontal displacement of the TSOP/PCB assembly is always smaller than $U_{\text{max}}^{\text{rel}}$ because of the mutual restraint [7–9]. Consequently, these results led to the forecast

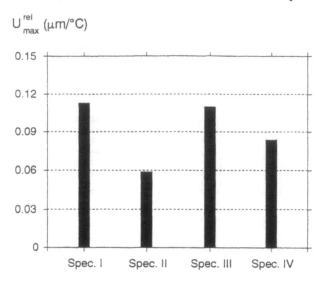

Figure 4.1.3. Maximum relative horizontal displacement per unit temperature change between the TSOP and PCB.

that specimen II would produce the smallest strains at the solder joint and, thus, provide the longest fatigue life of the TSOP/PCB assembly.

4.1.5   Application II: plastic ball grid array package

A CTE distribution of a 255I/O PBGA package, figure 4.1.4(*a*), was determined to estimate the local CTE mismatch between the package and the PCB. To form a final assembly, the package is connected mechanically and electrically to an FR4/glass PCB by the solder ball array. The deformation of the solder balls in the assembly were caused by the CTE mismatch not only between the module and the PCB (global CTE mismatch) but also between the solder and the module (or the PCB) at the interfaces (local CTE mismatch) [6, 7]. Consequently, determination of both global and local CTE was required to assess the solder joint reliability of the BGA package assembly accurately.

In the experiment, the specimen grating was replicated at 82 °C on the bottom surface of the basic package shown in figure 4.1.4(*a*). The deformation was measured at room temperature ($\Delta T = -60$ °C).

The fringe pattern shown in figure 4.1.4(*b*) represents the U displacement field of the package. The CTE distribution along the centre line AA' is determined from the pattern and plotted in figure 4.1.4(*c*). As expected for this thin PCB substrate (0.61 mm), the chip area had a much smaller CTE than the average CTE of the entire module. For the complete PBGA package assembly, the local CTE mismatch under the chip area dominates over the global CTE mismatch and,

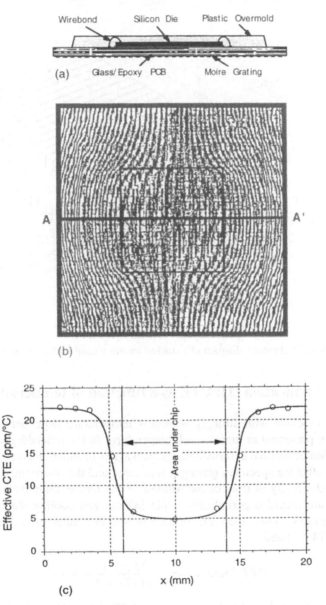

Figure 4.1.4. (*a*) Schematic diagram of a PBGA package, (*b*) *U* displacement field of the bottom for the package, induced by thermal loading of $\Delta T = -60\,^{\circ}\text{C}$ and (*c*) CTE distribution along the centre line of the package, AA′.

consequently, the maximum strain occurs at the solder ball located near the edge of the chip [7, 10].

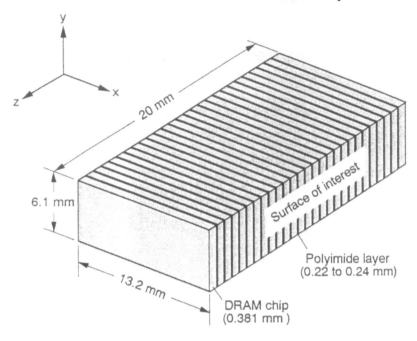

Figure 4.1.5. Schematic diagram of a stacked memory cube with relevant dimensions.

4.1.6 Application III: CTE as a function of temperature

For the packages with strongly nonlinear temperature dependency, the CTE values are presented as functions of temperature. In the procedure for real-time observation of thermal deformation, the package is installed in an environmental chamber after the specimen grating is replicated and the moiré interferometer is positioned directly in front of the window of the temperature chamber. Fringe patterns are recorded periodically while heating and cooling the specimen at moderate rates. Then, the CTE for each temperature interval is determined by equation (4.1.1) and

$$CTE \, (\text{ppm} \, ^{\circ}C) = \frac{N_2 - N_1}{f \Delta L (T_2 - T_1)} \times 10^6. \qquad (4.1.4)$$

The value obtained is taken to represent the CTE at $T = (T_2 - T_1)/2$.

The CTE of a stacked memory cube was determined by this procedure. The memory cube offers high performance within a small space and it is used in small portable computers as well as large system computers. The package is illustrated in figure 4.1.5 with relevant dimensions. It is an assembly of successive layers of memory chips, metal conductors and polyimide insulators, cemented together. The polyimide material has a very high CTE but a very low modulus compared to the adjacent silicon chips. In spite of its relatively high volume fraction

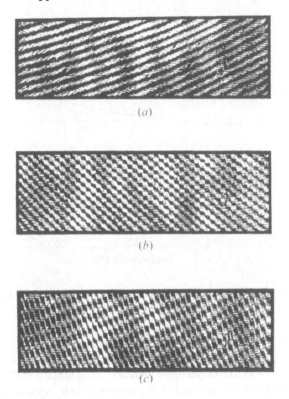

(*a*)

(*b*)

(*c*)

Figure 4.1.6. *Z* displacement fields of a memory cube at (*a*) $T = 110\,°C$, (*b*) $T = 20\,°C$ and (*c*) $T = -20\,°C$.

(approximately 38%), the polyimide layer has very small effect on the CTE of the cube in the x and y direction, because of its low modulus [11]. However, the polyimide layer expands or contracts more freely in the z direction, which results in a much higher CTE in the z direction. The whole-field measurement was required because of the multi-layer composite nature of the cube. In addition, the real-time measurement was required because of the temperature-dependent CTE of the polyimide material.

In the experiment, a specimen grating was replicated on the side of the stack at $80\,°C$. The specimen was installed in an environmental chamber and it was heated to the grating replication temperature. The moiré interferometer was adjusted to produce the minimum number of fringes at this temperature. The specimen was then heated to the maximum temperature of the thermal cycle ($140\,°C$), and the deformation in the z direction was documented periodically during the cooling cycle to $-20\,°C$. Figure 4.1.6 shows the z direction fringe patterns at (*a*) $110\,°C$, (*b*) $20\,°C$, and (*c*) $-20\,°C$. Again, the zigzag fringes in the patterns are caused by the laminate nature of the cube. The CTE was calculated

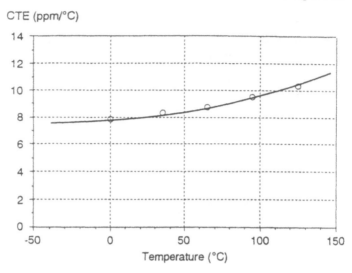

CTE (ppm/°C)

Figure 4.1.7. CTE of a stacked memory cube as a function of temperature.

as a function of temperature ad the results are plotted in figure 4.1.7. As expected from the temperature-dependent CTE of the polyimide layer, the CTE of the cube gradually increases as the temperature increases. The highest CTE at the elevated temperature was greater than that at room temperature by approximately 30%.

4.1.7   Conclusions

Moiré interferometry was employed to determine the effective CTE of electronic packaging components. With this method, the overall CTE as well as the local CTE were determined across the entire component with high accuracy. The method was compatible with a broad range of components. Two experimental procedures were employed: (1) CTE in a fixed temperature range and (2) CTE as a function of temperature. The procedures were implemented for various TSOPs a PBGA package and a stacked memory cube. The results emphasized the importance of a whole-field measurement technique. The measurement accuracy was approximately 0.1 μm.

References

[1] Han B and Guo Y 1996 Determination of an effective coefficient of thermal expansion of electronic packaging components: a whole field approach *IEEE Transa. Components, Packaging and Manufacturing Technol.* A **19**

[2] Guo Y 1995 Experimental determinations of effective coefficients of thermal expansion in electronic materials and components EEP vol 10-2 *Advances in Electronic Packaging (ASME)*

[3] Norris K C and Landzberg A H 1969 Reliability of controlled collapse interconnections *IBM J. Res. Development* **13** 266–71

[4] Guo Y, Ifju P, Boeman R and Dai F 1997 Formation of specimen gratings for moiré interferometry applications *Post Conf. Proc. 1997 SEM Spring Conf. on Experimental and Applied Mechanics* (Bethel, CT: SEM)

[5] Post D and Wood J 1989 Determination of thermal strains by moiré interferometry *Exp. Mech.* **29** 318–22

[6] Guo Y, Lim C K, Chen W T and Woychik C G 1993 Solder ball connect (SBC) assemblies under thermal loading: I, deformation measurement via moiré interferometry, and its interpretation *IBM J. Res. Development* **37** 635–47

[7] Han B and Guo Y 1995 Thermal deformation analysis of various electronic packaging product by moiré and microscopic moiré interferometry *J. Electron. Packaging, Trans. ASME* **117** 185–91

[8] Lau J W, Golwalkar S, Erasmus S, Surratt R and Boysan P 1992 Experimental and analytical studies of 28 pin thin small outline package (TSOP) solder joint reliability *J. Electron. Packaging, Transaction of ASME* **114** 169–76

[9] Hines L L 1987 SOT-23 surface mount attachment reliability study *Proc. 7th Ann. Int. Electron. Packaging Conf. (IEPS)* (Washington, DC: IMAP) pp 613–29

[10] Ejim T I, Clech J-P, Dudderar T D, Guo Y and Han B 1995 Measurement of thermomechanical deformations in plastic ball grid array electronic assemblies using moiré interferometry *Proc. 1995 SEM Spring Conf. Experimental Mechanics* (Bethel, CT: SEM) pp 121–5

[11] Wu T Y, Guo Y and Chen W T 1993 Thermal-mechanical strain characterization for printed wiring boards *IBM J. Res. Development* **37** 621–34

Chapter 4.2

Determinations of thermal strains in solder joints: an application in electric packaging

Yifan Guo
Motorola Semiconductor Products Sector, Tempe, USA

4.2.1 Introduction

Mechanical analysis in electronic packaging has a significant impact on the system reliability of an electronic device, and therefore, is critical in packaging design and manufacturing. The main task of mechanical analyses is to study the thermal strain/stress, interfacial strength/failures, moisture effects and the fatigue life of components and interconnections, such that the reliability of a package satisfies the manufacturing process requirements and the life expectancy under application conditions.

Electronic packages are complex material systems operating under thermal and mechanical loading conditions. Many components are organic materials which have highly nonlinear properties and are process and scale dependent. The properties of some materials are also temperate and strain rate dependent. Because of the developments and economics in semiconductor process technologies, the dimensions of electronic devices and interconnections are becoming smaller and smaller. Electronic components usually consist of multi-materials with bimaterial interfaces and are highly compact. The strain/stress concentrations in the structures are frequently large and localized in very tiny zones [1–3].

Conventional experimental tools such as strain gauges and extensometers can only measure point displacements and are not capable of determining deformations and strain distributions at the interfaces and in non-uniform media. Experimental techniques with high sensitivity and resolution are required to characterize the material properties and measure the local strains and stresses. Recently, optical techniques for *in situ* and whole-field measurements have become more mature and available [4–7]. These optical techniques have

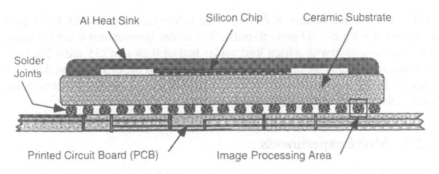

Figure 4.2.1. A schematic diagram of the ceramic ball grid array (CBGA) package.

been used for material property determinations, experimental stress–strain analysis, on-line inspections, simulation validations and hybrid methods in many packaging areas [8–11]. The information obtained is important in assessing the dependence of the system reliability on various structural and material configurations. Together with other analytical tools, the structural mechanics analysis is conducted at the design and qualification stage, so that the reliability can be estimated and optimized before the packages are put into production. The implementation of these advanced techniques has led to increased efficiency, improved reliability and reduced design cycle time, all of which play a critical role in determining the competitiveness of electronic products. In this section, the application of moiré interferometry to the reliability of solder joints is introduced.

4.2.2   CBGA packages

Ceramic Ball Grid Array (CBGA) is an area array surface mount technology used for high density and high performance packages [12, 13]. The mechanical reliability of the BGA (Ball Grid Array) solder joints is the key factor in determining the reliability of the whole system. The major concern regarding the solder joint reliability is the solder fatigue during system operation. Since the solder fatigue life is related to the thermal-mechanical strains, the mechanical reliability analysis of the solder joints can be conducted through the analysis of the thermal-mechanical strains. The application in the thermal strain measurements of a CBGA package is a typical example of using moiré interferometry for reliability analysis.

Figure 4.2.1 shows a typical CBGA package. In this package, a flip-ship is attached to a ceramic substrate which is then connected to a printed circuit board (PCB) by a grid array of solder joints. The ceramic substrate has a much lower effective coefficient of thermal expansion (CTE) than the printed circuit board. The CTE mismatch in the package induces thermal strain and stress in the solder joints during thermal cycles and causes solder fatigue failures. The dimensions of the substrate are 25 mm × 25 mm. The solder balls are in a 19 by 19 grid array with a pitch of 1.27 mm (50 mils). The solder interconnect is composed of a two-phase solder with a high lead solder ball of 0.89 mm (35 mils) diameter and eutectic solder at the interfaces of the substrate and the PCB. Due to the small scale of the local joint structure, it is a great challenge to determine the accurate strain values in the solder joints experimentally.

4.2.3   Moiré experiments

In thermal strain measurements using moiré interferometry, thermal loading can be applied by placing a uniform and known frequency grating (undeformed grating) on the specimen at an elevated temperature and measuring the deformed grating at room temperature. In order to have an undeformed grating at the elevated temperature, a grating mould with a ULE (ultra low expansion) glass substrate is used. The CTE of the ULE substrate is about 1×10^{-8} in the temperature range 0–100 °C. It does not experience any significant dimension change during the temperature change. In order to minimize the grating thickness, a very flat specimen cross section should be prepared. In the process, a thin and uniform layer of epoxy is applied to the ULE grating mould at an elevated temperature. The specimen and the ULE grating are then pressed together and maintained at the elevated temperature until the epoxy is cured. By separating the specimen from the ULE grating mould, a phase grating mould is transferred to the specimen surface. The specimen is then cooled to room temperature and measurements are conducted.

In this experiment, a specimen grating was replicated on the specimen cross section at 82 °C. After the epoxy cured, the specimen was cooled to room temperature (22 °C). The deformations measured on the specimen grating represent the thermal deformations of the specimen induced by the temperature change of 60 °C. The detailed procedures for applying gratings in the thermal strain measurements can be found in [14] and [15]. Moiré interferometry with a standard displacement sensitivity (0.417 μm per fringe order) was used in this investigation, which provided accurate relative displacements in the vicinity of a solder joint and the global deformations of the whole package assembly. The local strain distribution and the location of the maximum strain in a particular solder joint was determined by an image-processing program which transformed the displacement fields to strain fields.

4.2.4   Thermal deformations in CBGA

Figure 4.2.2 shows the fringe patterns of the U and V displacement fields captured from a cross section of the testing specimen. The fringes are contour lines of displacements with a contour interval of 0.417 μm per fringe order. They represent thermal deformation induced by 60 °C change of temperature. When

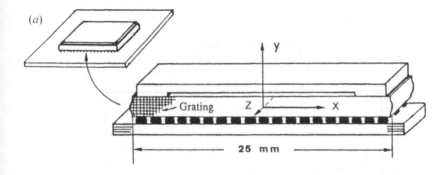

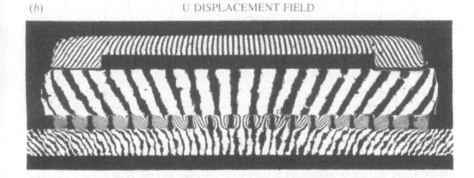

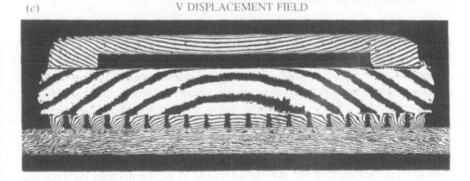

Figure 4.2.2. Fringe patterns of U and V displacement fields of the CBGA package.

the temperature decreases, all of the components contract in both the x and y directions. The thermal deformations in the ceramic substrate are very small (indicated by the low fringe density) because the material has a low CTE and high Young's modulus. In the PCB, the y-direction contraction is much greater than the x-direction contraction (shown by the higher fringe gradient in the V displacement field), indicating the anisotropic material property of the PCB. In

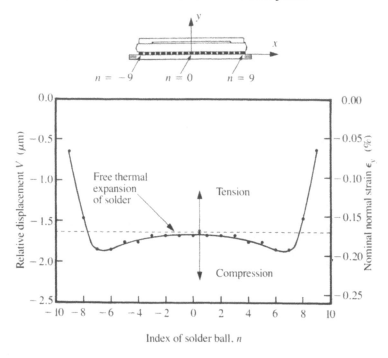

Figure 4.2.3. Normal displacement between the package and the PCB in the *y* direction and the resulting average normal strain in each solder joint.

the *x*-direction, the relative displacement between the substrate and the PCB is a function of the distance from the displacement neutral point (DNP) which is the geometric centre of the assembly. The relative displacement and, therefore, the stress and strain, reaches a maximum value where the DNP is the greatest. The increase in fringe density in the solder joints with increased distance from the DNP is readily observed in figure 4.2.2(*a*).

Under the applied thermal loading, the substrate and the PCB bend into different curvatures in the *x–y* plane (figure 4.2.2(*b*)). Because the bending curvatures are different in the substrate and the PCB, the distance between the bottom surface of the substrate and the top surface of the PCB varies. The distance change between the two surfaces is measured by the *y* direction relative displacement shown in figure 4.2.3, where the index number of solder joints, *n*, is used to label the horizontal axis. In this case (temperature decrease), the relative displacements are negative, implying that the two surfaces moved closer together everywhere. The relative displacement also describes the *y*-direction deformation of each solder joint because the solder height equals the distance between the two surfaces. The nominal normal strain (*y* direction) in each solder joint is calculated by dividing the relative displacement at that position (the centreline of the solder joint) by the height of the solder joint. The strain values are also

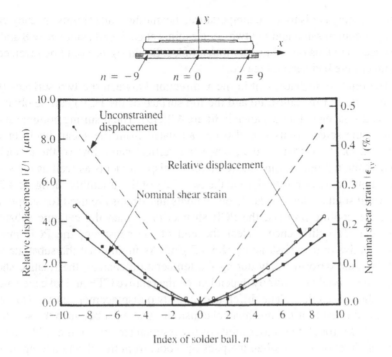

Figure 4.2.4. Shear displacement (absolute values) between the package and the PCB in the *x* direction, and the resulting average shear strain in each solder joint. The broken line shows the relative displacement that would be in the assembly if the assembly were free of the mechanical constraint from the solder joints.

given in figure 4.2.3 with the same curve but using the scale at the right, which is created by dividing the relative displacement value shown by the scale at the left by the nominal solder height.

The solder strains shown in figure 4.2.3 and 4.2.4 are the total strains which include: the thermal strain due to free thermal expansion of the solder and the mechanical strain from mechanical constrains. The relation of these strains can be expressed as

$$\varepsilon_{total} = \varepsilon_{mech} + \alpha \Delta T$$

where ε_{mech} is the mechanical strain, α is the CTE value and $\alpha \Delta T$ is the strain caused by the free thermal expansion which can be obtained from the material properties. The free thermal expansion of the solder is 28 ppm/°C (10^{-6}/°C) which corresponds to -0.17% strain for a $-60\,°C$ temperature change: it is shown as the horizontal broken line in figure 4.2.3. The mechanical strain is actually equal to the difference between the total strain (full curve) and the free thermal expansion of the solder (broken line). This is equivalent to shifting the curve so that the broken line becomes the horizontal axis of zero displacement and strain.

As the assembly cools to room temperature, the middle solder joints are subjected to a slight compression and the two solder joints at each end (index $n = 8$ and 9) are subjected to tensions of greater magnitudes. Strain signs would be reversed if the temperature had been increased.

The relative displacement in the x direction between the two surfaces (the bottom surface of the substrate and the top surface of the PCB) can be obtained from the displacement field shown in figure 4.2.2(*a*). The resulting nominal shear strains in the solder joints can also be calculated from the shear displacement (figure 4.2.4). It is obvious that, when the vertical centreline of the assembly is used as the reference point, the x-direction displacements as well as the shear strains should have opposite signs at the two sides of the assembly. In figure 4.2.4, however, absolute values of the displacements and strains are used for simplicity. During the cooling process, the PCB shrinks more than the ceramic substrate due to the CTE difference. Near the end of the substrate, the PCB moves inward relative to the substrate by about 5 μm. Assuming that the substrate and the PCB could deform freely during the temperature change, the relative shear displacement would be entirely dependent on the relative CTE and take the values give by the broken lines. The difference in the two displacement curves (full and broken) is the result of the mechanical constrains that exist between the substrate and the PCB through the solder joints or the residual stresses in the solder joints. A similar situation exists when the package cools from the elevated temperature used in an assembly process: residual stresses remain in the solder joints and the structure. Because free thermal expansion does not induce shear strain, the total shear strain in the solder joints is equal to the mechanical strain.

4.2.5 Thermal strain analysis

Fringe patterns obtained from the moiré interferometry technique are displacement contours where the displacement data are usually extracted at the fringe centre lines. From the fringes, strain contour maps can be obtained from image-processing software. For applications to electronic packaging, H Choi *et al* have developed a technique using Fourier transformation to calculate strain fields from the moiré interferometry fringe patterns [16]. In the so called Computational Fourier Transform Moiré (CFTM) technique, the essential principle is the application of the digital Fourier transform process. With this technique whole-field strain distribution can be obtained. The fringe patterns of the U and V displacement fields were obtained from the CBGA specimen with a displacement sensitivity of 0.417 μm per fringe order as shown in figure 4.2.2 The fringe patterns were captured by a CCD camera and were input to a desktop computer with image-processing software. The displacement and the strain distributions in the solder joints shown in the box of figure 4.2.1 were studied by the CFTM method. In order to use the image-processing software effectively, carrier fringes were added when the fringe patterns were captured. Figure 4.2.5

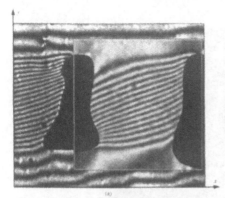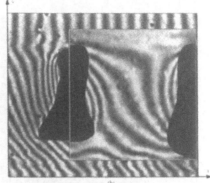

Figure 4.2.5. The phase maps of the displacement fields obtained by the image processing software for a particular solder joint shown in the box of figure 4.2.1. The backgrounds are the fringe patterns of the carrier-frequency-induced fringe patterns captured for the image processing. (*a*) *U* displacement field, (*b*) *V* displacement field. After [16].

shows the phase maps of the displacement fields obtained by the image processing software for a particular solder joint shown in the box of figure 4.2.1. The background is the original carrier-frequency-induced fringe patterns captured for the image processing. In the image processing for the phase maps, the carrier frequency was subtracted, so that the phase maps resemble the load-induced fringe patterns without the carrier frequency. Figure 4.2.6 shows the strain distributions in the solder joint, the same area as shown in figure 4.2.5 for the phase maps.

4.2.6 Conclusions

Electronic packaging is a relatively new and multi-disciplinary field where many exciting developments have happened in the last few years. The development of novel experimental techniques and methodologies for mechanics analyses of electronic packages is one of them. Most of the reliability models used in electronic packaging analyses employ thermal strains to predict the package fatigue life. Therefore, in order to achieve good reliability predictions, the thermal strains need to be determined with high accuracy. The moiré interferometry technique was able to provide strain distributions and the maximum strain in the solder joint, such that the reliability can be estimated. This whole-field technique is also ideal for simulation validations and for use in hybrid methods for providing displacement boundary conditions.

Because of the complexity of packaging structures and material properties, validation of results produced by numerical methods is often needed. It is difficult for the moiré interferometry technique with standard displacement sensitivity

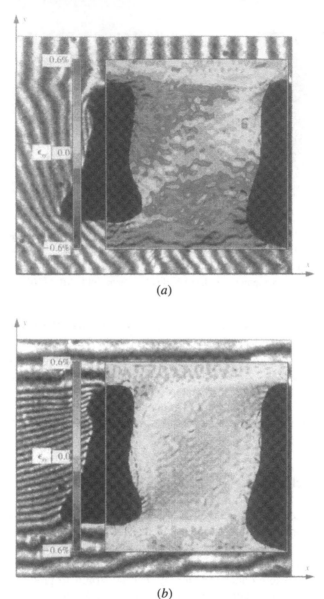

Figure 4.2.6. The strain distributions in the solder joint, the same area as shown in the figure 4.2.5 for the phase maps: (*a*) normal strain (ε_y) (*b*) shear strain ε_{xy}. After [16].

(0.417 μm per fringe order) to resolve the strain concentration and the maximum strain value in a localized area of a small structure such as a solder joint accurately. A micro moiré with higher displacement sensitivity [17] or a hybrid method using a local finite element method [10] which precisely models the local joint geometry with fine meshes, can offer much better resolution and accuracy for local strain concentration of even smaller solder joints and packaging structures.

References

[1] Norris C and Landzberg A H 1964 Reliability of controlled collapse interconnections *IBM J. Res. Development* **13** 266–71

[2] Lau J (ed) 1991 *Solder Joint Reliability, Theory and Applications* (New York: Van Nostrand-Reinhold)

[3] Lau J (ed) 1995 *Handbook of Ball Grid Array Technologies* (New York: McGraw-Hill)

[4] Guo Y and Liu S 1998 Development in optical methods for reliability analysis in electronic packaging applications *ASME J. Electron. Packaging* **120** 186–91

[5] Han B and Guo Y 1995 Thermal deformation analysis of various electronic packaging products by moiré and microscopic moiré interferometry *ASME J. Electron. Packaging* **117** 185–91

[6] Bastawros A and Voloshin A 1990 Thermal strain measurements in electronic packages through fractional fringe moiré interferometry *ASME J. Electron. Packaging* **112** 303–8

[7] Guo Y and Woychik C G 1992 Thermal strain measurements of solder joints in second level interconnections using moiré interferometry *ASME J. Electron. Packaging* **114** 88–92

[8] Han B, Guo Y, Lim C K and Caletka D 1996 Verification of numerical models used in microelectronic packaging design by interferometric displacement measurement methods *J. Electron. Packaging, Trans. ASME* **118** 157–63

[9] Dai X, Kim C, Willecke R and Ho P 1997 In-situ mapping and modelling verification of thermomechanical deformation in underfilled flip-chip packaging using moiré interferometry *Materials Research Society Symposium Proceedings* **445** 167–177

[10] Guo Y and Li L 1996 Hybrid method for local strain determinations in PBGA solder joints *SEM Experimental/Numerical Methods on Electronic Packaging* **1** (Bethel, CT: SEM) pp 22–30

[11] Teng S, Guo Y and Sarihan V 1997 Assessment of chip-scale package design options *ASME EEP* **21** 15–20

[12] Guo Y and Corbin J 1995 Reliability of ceramic ball grid array assembly *Handbook of Ball Grid Array Technologies* ed J H Lau (New York: McGraw-Hill) ch 8

[13] Guo Y, Lim C K, Chen W T and Woychik C G 1993 Solder ball connect (SBC) assemblies under thermal loading: I. deformation measurement via moiré interferometry, and its interpretation *IBM J. Res. Development* **37** 621–34

[14] Post D and Wood J 1989 Determination of thermal strains by moiré interferometry *Exp. Mech.* **29** 318–22

[15] Guo Y, Ifju P, Boeman R and Dai F 1997 Formation of specimen gratings for moiré interferometry applications *Post Conf. Proc. 1997 SEM Spring Conf. Exp. Appl. Mech. (Bellevue, WA)* (Bethel, CT: SEM)

[16] Choi H C, Guo Y, Lafontaine W and Lim C K 1993 Solder ball connect (SBC) assemblies under thermal loading: II. strain analysis via image processing, and reliability considerations *IBM J. Res. Development* **37** 649–59

[17] Post D, Han B and Ifju P 1994 *High Sensitivity Moiré* (New York: Springer)

Chapter 4.3

Thermomechanical behaviour of two-phase solder column connection under highly accelerated thermal cycle

Bongtae Han
Clemson University, USA

4.3.1 Specimen geometry

The ceramic column grid array (CCGA) package is an extension of the ceramic ball grid array (CBGA) package but it offers enhanced solder joint fatigue reliability through a compliant solder column [1]. The specimen was a CCGA package assembled to a FR-4 printed circuit board (PCB). The CCGA package was a 44 mm square module with 1089 I/O interconnections (33 by 33). A specimen with a strip array configuration was prepared from the assembly, containing four central rows of solder columns. A schematic diagram of the CCGA assembly after reflow is shown in figure 4.3.1 with relevant dimensions. The assembly contained cast-in-place columns, where wire solder columns were cast directly on the ceramic substrate and subsequently connected to a PCB using eutectic solder. The wire solder column is made of a high melting point solder alloy (90%Pb/10%Sn). The wire column does not reflow during the assembly process, which provides a consistent and reproducible standoff between the ceramic package and the PCB. One side of the specimen was ground flat to expose the largest cross section of solder columns for specimen grating replication.

4.3.2 Experimental procedure and fringe patterns

A special implementation of moiré interferometry was utilized to determine (1) the initial deformation associated with a reflow process and (2) the accumulated permanent deformations of the solder column during thermal cycling [2].

145

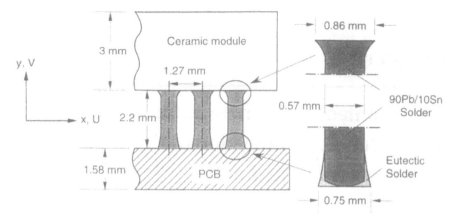

Figure 4.3.1. Schematic diagram of the CCGA package assembly after reflow.

In the experiment, a uniform cross-line grating was first produced on the specimen at an elevated temperature of 80 °C. The specimen was cooled down to room temperature and the deformation induced by an isothermal loading of $\Delta T = -60$ °C was documented by moiré interferometry. Figure 4.3.2 shows the U and V displacement fields for several columns near the left-hand side of the assembly.

The U and V displacement fields of the leftmost solder column are shown in figure 4.3.3(*a*). The displacement fields shown in figure 4.3.3(*a*) represent the total displacements of the column, which include the free thermal contraction part of displacement and the stress-induced part of displacement. In order to determine only the stress-induced part of displacement for strain analysis, the following procedure was implemented to optically subtract the free thermal contraction part of the deformation. The symmetric point of the assembly lies at the centre column, where the effect of the global mismatch of the coefficient of thermal expansion (CTE) between the module and the PCB is zero. In addition, it is safe to assume that the influence of the local CTE mismatch (the CTE mismatch between the solder and the module or PCB) would diminish completely at the middle of the column because of the high aspect ratio of the column. Consequently, the middle of the centre column contains only the free thermal contraction part of deformation. In the experimental procedure, carrier fringes of extension were applied until the U and V displacement fields of the middle of the centre column became devoid of fringes. Subsequently, the leftmost column was observed and the displacement patterns were recorded. The resultant stress-induced displacement fields of the leftmost column are shown in figure 4.3.3(*b*), where carrier fringes of rotation were applied to reduce the fringes of rigid body rotation. Mathematically, the total strain of the column is

$$\varepsilon^T = \varepsilon^\alpha + \varepsilon^\sigma = \alpha \Delta T + \varepsilon^\sigma \tag{4.3.1}$$

(a)

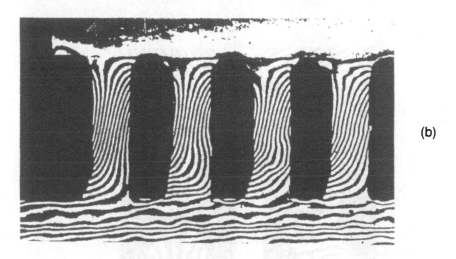

(b)

Figure 4.3.2. (*a*) U and (*b*) V displacement fields for several columns near the left-hand side, induced by an initial bithermal loading of $\Delta T = -60\,°\mathrm{C}$.

where ε^T is the total strain, ε^α is the free thermal contraction part, ε^σ is the stress-induced part and α is the CTE of the column. This procedure ensured optical subtraction of ε^α and, thus, produced ε^σ only. Note that this procedure does not require knowledge of the CTE of column material.

The same specimen was subjected to thermal cycles and the accumulated inelastic deformation was documented at room temperature. The maximum and minimum temperatures of the thermal cycles were 125 and $-40\,°\mathrm{C}$, respectively. The heating and cooling rate was approximately $16.5\,°\mathrm{C}\ \mathrm{min}^{-1}$ and a dwell time

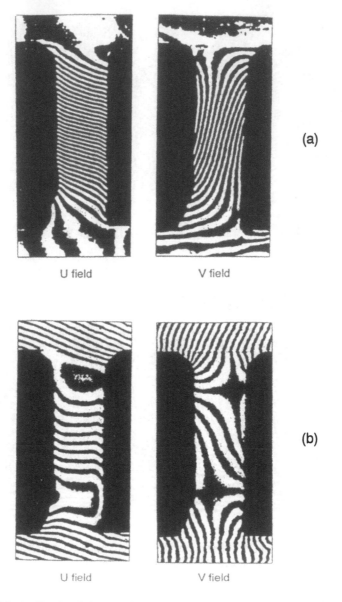

Figure 4.3.3. (*a*) Total and (*b*) stress induced U and V displacement fields of the leftmost solder column, induced by the initial bithermal loading.

of 5 min was used at each peak temperature, which resulted in 30 min per cycle. The total U and V displacements of the leftmost column after four thermal cycles of room–125 to $-40\,^\circ$C–room are shown in figure 4.3.4(*a*). This procedure to

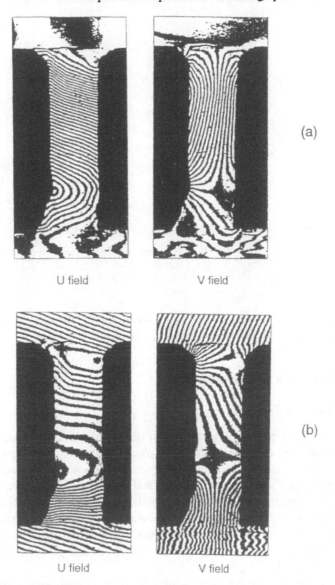

U field   V field

U field   V field

Figure 4.3.4. (*a*) Total and (*b*) stress induced *U* and *V* displacement fields of the leftmost solder column after four complete thermal columns.

subtract off the free thermal contraction part of deformation was repeated. The resultant fringe patterns are shown in figure 4.3.4(*b*).

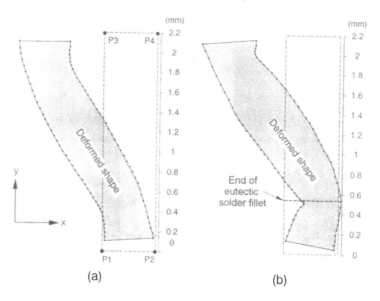

Figure 4.3.5. Deformed shape of the leftmost solder column, induced by (*a*) the initial bithermal loading and (*b*) four thermal cycles of −40 to 125 °C.

4.3.3   Analysis and discussions

4.3.3.1   Deformed shape

In order to understand the global behaviour of the column, the deformed shapes of the column were evaluated from the displacement fields. They were obtained from the fringe patterns in figures 4.3.3(*a*) and 4.3.4(*a*). representing the total displacements after the initial isothermal loading and after thermal cycles, respectively. The fringe patterns were marked at uniform intervals along the column boundary and the (fractional) fringe orders were determined at each mark. Then figure 4.3.5 was plotted by displacing each boundary point by an amount proportional to the *U* and *V* fringe orders. The deformations in figure 4.3.5 are greatly exaggerated.

It is interesting to note that while the relative horizontal displacement between P1 and P3 are nearly the same, the deformed shapes are quite different around the region with the eutectic solder fillet. Eutectic solder has a much lower melting point and, thus, it is more susceptible to deformation at the elevated temperature during the thermal cycles (smaller Young's modulus and higher creep rate), as compared to the higher melting solder. After thermal cycles, the column surrounded by the eutectic solder fillet experienced much more severe permanent deformation and, thus, is could not return to its previous geometry [2].

Figure 4.3.6 depicts the cross section of two leftmost solder columns that failed during an actual ATC test, where the columns failed after 1783 thermal

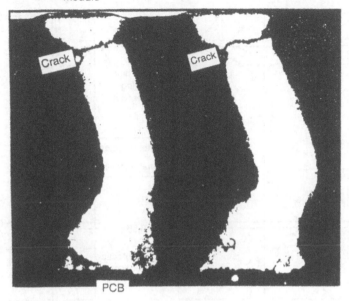

Figure 4.3.6. Cross-sectional view of two leftmost solder columns failed after 1783 thermal cycles of −55 to 125 °C.

cycles of −55 to 125 °C. The shape of the failed column is nearly identical to the deformed shape after four cycles, shown in figure 4.3.5.

4.3.3.2 Accumulated permanent deformation

The deformation accumulated during thermal cycles can be quantified by subtracting the deformation induced by the initial isothermal loading from the total deformation after thermal cycles. The deformation of the column was dominated by bending and the normal strain along the column height (ε_y) was much greater than the other in-plane strain components. The accumulated permanent normal strains, ε_y^{ac}, along the left- and right-hand side of the column, induced by the thermal cycles were determined from the fringe patterns and they are plotted in figure 4.3.7. The distribution of ε_y^{ac} along the left-hand side shows a significant strain concentration. The maximum accumulated tensile strain occurred at the end of the high melting solder fillet: its magnitude was 0.92%. This strain concentration was related to the fillet geometry. Without the fillet, a different distribution of ε_y^{ac} would be expected, suggesting that modifications of the geometry should be investigated. In the cross section shown in figure 4.3.6, the fatigue crack initiated at this strain concentration location and it propagated through the width of the column.

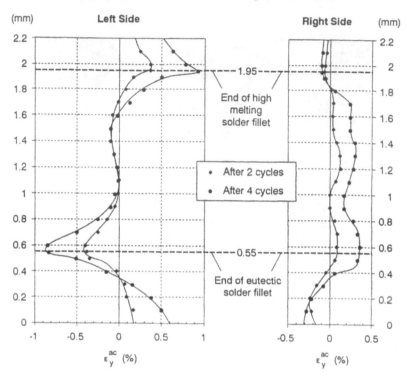

Figure 4.3.7. Distributions of strains accumulated during thermal cycles, ϵ_y^{ac}, along the left and right-hand sides of the column.

References

[1] Pullitz K J and Shutler W F 1994 C4/BGA comparison with other MLC single chip package alternatives *Proc. ECTC* (Piscataway, NJ: IEEE) pp 16–24
[2] Han B 1997 Deformation mechanism of two-phase solder column interconnections under highly accelerated thermal cycling condition: an experimental study *J. Electronic Packaging, Trans. ASME* **119** 189–96

Chapter 4.4

Verification of numerical models used in microelectronics product development

Bongtae Han
Clemson University, USA

Microelectronics products are part of every modern computer system, from laptops to mainframes. It is, therefore, imperative that the performance, reliability, development cycle time and cost be optimized. One of the key elements in accomplishing these goals involves the role of modelling/simulation of the physical design at the conceptual stage of a new technology product.

With advanced computational mechanics techniques and computing hardware, one can model/stimulate the performance of almost any kind of electronic packaging product under complex loading and boundary conditions. Then, what makes modelling predictions uncertain? One of the most critical reasons is nonlinear mechanical behaviour of the materials used in electronic packaging, especially solders, adhesives and other polymers or organic materials.

A compelling example is illustrated in figure 4.4.1 [1]. The specimen was an active silicon chip with an aluminum heat sink adhesively bonded to its lower surface. Moiré fringe patterns induced by isothermal loading are shown in figures 4.4.1(*a*) and (*b*), which represent vertical displacement fields of the assembly with a contour interval of 417 nm/fringe. In (*a*), the temperature excursion was $\Delta T = -60\,°C$, from 80 to 20 °C. In (*b*), $\Delta T = -120\,°C$, from 80 to $-40\,°C$. If all the materials had constant properties, independent of temperature, the deformation in (*b*) would be twice that in (*a*). However, the bending deformation of the chip was almost an order of magnitude larger in (*b*), which was indicated by the higher fringe order at the edge of the chip. The strong nonlinearity of the structural behaviour was ascribed to the temperature-dependent properties of the adhesive layer. Below 0 °C, Young's modulus and the shear modulus of the adhesive were an order of magnitude greater than at room temperature. Consequently, the coupling and bending were greatly increased at the low temperature.

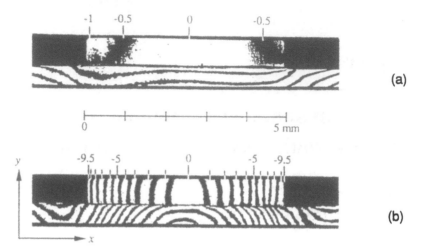

Figure 4.4.1. Illustration of the effect of temperature dependent material. Moiré fringe patterns represent the vertical displacement fields of a chip/heat sink assembly, obtained by cooling (*a*) from 80 °C to room and (*b*) 80 to −40 °C. In (*b*), the vertical displacement is nearly an order of magnitude greater (instead of twice), proving strong nonlinearity.

Numerical models predict package performance, i.e. the deformation, strain and stress distributions. If a package for a given model is fabricated and verified experimentally or if the numerical model is refined to achieve verification, the subsequent predictions associated with variations of the model can be predicted with a much higher confidence level. The verification would be even more critical if the predicted stresses are to be correlated with fracture parameters. In this chapter, moiré and microscopic moiré interferometry are presented as tools to verify the validity of numerical models used in microelectronics product development.

4.4.1   Ceramic ball grid array package assembly

The ceramic ball grid array (CBGA) package is comprised of a CBGA module attached to a printed circuit boar (PCB) through solder ball interconnections. Thermal deformations of the solder ball interconnections are caused by a coefficient of thermal expansion (CTE) mismatch between the ceramic module and the PCB.

The assembly used in the verification was a 25 mm CBGA module with 324 I/Os (18 × 18 solder ball array) assembled to a FR-4 PCB. A specimen with a strip array configuration [2] was prepared from the assembly, containing five central rows of solder balls, as illustrated schematically in figure 4.4.2. The specimen was subjected to an isothermal loading of $\Delta T = -60$ °C (cooling the specimen obtained from 82 °C to room temperature). The U and V displacement

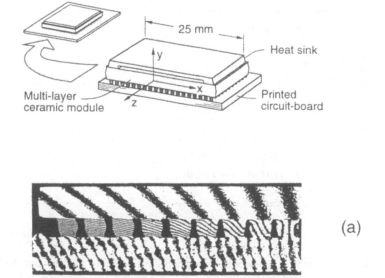

Figure 4.4.2. (*a*) *U* and (*b*) *V* displacements of the left half of a ceramic ball grid array package assembly, induced by thermal loading of $\Delta T = -60\,°C$.

fields of the left half of the specimen obtained from moiré interferometry are presented in figures 4.4.2(*a*) and (*b*).

Although a full array three-dimensional (3D) model was used in the final numerical analysis, a 3D model equivalent to the strip array specimen was analysed for the purpose of verification. The relative displacements between the module and the PCB obtained from the FEM model were compared with the experimental data. Figure 4.4.3 shows the relative displacements between the centre line of the module and that of the PCB. The numerical prediction matched the experimental data very well in magnitude as well as in distribution. At this point, numerical parametric studies with variations of the module and variations of the PCB thickness proceeded with a higher level of confidence. Note that important comparisons could be made for displacement fields, so determination of strains was unnecessary.

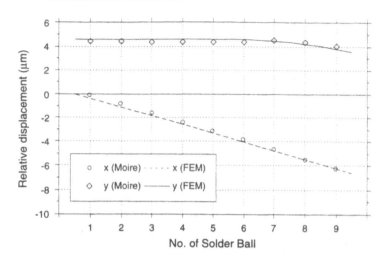

Figure 4.4.3. Relative displacement components between the module and the PCB obtained from the FEM analysis are compared with the experimental data. The experimental data were obtained from the fringe patterns in figure 4.4.2.

4.4.2   Bleed layer on power laminate

Several layers of glass/epoxy composite sandwiched by two copper sheets is called a power laminate. The copper sheets are etched initially to provide clearance holes at the locations of plated through holes (PTH). The holes are filled with epoxy in the lamination process. The usual procedure produces a thin epoxy layer on the top and bottom, of the power laminate. This thin layer of epoxy is called a *bleed layer*. While the assembly is cooled from the lamination temperature, the epoxy in the hole shrinks more than the adjacent copper, as illustrated schematically in figure 4.4.4(*a*), where the thickness of the bleed layer in figure 4.4.4(*a*) is greatly exaggerated. As a result, the bleed layer is subject to a very large tensile stress at the boundary of the clearance hole. If a crack is produced in the bleed layer at this stage, it could propagate through the adjacent layers during subsequent processing to fabricate a multi-layer PCB. In order to conduct parametric failure analysis with different thickness of bleed layers, a verified numerical model was required for quantitative stress determination.

In normal practice, the finite element method (FEM) model produces the total displacement and the stress-induced strain, whereas the moiré methods produce the total displacement and the total strain which is the stress-induced strain plus the free thermal expansion strain. Although the total displacement could be used for verification, the stress—induced strain of the bleed layer was sought to compare the strain distribution as well as the maximum tensile strain values.

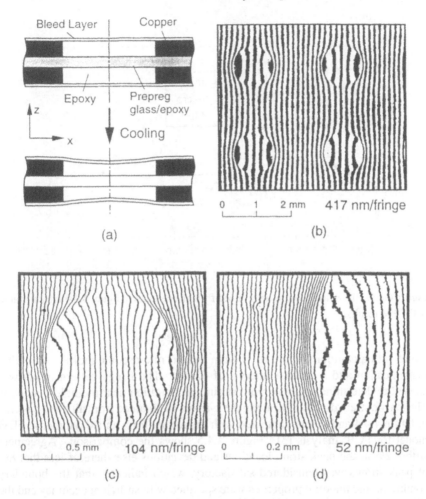

Figure 4.4.4. (*a*) Bleed layer deformation during cooling. Stress induced part of the *U* displacement field for the bleed layer, by (*b*) moiré and (*c*), (*d*) microscopic moiré interferometry.

In the experiment, a uniform specimen grating was produced on top of a power laminate at an elevated temperature of 82 °C and the deformation induced by cooling was documented at room temperature ($\Delta T = -60$ °C) by moiré and microscopic moiré interferometry. In order to obtain the stress-induced strain alone, the same specimen grating was also replicated on a separate piece of epoxy and the deformed grating on the epoxy piece was used as a reference to subtract optically the free thermal expansion part of the strain.

The results are depicted in figure 4.4.4, where the fringe patterns were obtained from (*b*) moiré with a contour interval of 417 nm, (*c*) microscopic

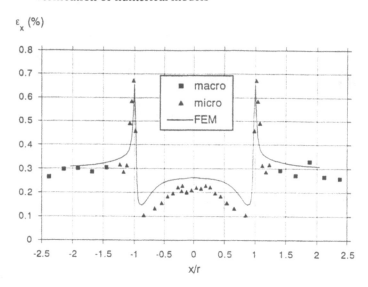

Figure 4.4.5. Bleed layer strains determined from the patterns in figure 4.4.4 are compared with FEM results.

moiré with a contour interval of 104 nm and (*d*) microscopic moiré with higher magnification and a contour interval of 52 nm. The fringe pattern in (*d*) was needed to determine the maximum strain concentration near the hole boundary. The tensile strain distribution along the x-axis was obtained from all three patterns in figure 4.4.4 and it was compared with the results from an axisymmetric FEM model for PCB analysis [3]. Figure 4.4.5 shows the comparison. Agreement within 2% at the peak strain locations and deviations elsewhere of less that 8% of peak strains was considered satisfactory, which indicates that the boundary conditions and material properties were specified with sufficient accuracy and the element sizes were appropriate.

The maximum tensile strain in the bleed layer occurred just above the copper/epoxy interface, where the fringes were spaced most closely. Note that fringes in this region had almost the same spacing, which indicated a finite width of uniform strain. The width was used to guide the FEM model on the smallest mesh size to be used at the copper/epoxy interface.

4.4.3 Discussion

Moiré and microscopic moiré interferometry provide the displacement fields for the specimen surface. Usually, the displacement field itself is sufficient to verify numerical models. What degree of correlation should be sought between the experimental results and numerical predictions?

Experimentally, the accuracy of displacement measurements on a specimen can realistically approach 100%. Uncertainties that may exist stem from variations within a production run from which the specimen is taken, e.g. dimensional and material variations that fall within the limits of quality control. Experimental accuracy is also determined by the correct application of the intended loading conditions, e.g. the accurate measurement of temperatures in thermal loading, and the accurate extraction of data from the contour maps of displacements. With careful work, such errors should be, cumulatively, only 1% of 2%.

The accuracy of numerical modelling is subject to the variations inherent in production, just as experimental analysis. Its accuracy also depends upon authentic modelling of the geometry and loading conditions, and accurate prescription of all the material properties. For isotropic elastic linear materials and well-defined geometry and dimensions, numerical modelling analyses can realistically approach 100% accuracy, too. In cases where the material properties are not fully quantified, especially for nonlinear material properties that vary with temperature and time (viscoelasticity, relaxation, creep) lower predictability is normal. Other sources of error may include element sizes that are too large and purposeful approximation of material properties (e.g. linearization of nonlinear properties): such steps are often taken to make massive computational problems more tractable.

Accordingly, the expected match between displacement fields determined by experimental measurements and numerical modelling depends upon the complexity of the package and the degree of approximation accepted in the model. Correlation within 10% are frequently considered realistic verification of a model, whereupon the model can be used for parametric studies and prototype predictability.

References

[1] Han B, Guo Y, Lim C K and Caletka D 1996 Verification of numerical models used in microelectronics packaging design by interferometric displacement measurement methods *J. Electron. Packaging, Trans. ASME* **118** 157–63

[2] Han B, Caletka D and Guo Y 1995 On the effect of moiré specimen preparation on solder ball strains of ball grid array package assembly *Proc. 1995 SEM Spring Conf. Experimental Mechanics (Grand Rapid, MI)* (Bethel, CT: SEM)

[3] Ramakrishna K and Han B 1995 Predetermined boundary conditions for the unit cell model used in stress analysis of plated through hole *1995 Int. Mechanical Engineering Congress and Exposition (95-WA/EEP-1, ASME) (San Francisco, CA)* (New York: ASME)

Chapter 4.5

Analysis of residual stresses in a die-attaching adhesive

A S Voloshin
Lehigh University, Bethlehem, USA

4.5.1   Introduction

The contemporary microelectronics package is a very sophisticated composite structure constructed from a large number of various materials with diversified thermomechanical properties, which very often are not well defined.   This precludes exact prediction of the residual stresses built into the package during the production process.   In many packages inorganic adhesives such as Au–Si eutectic solder [1] and silver-filled glass [2, 3] are used to attach the silicon die to the substrate. To use an Au–Si eutectic as the die-attaching adhesive, the die normally needs to be coated with an Au layer on its back side and then bonded onto a lead frame or substrate with Au plating on the die-attaching paddle. The thin layer of eutectic, which can be considered rigid, transmits to the die the stress induced by the mismatch in the coefficient of thermal expansions (CTES) of the die, solder and lead frame. This stress can lead to vertical or horizontal die cracking especially when the die size increases [4, 5]. Thus, alternate die-attaching adhesives are needed to improve the reliability of the die attachment.

Organic adhesives, polyimide or epoxy filled with 70–80 wt% metal particles to improve electrical or thermal conductivity, are widely used materials for the die attachment [6]. They offer the advantage of lower processing temperatures, lower cost compared to the Au–Si eutectic and lower manufacturing-induced stresses in the silicon die [7]. However, residual stresses still develop in the structure during curing. These stresses can reduce bond integrity and, as the die size increases, they may cause delamination. Thus, it is the objective of this work to investigate the *in situ* development of the residual stresses in organic adhesives during curing and cooling processes. A typical adhesive, alumina-filled epoxy (EPO-TEK H65-

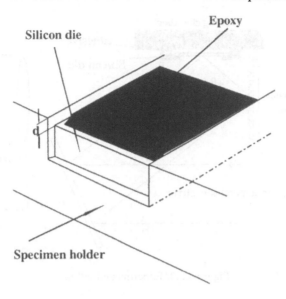

Figure 4.5.1. Schematic diagram of the fixture for sample preparation.

175MP), was studied here for evaluation of the curing stress developed during polymerization of the adhesive.

An experimental technique of moiré interferometry [8] was chosen as a tool for measuring the *in situ* out-of-plane displacement of the sample's surface during curing process. This technique is well developed and it was used in various applications [9], where a high resolution was needed. Its resolution is about 67% greater than the $\lambda/2$ of most interferometric methods for out-of-plane measurements, including holographic interferometry.

4.5.2 Experiment

Due to the shrinkage of the organic adhesive during cure and the CTE mismatch of the adhesive and silicon substrate during the cooling down process, the topology of the substrate's surface changes continuously during entire process. This change in curvature can also be used to yield information on the curing stresses developed in the adhesive. The approach presented here utilized the digital image analysis enhanced moiré interferometry (DIAEMI) to monitor the surface topology of the silicon substrate throughout the entire curing process of the epoxy and cooling down of the die.

A one half millimetre thick, single-side polished silicon wafer was diced into eight 12.5 mm × 12.5 mm test dies. The dies were cleaned to remove the organic and inorganic substance from the surfaces. A layer of the organic adhesive in different thickness (0.047 mm, 0.069 mm, 0.138 mm and 0.254 mm) was then

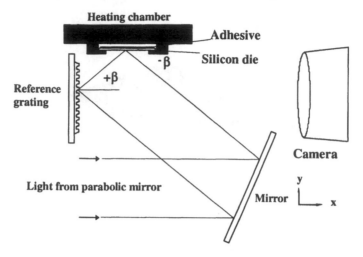

Figure 4.5.2. Experimental set-up.

applied to the unpolished surface of the die. The thickness 'd' of the layer was controlled by the depth of the slot in the processing fixture (figure 4.5.1). A modified optical system for moiré interferometry was utilized for *in situ* curvature measurements during the curing process (figure 4.5.2). The polished surface of the sample was placed faced-down. The flat mirror reflected the collimated laser ray up to the reference grating and the sample's surface at angle $-\beta$. The reflection of the sample's surface projected onto the reference grating at angle around $+\beta$ which depends on the curvature of the surface. The interference pattern (moiré pattern) was then created and captured by a CCD video camera.

The initial topology of the die's surface was recorded before the layer of the epoxy was applied. The die was inserted into a custom-built heating chamber which was located in the moiré interferometer. The apparatus was adjusted to produce a minimum fringe number pattern at room temperature and the initial moiré pattern, representing the curvature of the die surface, was recorded (figure 4.5.3(a)). The die then was removed from the chamber and the layer of the epoxy was applied onto the back side of the die. Since the curing schedule of the epoxy requires 180 °C for 1 hr, the chamber was preheated to 180 °C and after this the sample was placed into it. The moiré patterns were recorded continuously, using a video recorder, during the 1 hr cure cycle and during the consequent cooling cycle to room temperature. Figure 4.5.3(b) shows a moiré pattern recorded at the end of the curing process. The recorded patterns contain information on the time and temperature-dependent curvature of the sample. All moiré patterns were analysed off-line, after the end of the curing process.

It was also found that the curvature of the samples changed significantly after they were kept in an open environment at room temperature. It was assumed that this change was due to the moisture absorption. In order to prove this assumption,

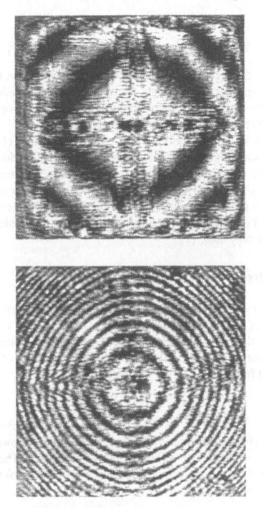

Figure 4.5.3. The moiré fringe patterns of the sample (*a*) before (*b*) after the curing process. The epoxy layer is 0.138 mm thick.

the samples were dried in an oven at 110 °C for 1 hr, cooled down to the room temperature in a desiccator and then placed into a moiré interferometer. The moiré patterns were recorded at eight different times within 24 hr.

Moiré pattern characterizes the out-of-plane displacement at every point on the specimen surface. By using a digital image analyser and the fractional fringe analysis technique [10], the displacement can be computed at every location throughout the moiré pattern. This is accomplished by using the basic optical law [11] which relates light intensities in a moiré field to the corresponding displacements:

$$W(x, y) = W_0 + \tfrac{1}{4}\pi f \cos[(I(x, y) - I_0)/I_1] \qquad (4.5.1)$$

where W_0 is the displacement at some starting point, f is the frequency of the diffraction grating, $I(x, y)$ is the light intensity at the point under consideration, I_1 is the intensity amplitude of the first harmonic term in the optical law for the field and I_0 is the average background intensity.

Equation (4.5.1) utilizes only the first term in the series [11], since here one is dealing with nearly pure cosinusoidal intensity distribution due to the nature of the moiré grating. The values of I_1 and I_0 are determined by the image analyser for each half fringe separately. Since the digital image analyser has light intensity resolution of 256 grey levels, one can hope for effective fringe multiplication of 512; however, in practice multiplication of 20 are more reliable due to inherent optical and electronic noise [12].

4.5.3   Results

The density of moiré fringes, which represents the topology of the sample's surface, increases significantly during the curing and cooling of the epoxy. The obtained out-of-plane displacement data of the sample's surface were smoothed by a second-order least-squares fitting, and the curvature, κ, of the surface was then determined from the following relation:

$$\kappa = \frac{\partial^2 W}{\partial x^2}. \qquad (4.5.2)$$

The curvature changes during the curing and cooling down process were calculated for all samples with different epoxy thicknesses. The induced tensile stress σ_e (figure 4.5.4) in the epoxy layer was calculated from the beam theory assuming $E_{epoxy} \ll E_{silicon}$:

$$\sigma_e = \frac{\kappa E_s^0 t_s}{6(1 + t_c/t_s)t_c/t_s} \qquad (4.5.3)$$

where t_s and t_c are the thickness of the substrate (die) and the coating (epoxy) and $E_s^0 = E_s/(1 - v)$ is the biaxial modulus of the die.

Each data point represents average of two samples and horizontal bars represent high and low values. The induced stresses in the epoxy due to polymerization became almost constant (~ 2 MPa) after 40 min in 180 °C . This indicates that the epoxy had been fully cured during the process. The total induced stresses by curing and cooling were found from equation (4.5.3) and are shown in figure 4.5.5. It is clear that most of the residual stresses are attributed by the mismatch in coefficient of thermal expansion of the epoxy and die.

It is well known [13] that the calculation of the simply supported plate's deflection at the cross section far from the edges reduces to the equation identical

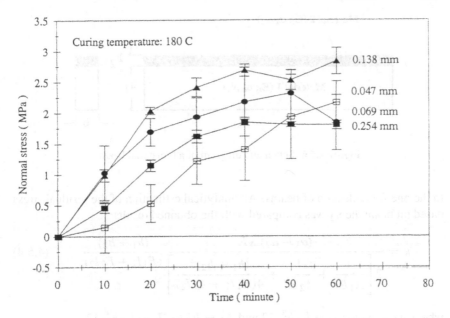

Figure 4.5.4. The induced normal stresses in the epoxy during cure.

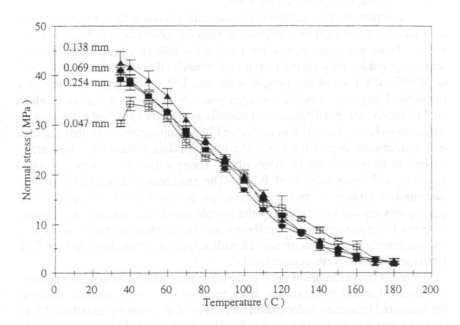

Figure 4.5.5. The induced normal stresses in the epoxy during the cool-down process.

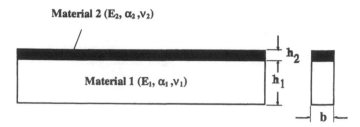

Figure 4.5.6. Schematic diagram of a bimaterial beam.

to the one for deflection of beams. An analytical estimation of the residual stress based on beam theory was compared with the obtained results.

$$\kappa = \frac{(\alpha_1 - \alpha_2)\Delta T}{2\left[\dfrac{1}{A_1 E_1^0} + \dfrac{1}{A_2 E_2^0} + \dfrac{(h_1 + h_2)^2}{4(E_1 I_1 + E_2 I_2)}\right]} \frac{(h_1 + h_2)}{(E_1 I_1 + E_2 I_2)} \qquad (4.5.4)$$

where $A_1 = b \cdot h_1$; $I_1 = b \cdot h_1^3/12$ and $A_2 = b \cdot h_2$; $I_2 = b \cdot h_2^3/12$.

The curvature change due to CTE mismatch in a bimaterial beam can be described by this expression (figure 4.5.6).

To compare the theoretical and experimental curvature data, it was necessary to determine the temperature-dependent material properties (E and α) of the epoxy. Those properties were measured from a bulk specimen. Both samples were prepared by fully curing epoxy in a mould, 100 mm × 6 mm × 3 mm³, at 180 °C. The ends of the sample were attached to the grips of a dynamic mechanical analyser. One of the grips was connected to an activator, which applied torsion and had displacement transducer, the other grip had a strain gauge torque transducer. The test was performed in a temperature-controlled chamber in a temperature range 20–200 °C. The obtained data yielded the temperature-dependent shear modulus G, from which Young's modulus E was calculated assuming a Poisson ratio ν of 0.33. The temperature-dependent CTE was measured on a 6 mm × 3 mm × 1 mm sample prepared from bulk epoxy. Cross-grating was applied on one side of the sample which was placed in the heating chamber located in the moiré interferometer. The in-plane displacement field of the specimen's surface was monitored with respect to temperature and the CTE *versus* temperature curve as calculated.

The curvature change of the bimaterial samples due to temperature changes was then calculated incrementally from the cure to room temperature utilizing the measured temperature-dependent behaviour of E and α by equation (4.5.4). Good agreement was found between the calculated results and the experimental data (figure 4.5.7). This means that one can use traditional analytical expression for calculation of residual stresses for this particular type of epoxy, assuming that temperature-dependent properties are used in the calculations. The results also

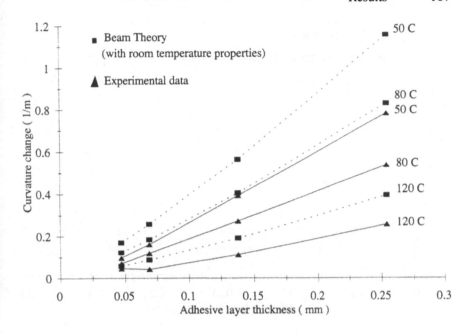

Figure 4.5.7. The curvature changes of the samples during the cooling process.

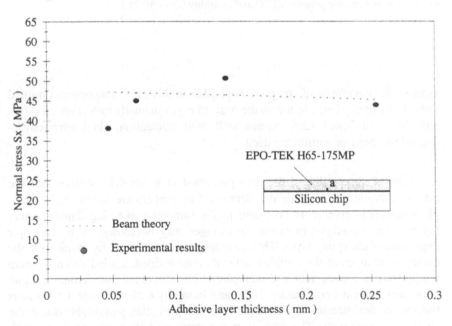

Figure 4.5.8. Normal stress in the epoxy layer after curing.

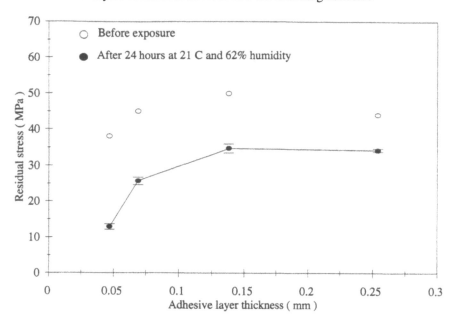

Figure 4.5.9. Residual stresses in the epoxy layer after the sample had been placed in the given environment (temperature 21 °C and humidity 62%) for 24 h.

suggest that the effects of creep are negligible in this case. The residual normal stress in the epoxy layer, found on the basis of experimentally measured curvature and shown in figure 4.5.8, agrees well with calculation when temperature-dependent epoxy properties are used.

Careful observation of the data presented in figure 4.5.7 reveals that the rate of curvature change slightly decreased at temperature below 50 °C. This phenomenon is even more profound in the samples containing thinner epoxy layers. The slowdown in curvature changes may be caused by hydroscopic expansion of the epoxy layer. This assumption was evaluated by monitoring the curvature changes of the samples, after they were dried, cooled down to room temperature in a desiccator and then placed into the moiré interferometer in an open environment (temperature 21 °C and humidity 62%). Figure 4.5.9 shows that the residual stresses in the thinner layer samples relax presumably due to the shorter diffusion path. The amount of the stress relaxation aslo decreases with the increase in the thickness of the epoxy, perhaps attributable to the fact that the moisture absorption by the outside surface of the epoxy is much easier than the inside layer.

4.5.4   Conclusions

Formation of the residual stresses in the organic adhesive, EPO-TEK H65-175MP, was investigated experimentally. The obtained results indicate that the mismatch in coefficients of thermal expansion of the epoxy and the Si die plays an important role in the development of this stress. The experimental results were compared with analytical solutions based on beam theory by using temperature-dependent bulk material properties. Good agreement was found for all samples with various thickness of the epoxy layer. Therefore, it can be concluded that there is no thickness effect of the epoxy layer on residual stress for the thickness range used in this study. The obtained results indicate that one can use analytical solution with temperature-dependent bulk material properties for the analysis of residual stress due to epoxy curing with the layer thickness over 50 μm.

Moisture can cause the formation of a water rich layer [14] at the interfaces in the die-attaching assembly (die, adhesive and lead frame). This will reduce the strength of the interface and degrade the bond integrity and even may cause a 'popcorn effect' when the device is under thermal loading. The residual stresses may act as the driving force and cause the delamination while the strength of the interface was weakened by moisture. The results obtained show that moisture absorption by the epoxy reduces the residual stresses, more so, in the thinner adhesive layer than in the thicker layer. The effects of moisture absorption by the epoxy on the bond integrity should be further studied and taken into account in an analysis of the reliability of the die attachment.

References

[1] Mencinger N P, Carthy M P and McDonald R C 1985 Use of wetting angle measurements in reliability evaluations of Au–Si eutectic die attach *Proc. 23rd Annual Reliability Physics Conf.* (New York: IEEE) pp 173–8

[2] Dietz R and Winder L 1982 An innovation in materials for attaching silicon device *Int. J. Hybrid Microelectron.* **5** 480–6

[3] Moghadam F K 1984 Development of adhesive die attach technology in cerdip packages; material issues *Semicond. Sci. Technol.* **27** 149–54

[4] Kessel C G M van, Gee S A and Murphy J J 1983 The quality of die-attachment and its relationship to stresses and vertical die-cracking *Proc. 33rd Electronic Components* (New York: IEEE) pp 237–41

[5] Chiang S S and Shukla R K 1984 Failure mechanism of die cracking due to imperfect die attachment *Proc. 34th Electronic Components Conf.* (New York: IEEE) pp 195–200

[6] Bjorneklett A 1993 Characterization of silver filled adhesives for attachment of microelectronics chips to substrates *Polymers and Polymer Composites* **4** 275–82

[7] Minges M L 1989 *Electronic Materials Handbook: Packaging* (Cleveland, OH: ASM International)

[8] Post D 1987 Moiré interfereometry *Handbook on Experimental Mechanics* ch 7 (Englewoods Cliff, NJ: Prentice-Hall) pp 378–80

[9] Asundi A, Cheung M T and Lee C S 1989 Moiré interferometry for simultaneouse measurement of U, V, W *Exp. Mech.* **29** 258–60

[10] Voloshin A S, Burger C P, Rowland R E and Richard T S 1986 Fractional moiré strain analysis using digital imaging technique *Exp. Mech.* **26** 254–8

[11] Sciammarella C A 1965 Basic optical law in the interpretation of moiré pattern applied to the analysis of strain—part I *Exp. Mech.* **5** 154–60

[12] Bastawros A F and Voloshin A S 1990 Thermal strain measurements in electronic packages through fractional moiré interferometry *J. Electron. Packaging* **112** 303–8

[13] Timoshenko S and Woinowsky-Kreiger S 1959 *Theory of Plates and Shells* 2nd edn (New York: McGraw-Hill) ch 1

[14] Lupinski J H and Moore R S 1989 Polymeric materials for electronics packaging and interconnection *ACS Symposium Series* **407** 1–100

Chapter 4.6

Characterization of thermomechanical behaviour of a flip-chip plastic ball grid array package assembly

Bongtae Han
University of Maryland, USA

Ball grid array (BGA) packaging technology has matured rapidly in the past decade. It allows the attachment of high I/O single- or multi-chip modules directly to a standard printed circuit board (PCB). This investigation contributed to the understanding of the thermomechanical behaviour of flip-chip plastic ball grid array (FC-PBGA) package assemblies, with an emphasis on the effect of substrate deformation on solder ball strains.

4.6.1 Flip-chip plastic ball grid assembly

A cross-sectional view of the FC-PBGA package assembly tested in the experiment is shown schematically in figure 4.6.1. In the assembly, a silicon die (6.8 mm × 6.1 mm × 1.2 mm) was attached to a printed circuit board (PCB) substrate through eutectic C4 solder bumps. The typical C4 solder bump height was 80 μm. The gap between the chip and the substrate was filled with an epoxy underfill to reduce the strains of the C4 connection. The final dimension of the package body was 21 mm by 25 mm, which provided 304 I/O connections (16 by 19). The package was surface-mounted to a typical FR4 PCB. The solder interconnection of the package assembly consisted of a high melting point solder alloy (90%Pb/10%Sn) and an eutectic solder fillet (63%Pb/37%Sn).

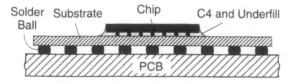

Figure 4.6.1. Cross-sectional view of a flip-chip PBGA package assembly.

4.6.2 Grating replication for complex geometry

A special technique is required for replicating a specimen grating on the cross sections of microelectronics devices because they usually have such a tiny and complex geometry that the excess epoxy produced by the usual grating replication procedure [1] cannot be swabbed away. The excess epoxy is critical since it could reinforce the specimen and change the local strain distribution.

An effective replication technique was developed to circumvent the problem [2, 3]. First, a tiny amount of liquid epoxy is dropped onto the grating mould; the viscosity of the epoxy should be extremely low at the replication temperature. Then, a lintless optical tissue (a lens tissue) is dragged over the surface of the mould, as illustrated in figure 4.6.2. The tissue spreads the epoxy to produce a very thin layer of epoxy on the mould. The specimen is pressed gently into the epoxy and it is pried off after the epoxy has polymerized. Before polymerization, the surface tension of the epoxy pulls the excess epoxy away from the edges of the specimen. The result is a specimen grating with a very clean edge. The specimen must be made very flat and smooth to be compatible with the thin film of epoxy.

4.6.3 Bithermal loading

Thermal deformations can be analysed by room temperature observations. In this technique, the specimen grating is applied at an elevated temperature and it is allowed to cool to room temperature before it is observed in the moiré interferometer. Thus, the deformation incurred by the temperature increment is locked into the grating and recorded at room temperature [4].

A typical temperature increment is 60 °C, whereby the grating is applied at about 82 °C and observed at about 22 °C. An adhesive that cures at elevated temperature is used, usually an epoxy. The specimen and mould are preheated to the application temperature, the adhesive is applied and it is allowed to cure at the elevated temperature. The mould is a grating on a zero expansion substrate, so its frequency is the same at elevated and room temperatures. Otherwise, a correction is required for the thermal expansion of the mould.

This technique had come to be known as *isothermal loading* in recent years but that name implies constant temperature and, therefore, it is not descriptive.

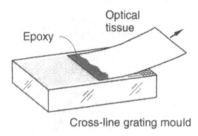

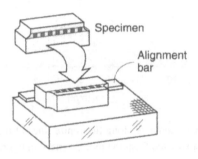

Figure 4.6.2. Procedure to replicate a specimen grating on a specimen with a complex geometry.

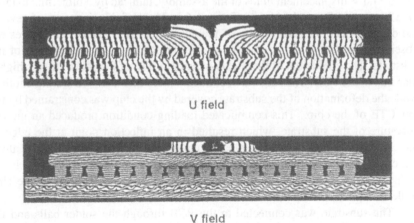

Figure 4.6.3. *U* and *V* displacement fields of the FC-PBGA package assembly, induced by thermal loading of $\Delta T = -60\,^{\circ}\text{C}$.

It was proposed to call the technique *bithermal loading*, implying two discrete temperatures [5, 6].

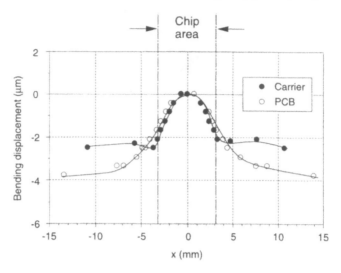

Figure 4.6.4. Vertical displacements along the centre lines of the substrate and the PCB. The displacements were extracted from the moiré pattern in figure 4.6.3.

4.6.4   Result and implication

The U and V displacement fields of the assembly, induced by a bithermal loading of $\Delta T = -60\,^{\circ}$C, are depicted in figure 4.6.3. The V-field fringe pattern reveals the detailed bending deformation of the substrate. Two distinct curvatures are observed: one in the area connected to the chip and the other in the rest of the substrate. The coefficient of thermal expansion (CTE) of the substrate was higher than that of the PCB. The substrate contracted more than the PCB during cooling, while the deformation of the substrate covered by the chip was constrained by the low CTE of the chip. This complicated loading condition produced an uneven curvature of the substrate, which resulted in an inflection point at the edge of the chip. The actual vertical displacements of the substrate and the PCB along the centre lines were determined from figure 4.6.3 and the results are plotted in figure 4.6.4. The curvature change of the substrate at the edge of the chip (inflection point) is evident.

The substrate was connected to the PCB through the solder balls and the difference of curvature between the substrate and the PCB must be accommodated by the deformation of the solder balls. The load-induced normal and shear strains (averaged along the vertical centreline) at each solder ball was calculated from the fringe patterns and they are plotted in figure 4.6.5. The largest of these normal strains occurred in the solder ball located at the edge of the chip and its magnitude was nearly four times greater than the largest shear strain. This result is entirely different from the results of the previous study on a ceramic ball grid array package assembly, where the shear strain was a dominant deformation and it

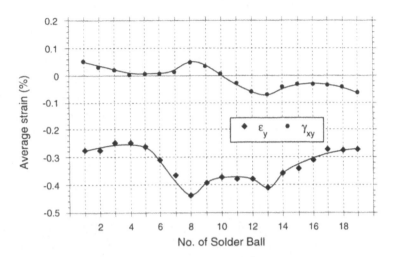

Figure 4.6.5. Distribution of normal and shear strains (averaged along the vertical centre line) at each solder ball.

was linearly proportional to the distance from the neutral point [2]. These results guided and validated a subsequent numerical parametric study that quantified the effect of the substrate CTE on the solder ball strains. The study indicated that the substrate CTE was one of the most critical design parameters for optimum solder joint reliability [7]. The experimental evidence provided by moiré interferometry was essential to revealing this important design parameter.

References

[1] Post D, Han B and Ifju P 1994 *High Sensitivity Moiré: Experimental Analysis for Mechanics and Materials* (New York: Springer)

[2] Guo Y, Lim C K, Chen W T and Woychik C G 1993 Solder ball connect (SBC) assemblies under thermal loading: I. deformation measurement via moiré interferometry, and its interpretation *IBM J. Res. Development* **37** 635–48

[3] Han B 1998 Recent advancement of moiré and microscopic moiré interferometry for thermal deformation analyses of microelectronics devices *Exp. Mech.* **38** 278–88

[4] Post D and Wood J 1989 Determination of thermal strains by moiré interferometry *Exp. Mech.* **29** 318–22

[5] Han B, Post D and Ifju P 2001 Moiré interferometry for engineering mechanics: current practice and future development *J. Strain Analysis* **36** 101–17

[6] Post D, Han B and Ifju P 2000 Moiré methods for engineering and science— moiré interferometry and shadow moiré *Photomechanics for Engineers* ed Pramod Rastogi (Berlin: Springer) ch 7

[7] Han B, Chopra M, Park S, Li L and Verma K 1996 Effect of substrate CTE on solder ball reliability of flip-chip pbga package assembly *J. Surf. Mount Technol.* **9** 43–52

CHAPTER 5

COMPLEX SHAPES

Chapter 5.1

Strain gauge verification by moiré

Peter G Ifju
University of Florida, USA

5.1.1 The objective

The objective was to measure the strain distribution in an electrical resistance strain gauge. It was a unique experiment, performed to verify the strain averaging capabilities of a strain gauge called the *shear gauge* [1, 2], illustrated in figure 5.1.1. The shear gauge is a specialized foil resistance strain gauge that is used on notched beam specimens such as the Iosipescu [3, 4] and compact shear [5] specimens. The shear gauge was designed to measure the average shear strain across the test section (the centre line connecting the notches). The gauge is a $\pm 45°$ strain gauge rosette which spans the entire test section of the specimen, over a finite width. It is available in two configurations, stacked and side-by-side.

In most case where strain gauges are used, the gauge length is selected to ensure that the peak strain is measured. In regions where the strain gradient is small, a large gauge can be used. Where the strain gradient is high, small gauges are typically used to determine the strain at a 'point'. However, in the case of the Iosipescu specimen, the shear strain distribution is highly non-uniform but it is the average shear strain that is desired. This is because the shear modulus of the material is determined as the ratio of the average shear stress to the average shear strain. The average shear stress is determined by the load divided by the cross-sectional area at the test section. The shear gauge was developed to provide an easy means of determining the average shear strain.

In addition, the shear strain distribution is a function of the loading conditions, notch geometry and the orthotropy [6] of the material being tested. In the past, small strain gauge rosettes, located in the centre of the test section, were used to measure the shear strain. Shear modulus calculations were obtained by dividing the *average* shear stress by the *local* shear strain. As a result, serious errors in modulus measurement occurred because the local shear strain

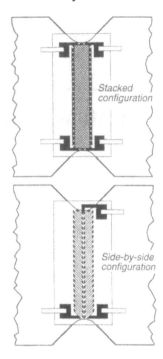

Figure 5.1.1. The shear gauges span the entire length of the test section, thus integrating shear strains and recording the average. Two shear gauge styles were tested: stacked and side-by-side.

differs from the average shear strain. Complicated procedures [7] were proposed to perform corrections by predicting the shear stress distributions but they rely partially on knowledge of the property that is to be measured. Although the idea of using a gauge to record the average normal strain over a highly varying field is not new, the application to average shear strain measurement by integration over the entire test section of a specimen is novel.

To verify that the shear gauge measures the average shear strain on the surface of the specimen, moiré interferometry [8] was used to measure the average shear strain independently. Moiré interferometry was selected since high measurement sensitivity, high spatial resolution and full-field capabilities were required. Since variability in the specimen materials exist, especially for the orthotropic composite materials of interest, it was necessary that the moiré interferometry test and the strain gauge test be implemented on the same specimen. Furthermore, specimen twist, which is unavoidable and random in nature, warranted that the two measurements be performed on the same side of the specimen. Therefore, the diffraction grating for the moiré experiment was replicated directly over the shear gauge and the two measurements were

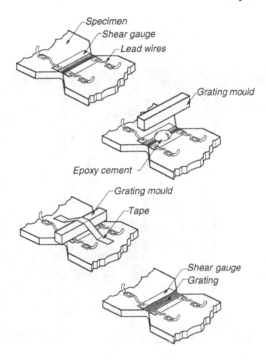

Figure 5.1.2. The diffraction grating for moiré interferometry was replicated directly over the shear gauge. The procedure is illustrated here.

performed over the same region of the specimen, for the same loads, and at the same time.

5.1.2   The experiment

A special procedure for replicating a diffraction grating on top of the shear gauge was developed, as illustrated in figure 5.1.2. It was desired that the grating be thin and provide minimal reinforcement. A cross-lined grating with 1200 lines/mm was used. The procedure starts by installing the shear gauge on the specimen surface using common techniques specified in the manufacturer's literature. A grating mould was prepared by slicing, with a diamond impregnated saw, a narrow strip (2.5 mm wide) of a photoresist grating on a glass substrate. The grating had to be narrow to allow for attachment of the strain gauge leads, as illustrated in figure 5.1.2. Precautions were taken to ensure that the saw cuts were aligned with the grating lines. The diffractive surface of the mould was then coated with a thin layer of aluminium by vacuum deposition. The grating mould was lowered, diffractive surface down, into a drop of epoxy (Measurements Group PC-10C) which was poured directly on top of the shear gauge. The edges of the grating

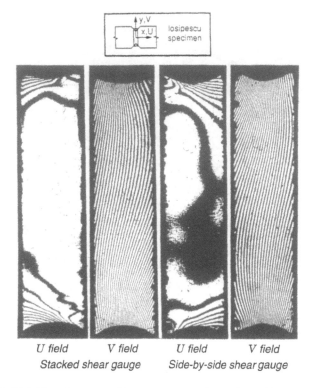

U field   V field   U field   V field
Stacked shear gauge   Side-by-side shear gauge

Figure 5.1.3. Moiré interferometry fringe patterns on the surface of the shear gauge for 0° fibre direction Iosipescu specimens.

were aligned with respect to the strain gauge and the grating was held stationary by a thin piece of transparent tape. The spew from the excess epoxy was absorbed with a cotton swab and the epoxy was left to cure. The grating mould was then removed leaving some photoresist and the replica of the diffractive surface. Acetone was used to wash away the leftover photoresist. The resulting grating spanned the entire test section: it was about 2.5 mm wide and less than 25 μm thick (significantly less than the thickness of the shear gauge itself).

This procedure was used to test a total of ten Iosipescu specimens and ten compact shear specimens. They were made from 0°, 90°, ±45°, 0/90 and quasi-isotropic graphite/epoxy (Hercules IM7G/2503). The shear gauges were made in two styles, side-by-side and stacked, and both styles were tested with moiré interferometry. Typical fringe patters over the gauge for the Iosipescu and compact specimens are shown in figures 5.1.3 and 5.1.4. At each load level, the shear strain from the shear gauge was recorded at the same time as the fringe patterns were photographed. The average shear strain from the fringe pattern was determined along a line connecting the centre of the notches. Since calculation of the shear strain requires the sum of the two cross-derivatives, the average $\partial V/\partial x$

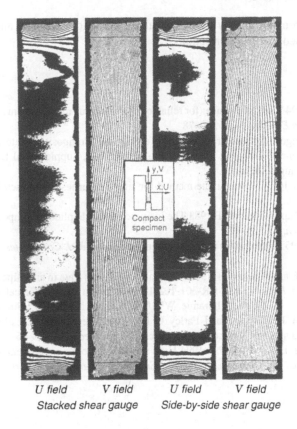

U field	*V field*	*U field*	*V field*
Stacked shear gauge		*Side-by-side shear gauge*	

Figure 5.1.4. Moiré interferometry fringe patterns on the surface of the shear gauge for 0° fibre direction compact shear specimens.

was determined by sampling the gradient at 21 locations along the vertical line and integrating.

The data from these tests showed that the average shear strain from the moiré tests were consistently higher than the shear gauge by, on average, 4% for the Iosipescu specimen and 2.5% for the compact shear specimen. This variation was attributed to the fact that the moiré data were taken on the centreline, while the shear gauge averages over a finite width. The shear strain in the transverse direction away from the centre drops off from the peak values at the centre line. These differences are small but they can be incorporated into the gauge factor of the strain gauge. There was no significant difference of results for the stacked and side-by-side shear gauge configurations. A comprehensive report is given in [1].

Moiré interferometry was used to confirm that the shear gauge performs its intended task. A small difference between the average shear strain measured by moiré interferometry and by the electrical shear gauge was found and the moiré data can be used to apply corrections.

References

[1] Ifju P G 1994 The shear gauge: for reliable shear modulus measurements of composite materials *Exp. Mech.* **34** 369–78

[2] Strain Gauges for Shear Modulus Testing of Composite Materials, Micro-Measurements Division, Measurements Group, Inc., Supplemental Data Sheet for Strain Gauge Catalogue.

[3] Iosipescu N 1967 New accurate method for single shear testing of metals *J. Mater.* **2** 537–66

[4] Adams D F and Walrath D E 1983 The Iosipescu test as applied to composite materials *Exp. Mech.* **23** 105–10

[5] Ifju P G 1995 Shear testing of textile composite materials *J. Composites Technol. Res.* **17**

[6] Walrath D E and Adams D F 1983 Analysis of the stress state in an Iosipescu shear test specimen *Report* UME-DR-301-102-1, Department of Mechanical Engineering, University of Wyoming Laramie, WY, NASA Grant NAG-1-272, June

[7] Ho H, Tsai J, Morton J and Farley G L 1991 Experimental investigation of the Iosipescu specimen for composite materials *Exp. Mech.* **31** 328–36

[8] Post D, Han B T and Ifju P G 1994 High sensitivity moiré: experimental analysis for mechanics and materials *Mechanical Engineering Series* (New York: Springer)

Chapter 5.2

A high-precision calibration of electrical resistance strain gauges by moiré interferometry

Robert B Watson
Measurement Groups Inc, Raleigh, USA

5.2.1   Introduction

Obtaining precise measurements with electrical resistance strain gages requires accurate knowledge of the gauge factor, which relates the change in the strain magnitude to the change in electrical resistance of the gauge. Strain gages are manufactured to a specific resistance with close tolerance. When bonded to a specimen and subjected to strain, the gauge changes resistance with close tolerance. When bonded to a specimen and subjected to strain, the gauge changes resistance predictably and linearly over the elastic range of engineering materials. Conveniently, resistance increases with tensile strain and decreases with compressive strain.

Measurement of the gauge factor requires a destructive test, since the gauge must be bonded to perform the test and it cannot be used again. Thus, the gauge factor is typically determined using a systematic, statistical approach. A sufficient number of gages of each geometric type and construction are tested to ensure an uncertainty in the gauge factor value of less than 0.5%. internationally recognized standards define methods for calibrating strain gages [1,2]. They specify applying a known strain to a test gauge and measuring the resultant resistance change.

High-accuracy resistance measurements are made in several different ways, with the most common technique employing a Wheatstone bridge circuit to resolve the micro-ohm changes in resistance associated with precise calibration. Measuring the applied strain with sufficient accuracy is a more difficult problem. In the past, high accurate extensometers were used to determine the calibration strain [3]. One such device is the Tuckerman optical extensometer. These were manufactured, calibrated and supported by a spin-off company from the National Bureau of Standards, providing the dual benefit of being accepted as a primary standard for strain measurement and being directly traceable to the national standards laboratory. These extensometers are no longer manufactured or supported, so currently there is no recognized standard for strain measurement.

In 1982, Watson and Post proposed the use of moiré interferometry as a method to determine reference strains [4]. More recently, the method was refined to allow easy adaptation to an existing load system, and to enhance its practical implementation. This refined approach is reported in detail in [5, ch 13], and described briefly here.

5.2.2   The challenge

Accurate determination of the gauge factor requires a precisely known applied strain. Gauge factor values are typically supplied with a 0.5% uncertainty at a 2σ confidence interval. This includes all variables of the measurement system and any gauge-to-gauge variations. Accounting for all possible system and gauge uncertainties, the allowable uncertainty for the applied strain is only 0.1%. Recognized standards specify a room temperature gauge factor calibration strain of 1000 μm m^{-1}. Therefore, achieving a 99.9% accuracy for the applied strain dictates that the strain measuring technique can have no more than a 1 μm m^{-1} total error, i.e. the error for change of length measurement cannot exceed one part per million.

5.2.3   The high-sensitivity moiré solution

The classical technique for strain measurement is to apply fiducial marks to a surface and measure the strain-induced change of length between the marks. Using moiré interferometry, this can now be accomplished with sub-wavelength sensitivity.

To provide the uniform strain field necessary for gauge factor measurement, a deflection-based, constant-stress cantilever beam test strand is used [1]. the active element of the test stand is illustrated in figure 5.2.1. For routine measurement of gauge factor, test gages are bonded along the length of the constant-stress section of the cantilever beam and the beam is deflected to produce approximately 1000 μm m^{-1} in the test section. The direct use of moiré interferometry as the technique to determine the applied strain accurately is impractical. Instead, master reference gages are installed on the beam, calibrated and periodically verified using high-sensitivity moiré interferometers. During calibration and verification, the output from the master gauge is correlated with the test section strain as established by the moiré measurements. Subsequently, the output from the test gages is compared to that of the master gauge. This scenario is identical to that of modern-day, high-precision strain-gauge-based load

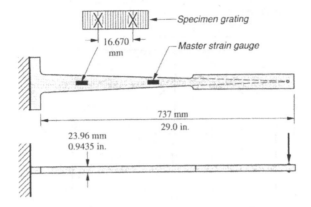

Figure 5.2.1. Constant strain cantilever beam.

cells, wherein a gaged spring element is calibrated in engineering units dictated by the intended use of the product. In this case, the unit of interest is strain.

A nominally 1200 lines/mm, linear specimen grating was replicated in epoxy on the cantilever beam in the constant-stress test section. The specimen grating was replicated from a photoresist mould. Teflon® shims were placed under the mould to produce a grating thickness typical of the installed thickness of test gages. To establish a gauge length (GL) for the moiré measurements, two **x** marks were scribed into the grating mould prior to replication. The marks were 16.670 mm apart. This distance was determined using a biaxial measuring machine manufactured by the David W Mann Company. Resolution of the machine is 1.23 μm or 74 parts per million (ppm) of the gauge length.

The frequency of the grating mould was measured by interferometric comparison with a known reference grating. The fused silica reference grating, manufactured by the Baush and Lomb Optical Company (B&L), has a traceable frequency of 1200.131 lines/mm with an uncertainty of 5 ppm. The measurements yielded a frequency for the mould of 1200.34 lines/mm with an uncertainty of 10 ppm.

Figure 5.2.2 shows a schematic diagram of the moiré interferometer used to interrogate the specimen grating. The upper grating, use primarily as a beam-splitter, is attached to a translation mechanism, which provides the capability of shifting the fringe pattern relative to the gauge length **x** marks. A top view of the complete optical system is illustrated in figure 5.2.3.

5.2.4   Assessment of measurement system accuracy

Potential errors from each part of the calibration system were quantified and used in conjunction with an experiment to determine if the desired 99.9% accuracy for measurement of the reference strain had been achieved.

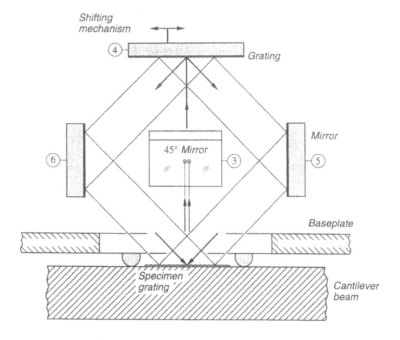

Figure 5.2.2. Moiré interferometer to interrogate the specimen grating.

The master gauge response was measured with a Model 3800 wide range strain gauge indicator [6], using an instrument gauge factor setting of 2.000. The indicator was calibrated with a Model 1550A strain indicator calibrator [7]. After calibration, the strain indicator had a National Institute of Standards and Technology (NIST) traceable accuracy (at 1000 μm m^{-1}) within 0.025% or reading or 250 ppm.

Angular alignment of the specimen grating was done by the method in [8]. Consistent with that method, the alignment accuracy was estimated to be within 0.2°. Using a Poisson's ratio of 0.289 for the steel cantilever beam, the error in the measured longitudinal strain due to misalignment is within the range 0 to −16 ppm.

The specimen grating has a thickness uncertainty of 3 μm, which influences the distance to the neutral axis of the beam by 0.025% or 250 ppm. This produces a corresponding uncertainty in the calibration factor.

Additional consideration was given to errors introduced from curvature of the beam during loading, slope of the specimen grating at load and temperature effects. Errors from beam curvature and grating slope are well below 5 ppm and, therefore, considered negligible. Effects from changing temperature are eliminated by calibrating the specimen grating mould frequency and gauge length distance at the same temperature. Subsequent temperature-induced deformation

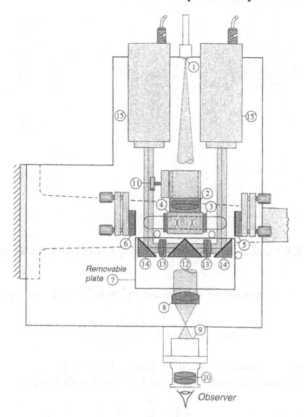

Figure 5.2.3. Optical system mounted on the cantilever beam: (1) optical fibre, (2) collimating lens, (3) 45° mirror, (4) grating beam-splitter, (5, 6) mirrors, (7) removable table, (8) focusing lens, (9) screen, (10) magnifier, (11) fringe shifter, (12, 14) prisms, (13) focusing lens and (15) video camera.

of the replicated specimen grating will equally affect the grating frequency and gauge length such that net sensitivity for the measurement is unaffected.

5.2.5 Experimental procedure

The following procedure was established to determine the calibration factor. Strain is induced in the cantilever beam by a deflection mechanism at the free end. The deflection is adjusted to produce an integral number of fringes, ΔN, in the gauge length between the **x** marks (e.g. $\Delta N_1 = -40$). The reading η_1 from the master strain gauge is recorded. The deflection is reversed and adjusted for an integral number of fringes (e.g. $\Delta N_2 = +40$) master gauge reading η_2 is recorded.

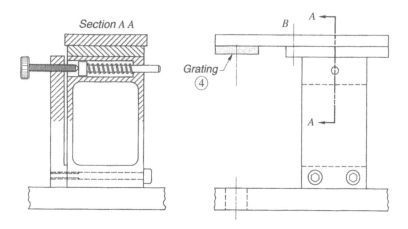

Figure 5.2.4. Flexural fixture for fringe shifting by tiny translation of grating (4).

The strain increment $\Delta\varepsilon$ determined by moiré interferometry is calculated by

$$\Delta\varepsilon = \varepsilon_2 - \varepsilon_1 = \frac{(\Delta N_2 - \Delta N_1)}{f(GL)} \qquad (5.2.1)$$

where f is equal to twice the specimen grating frequency.

The corresponding master gauge readings are totalled to give

$$\Delta\eta = \eta_2 - \eta_1.$$

Then, a calibration factor (CF) for the master gauge is calculated by

$$CF = \frac{\Delta\eta}{\Delta\varepsilon} = f(GL)\frac{(\eta_2 - \eta_1)}{(\Delta N_2 - \Delta N_1)}. \qquad (5.2.2)$$

An iterative process is used to adjust ΔN_1 (and ΔN_2) to an integral number. First, the beam deflection is adjusted to give a first approximation of the desired integral number of fringe between the x marks. Then, the fringes are shifted to centre a black fringe over the left-hand x mark. The deflection is adjusted and the fringes shifted again, iteratively, until black fringes are centred over both the left-hand and right-hand x marks. Fringe shift is accomplished by a tiny translation of the upper grating in figure 5.2.2, using the mechanism illustrated in figure 5.2.4. The flexural fixture is basically a hollow tube that deflects to the right when the thumbscrew depresses the internal spring. The displacement required to shift fringes by one full fringe order is only 1/2400 mm or 0.417 μm.

The fringe pattern can be viewed in two different ways. Referring to figure 5.2.3, when plate (7) is removed, an observer can see the entire fringe pattern. In this mode the observer makes adjustments to obtain roughly the desired integral number of fringes in the gauge length. When close to the desired count,

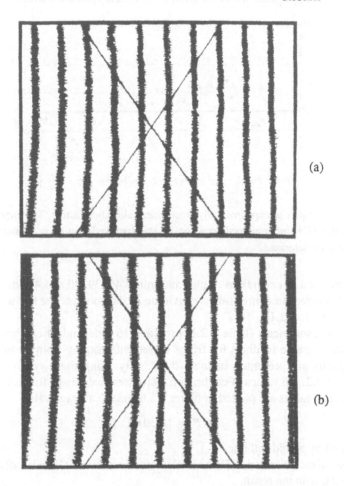

(a)

(b)

Figure 5.2.5. Images from the video camera (*a*) before alignment of a black fringe on a gauge length mark and (*b*) after estimated alignment.

plate (7) is installed to facilitate accurate fringe centring. The fringe pattern is split, with the area surrounding the left-hand **x** mark being diverted to the left camera and a similar area surrounding the right-hand **x** mark being diverted to the right-hand camera. The two images, viewed on separate video monitors, are magnified, allowing high precision in the centring operation. Figure 5.2.5 shows typical images for the misaligned and closely aligned conditions.

5.2.6   Results

Readings from the master gauge were taken at four different strain levels in both tension and compression. The strain was indicated by the moiré fringe count

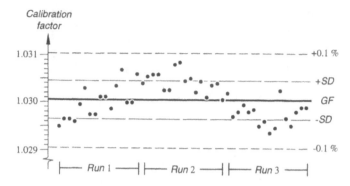

Figure 5.2.6. Scatter of experimentally determined calibration factors. The lines show the average value (CF) of combined error for ±1 standard deviation ($\alpha = 1$) and the range within the 0.1% tolerance.

which consisted of excursions to plus and minus 40, 39, 40 and 41 fringes. This sequence was repeated three times, providing 24 data points. The test sequence is outlined in table 5.2.1.

The measurements (table 5.2.1) provided 16 independent combinations of master strain gauge readings for fringe increments ranging from 78 to 82. The combinations are presented in table 5.2.2. By using these combinations in equation (5.2.2), 16 values of calibration factor were calculated for each run. The 48 calculated values are plotted in figure 5.2.6 and the average value

$$CF = 1.030\,06$$

is indicated by a bold full line.

A standard deviation, SD, was calculated based on a Gaussian distribution for the data, with the result,

$$SD = 0.000\,39$$

which is 380 ppm of the calibration factor. This uncertainty represents the error associated with centring black fringes on the x marks. To obtain an understanding of the overall system uncertainty, each individual contribution must be considered.

Table 5.2.3 lists significant (exceeding 10 ppm) uncertainties in the measurement system. The uncertainty associated with the strain indicator, SD_2, is embedded in the uncertainty of centring fringes, SD_1, and is not considered independently.

Statistically, the bounds for the combined error are predicted by [9]

$$e = \pm\alpha\sqrt{\sum_{i=3}^{5} SD_i^2 + \frac{SD_1^2}{48}} \qquad (5.2.3)$$

where α is the confidence interval in terms of standard deviations (i.e. for a 2σ confidence interval, $\alpha = 2$), i is the subscript for the cases of uncertainty listed in table 5.2.2 and 48 is the number of readings used to determine SD_1. Note that the first term under the radical comprises the systematic errors and the second term comprises the random errors.

A goal for this investigation was to determine a calibration value with high confidence, so a confidence interval of 3σ (99.7%) was used for equation (5.2.3) ($\alpha = 3$). This resulted in a calculated error band for the calibration value of ± 801 ppm, which translates into an absolute band of ± 0.00082 on the calibration factor.

In a practical sense, the specification of confidence to three significant figures is artificial. It stems from the idealized rules of statistics. The experiment was designed to measure the calibration factor of the strain standard within 0.1% tolerance and that accuracy was achieved with extremely high confidence. Rounded off to the 0.1% tolerance, the calibration factor is prescribed as

$$CF = 1.030 \pm 0.001.$$

5.2.7   Refinements

Since the calibration factor is determined periodically, refinements to the experimental apparatus are justified. An example is the manual screw adjustment for fringe shifting (part 11 in figure 5.2.3), which could be replaced by a remotely controlled electrical device, e.g. a piezoelectric actuator. Another example is the observing system, parts (8), (9), (10), which could be replaced by another video camera. The utility and convenience of the apparatus could be enhanced by these and other refinements.

References

[1] 1998 Standard Test Methods for Performance Characteristics of Metallic Bonded Resistance Strain Gages, Designation: E251-98 (West Conshohocken, PA: American Society for Testing and Materials)

[2] 1985 Performance Characteristics of Metallic Resistance Strain Gauges, International Recommendation No. 62, Organization Internationale de Metrologic Legale (OIML), Bureau Internationale de Metrologie Legale, 11 Rue Turgot, 75009 Paris, France

[3] Watson R B Calibration techniques for extensometry: possible standards of strain measurement *J. Testing Evaluation, ASTM* **21** 515–21

[4] Watson R B and Post D 1982 Precision strain standard by moiré interferometry for strain gauge calibration *Exp. Mech.* **22** 256–61

[5] Post D, Han B and Ifju P 1993 *High Sensitivity Moiré: Experimental Analysis for Mechanics and Materials* (Berlin: Springer)

[6] 1985 Model 3800 Wide Range Strain Indicator Instruction Manual, Instruments Division, Measurements Group, Inc., Raleigh, NC, USA

[7] 1985 Model 1550A Strain Indicator Calibrator Instruction Manual, Instruments Division, Measurements Group, Inc., Raleigh, NC, USA

[8] Joh D 1992 Optical-precision alignment of diffraction grating mould in moiré interferometry *Exp. Techniques* **16** 19–22

[9] Lipson C and Seth N J 1973 *Statistical Design and Analysis in Engineering Experiments* (New York: McGraw-Hill)

Chapter 5.3

Interlaminar deformation along the cylindrical surface of a hole in laminated composites

Raymond G Boeman
Oak Ridge National Laboratory, USA

5.3.1 Introduction

The free-edge problem has attracted considerable attention during the past three decades. The proliferation of analytical investigations is a direct result of experimental observations indicating that laminate strength and failures at free edges can be influenced significantly by the stacking sequence of the individual layers. Free-edge stresses exist in a boundary-layer region and they are not accounted for by classical lamination theory. Singular behaviour is exhibited by some of these stresses, making laminates susceptible to delamination at the free edge. A fundamental understanding of the interlaminar stress field associated with the free edge in laminated composites is critical in order to develop delamination prediction models. Consequently, numerous analytical and numerical approaches are described in the literature. For straight free edges, these include finite-difference and finite-element techniques, boundary-layer approaches, force balance methods, stress potentials, sub-structuring and others. As can be seen from figure 5.3.1, discrepancies exist between different solutions even to the extent if the sign of the local stress field [1].

Requirements for cutouts or access holes in composite laminates necessitate the determination of the interlaminar response at the curved free edge of a hole. The presence of a hole complicates the problem significantly. Unlike the straight free-edge problem, where the response is independent of the axial coordinate, this problem is inherently three-dimensional. Three-dimensionality increases the computational burden dramatically. For the curved free edge, the

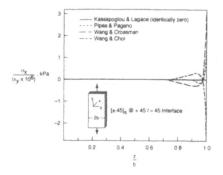

Figure 5.3.1. Comparison of different analytical solutions for the interlaminar normal stress distribution σ_x along a ±45 interface for a $[\pm45]_S$ laminate. The distance z is normalized by the half-width b (after [1]).

analytical approaches taken have included the finite-element method, boundary-layer approaches, force balance techniques, stress potentials and others.

The myriad of analytical and numerical approaches and their seemingly conflicting results suggests a clear need for experimental data to which the analytical solutions can be compared and validated. Experimental investigations reported in the literature are far less abundant than analytical investigations. Moreover, measurements are generally taken in the global sense. That is, the desired quantities are either measured on the face of the laminate, or the gauge-length of the measurement technique is too large (as in the case of strain gages) to measure ply-by-ply variations. Experimental approaches to obtain the global response have included geometric moiré, photoelastic coatings and strain gauge techniques.

Reported herein are partial results of an extensive experimental programme [2] that utilized high-sensitivity moiré interferometry to study free-edge effects along the cylindrical surface of a hole in thick composite laminates. The goal was to generate unique data that reveal the ply-by-ply distributions of the interlaminar normal and shear strains on the cylindrical surface of the hole. The data are intended to document the phenomenal behaviour and serve as a benchmark to which analytical and numerical studies can be prepared and verified.

5.3.2   Experimental procedures

Six composite laminates, 254 mm (10 in) long by 127 mm (5 in) wide, were machined from graphite-epoxy panels fabricated with two different material systems.   Specimen designations, thickness, material types and stacking sequences are given in figure 5.3.2. All specimens except B1 and B2 contained centrally located holes. A convenient hole size of 25.4 mm (1 in) was chosen to facilitate the experimental investigations on the surface of the hole. Specimens

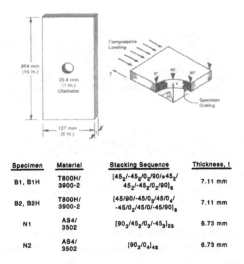

Specimen	Material	Stacking Sequence	Thickness, t
B1, B1H	T800H/ 3900-2	$[45_2/-45_2/0_2/90/\pm45_2/ \\ 45_2/-45_2/0_2/90]_s$	7.11 mm
B2, B2H	T800H/ 3900-2	$[45/90/-45/0_3/45/0_4/ \\ -45/0_2/45/0/-45/90]_s$	7.11 mm
N1	AS4/ 3502	$[90_3/45_3/0_3/-45_3]_{2s}$	6.73 mm
N2	AS4/ 3502	$[90_2/0_4]_{4s}$	6.73 mm

Figure 5.3.2. Specimen geometry, designation, material and stacking sequence.

B1, B1H, B2 and B2H were fabricated from Toray T800H/3900-2 graphite-epoxy, which incorporates toughened, compliant interleaves between layers to improve the damage tolerance.

Coordinate axes x, y, θ, as shown in figure 5.3.2, were chosen such that y is in the direction of loading, x is the thickness direction and θ is measured from the y-axis. The displacements at points along the straight free edge of the specimen, in the x and y directions, are given by U_x and U_y, respectively. At the cylindrical surface of the holes, U_x and U_θ denote the in-plane displacements in the x and θ directions.

The specimens were loaded in compression on a 267 kN (60 000 lbs) Tinius-Olsen electro-mechanical testing machine. The testing load was chosen for each specimen to ensure that the response is within the elastic range and that there are sufficient moiré fringes to permit an accurate ply-by-ply analysis. Relatively modest loads were required and anti-buckling restraints were not necessary.

Interlaminar deformations at the straight edges were obtained using standard techniques described in [3] and [4]. For this particular study, a moiré interferometer suitable for the testing machine environment [5] was used for the straight edges. Fringe patterns were recorded for the longitudinal (in the direction of loading) and the transverse (in the thickness direction) displacements. Strains ε_x (transverse), ε_x (longitudinal), and γ_{xy} were calculated and plotted as functions of the ply number.

To conduct the tests on the hole surface, specialized experimental techniques had to be developed. A flexible grating mould was used to produce the curved diffraction grating on the surface of the hole. To extract the deformation field from the curved specimen grating while it was under load, a replica of the

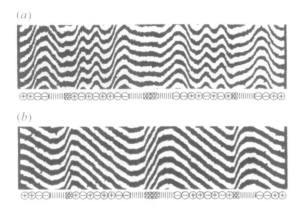

Figure 5.3.3. Moiré fringe patterns depicting the (*a*) longitudinal, U_y, and (*b*) transverse, U_x, deformations at the straight boundary for specimen B1. Contour interval = 417 nm. Load = 71.1 kN (16 000 lbs). The symbols define the fibre orientation in each ply as plus 45°, −45°, 0° and 90° from the direction of loading.

deformed specimen grating was made on a flexible substrate. The replica was removed and straightened for remote viewing in a standard interferometer such as those described in [3]. With these developments it was possible, for the first time, to determine the distributions of the interlaminar strains ε_x (transverse), ε_θ (tangential) and $\gamma_{x\theta}$ on the cylindrical surface of the hole. Details of the experimental procedures can be found in [2].

5.3.3   Results—straight boundary free edge

To serve as a baseline for comparisons, interlaminar strain distributions at the straight free edge were determined. Figure 5.3.3 displays the moiré fringe patterns obtained for specimen B1. Figure 5.3.3(*a*) is a contour map of the longitudinal displacement U_y. The constant fringe gradient in the *y* direction indicates that each ply experiences the same compressive strain ε_y. Figure 5.3.3(*b*) is a contour map of the corresponding transverse displacements U_x, transformed into a more tractable form for detailed data analysis by the addition of carrier fringes of rotation, i.e. a constant fringe gradient in the θ direction. Ply-by-ply variations of the normal strain ε_x are manifest as ply-by-ply variations of the slope of the fringes [3]. Figure 5.3.4(*a*) depicts the through-thickness distribution of the transverse strain ε_x normalized by the magnitude of the axial strain measured from the fringe pattern. Although the overall Poisson's effect is positive, the transverse strain exhibits sign reversals and changes rapidly and systematically in concert with the ply orientation. Variations of the interlaminar shear strain, γ_{xy}, are indicated by the undulating nature of the U_Y fringes. Figure 5.3.4(*b*) represents the corresponding graph of γ_{xy} and indicates that large values occur at

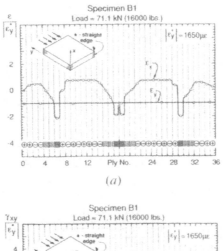

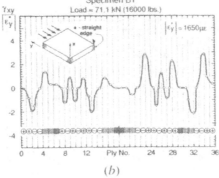

Figure 5.3.4. Interlaminar strain distributions at the straight boundary for specimens with and without holes.

interfaces between the +45° and −45° layers. Ideally, the curve should be anti-symmetrical about the centre plane. The variation or scatter in the magnitudes of the peak values is attributed to material variability, where nominally equal plies exhibit somewhat different responses. Figure 5.3.4 shows that the strains are essentially the same along the straight boundaries for specimens with and without holes. Accordingly, the strains at the straight boundary of specimens with holes are used as normalizing factors to report interlaminar strains as the holes.

5.3.4 Results—cylindrical surface of the hole

With the experimental techniques developed in this study, it is possible to obtain continuous deformation information over more than a quarter of the hole surface. A typical set of fringe patterns is displayed in figure 5.3.5. From symmetry, including the symmetrical ply sequence, this is sufficient to infer the distributions of the interlaminar strains ε_x (transverse), ε_θ (tangential), and $\gamma_{x\theta}$ on the entire cylindrical surface of the hole. As an illustration, figure 5.3.6 depicts the rapid

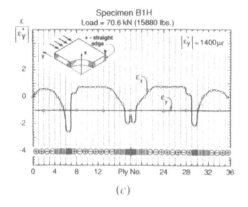

(c)

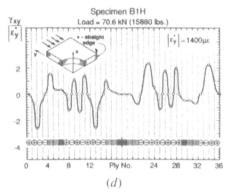

(d)

Figure 5.3.4. (Continued.)

variation of $\gamma_{x\theta}$ as a function of angular position for specimen N2. Note that the shear strain has been normalized by the applied axial strain at the straight edge, indicating that shear stain concentrations exceed a factor of 25. The remaining strain distributions reported herein were extracted at angular position $\theta = 90°$ with respect to the loading axis. Strain distributions at other angular locations can be found in [2].

5.3.4.1   Tangential strains in hole at 90°

Fringe patterns for the U_x and U_θ displacements were obtained from the hole surface. A representative fringe pattern for the 90° location is shown in figure 5.3.7 for specimen N1 illustrating again the rapid changes with angular position (including extremely rapid changes of shear stresses). Tangential strains, ε_θ, were obtained at the horizontal centre line, $\theta = 90°$. Figure 5.3.8 shows that the distribution of ε_θ at the hole differs dramatically from the corresponding ε_y strains at the straight boundary (figure 5.3.4), both in magnitude and distribution. At the straight edge, ε_y is constant throughout the thickness, whereas at the hole

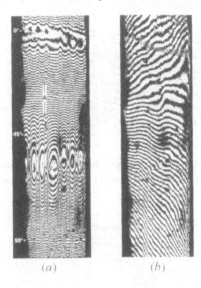

(*a*)   (*b*)

Figure 5.3.5. Moiré fringe patterns depicting the (*a*) tangential, U_θ and (*b*) the transverse, U_x deformations on the cylindrical surface of the hole for specimen N1. Contour interval = 417 nm. Compressive load = 52.9 kN (11 900 lbs).

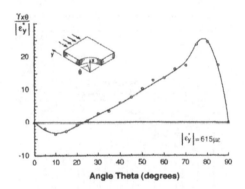

Figure 5.3.6. Interlaminar shear strain distribution at a representative $0°/90°$ interface as a function of angular position for specimen N2. Compressive load = 52.9 kN (11 900 lbs).

ε_θ is amplified on a ply-by-ply basis. Amplification factors can be large, reaching 7.5 at the ply interfaces for specimen N2.

5.3.4.2 Transverse strains in hole at $\theta = 90°$

Figure 5.3.9(*a*) shows that the transverse strain, ε_x at the $90°$ location exhibits a similar trend as that of the corresponding straight boundary free-edge distributions shown in figure 5.3.4. The magnitudes however, are significantly different. The

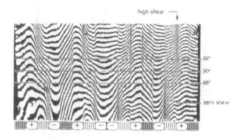

Figure 5.3.7. Moiré fringe patterns depicting the tangential deformation, U_θ, on the cylindrical surface of the hole for specimen N1. Only the region near the horizontal centre line is shown. Contour interval = 417 nm.

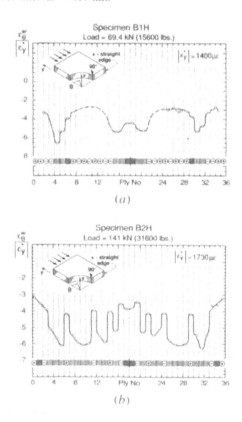

Figure 5.3.8. Tangential strain distributions at the horizontal centre line for the hole boundary. At the straight edge, ε_y^* is constant, independent of the fibre direction. Thus, the strains at the hole are amplified on a ply-by-ply basis.

other graphs in figure 5.3.9 show the ratio of ε_x at the 90° location on the hole boundary to ε_x in the same ply on the straight edge. If every ply exhibited the

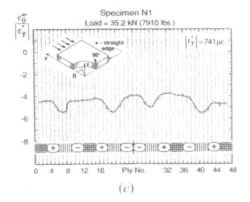

(c)

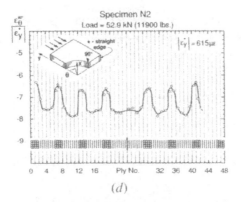

(d)

Figure 5.3.8. (Continued.)

same strain concentration factor, the graph would show a constant ratio. Instead, the ratio varies on a ply-by-ply basis for all the specimens, with the disparity in the range of a full order of magnitude. The transverse strains ε_x^{90} depend upon the fibre direction but also upon the direction and proximity of the neighbouring plies. Added to these variations, there are differences attributable to material variations in nominally equal plies.

5.3.5 Summary

Free-edge effects in laminated composite materials were studied experimentally using high-sensitivity moiré interferomety. Interlaminar deformations were measured on a ply-by-ply basis for straight and curved boundaries in laminated composite panels. The measurements on the curved boundary of the holes appear to be unique, representing the first experimental analyses on a ply-by-ply basis. Interlaminar strains were determined with high fidelity and plotted *versus* the ply number at the horizontal centreline. Results indicate that the strains at the hole

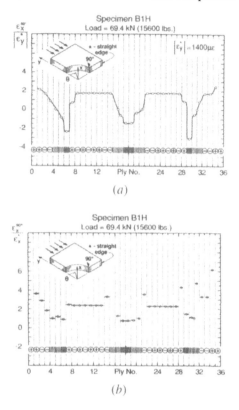

Figure 5.3.9. Transverse strain distributions at the horizontal centre line for the hole boundary. The strains have markedly different distributions at the hole and straight boundaries.

are dramatically different from the corresponding strains at the straight boundary in both magnitude and distribution. The tangential strains and transverse strains at the 90° location are not proportional, on a ply-by-ply basis, to those at the straight boundary. Instead, they are amplified by large factors, up to eight, that appear to be a function of the ply orientation and stacking sequence of each ply. Interlaminar shear strains vary very rapidly and reach very high magnitudes at some locations along the hole boundary. The results are reasonably systematic but exceptions are noted and attributed to variations in properties of nominally equivalent plies.

Acknowledgments

The experimental work described herein was conducted at the Photomechanics Laboratory at the Virginia Polytechnic Institute and State University. The research was supported by the Office of Naval Research (ONR) under grant N00014-90-

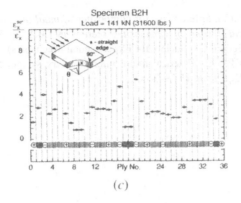

(c)

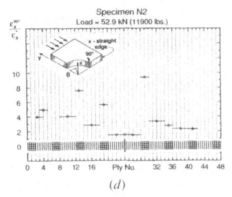

(d)

Figure 5.3.9. (Continued.)

J1688, The National Center for Composite Materials Research (Urbana Illinois) under University Research Initiation grant from ONR N0014-86-K0799, and the Boeing Commercial Airplane Group under contract TZ-689900-0A57N. Specimens N! and N2 were supplied by Dr Mark Shuart of NASA-Langley. Specimens B1, B1H, B2 and B2H were supplied by Dave Cooper of Boeing. This support is greatly appreciated and warmly acknowledged.

References

[1] Kassapoglou C and Lagace P A 1986 An efficient method for the calculation of interlaminar stresses in laminated composites *J. Appl. Mech.* **53** 744–50
[2] Boeman R G 1991 Interlaminar deformations on the cylindrical surface of a hole in laminated composites: an experimental study *Center for Composite Materials and Structures Report* Virginia Polytechnic Institute and State University, CCMS-91-07
[3] Post D, Han B and Ifju P 1994 *High Sensitivity Moiré—Experimental Analysis for Mechanics and Materials* (New York: Springer)

[4] Post D 1989 Moiré interferometry for composites *Manual on Experimental Methods for Mechanical Testing of Composites* section IV A-2, ed R L Pendleton and M E Tuttle (Bethel, CT: Society for Experimental Mechanics)

[5] Guo Y and Ifju P 1991 Practical moiré interferometry system for testing machine applications *Exp. Tech.* **15** 29–31

Chapter 5.4

Interlaminar deformation along the cylindrical surface of a hole in laminated composite tensile specimens

David H Mollenhauer
Wright Patterson Air Force Base, Ohio, USA

5.4.1   Introduction

The free-edge effects in laminated composites were studied for curved surfaces, specifically the cylindrical surface of a hole in tensile specimens. The two-fold objectives of this work were to develop improved experimental techniques and to provide experimental measurements for validation of a new, spline-based numerical analysis method. The research focused on the development and use of moiré interferometry techniques applied to the open hole problem in laminated composites. Strains along the edge of the hole were expected to vary extremely rapidly near ply interfaces. Moiré interferometry was chosen because the excellent spatial resolution and full-field nature of the technique would allow these rapid strain variations across a laminate thickness to be recorded. Much of the research builds upon the excellent research accomplishments of [1]. A comprehensive report of the research can be found in [2] which is available for download on the World Wide Web.

5.4.2   Fringe patterns

Figure 5.4.1 shows two mosaics of fringe patterns from one quadrant of an open hole in a $[0_4^\circ/90_4^\circ]_{3s}$ IM7/5250-4 composite loaded in tension in the x direction. These mosaics are constructed in a piecewise manner with individual fringe patterns, each spanning 3° of arc on the cylindrical surface of the hole. Both the U_θ and U_z components of the displacement are shown. Similar fringe pattern

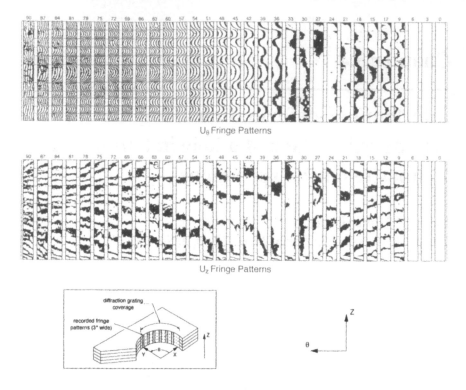

Figure 5.4.1. Mosaic of fringe patterns from the $[0^\circ_4/90^\circ_4]_{3s}$ laminate loaded at 50 MPa. Each fringe pattern is 3° wide. The horizontal lines between fringe patterns are the approximate ply boundaries.

mosaics from a $[+30^\circ_2/-30^\circ_2/90^\circ_4]_{3s}$ IM7/5250-4 composite specimen are shown in figure 5.4.2. Both composite specimens were 610 mm long and 73.3 mm wide, with a 25.4 mm diameter hole located at the specimen centre. The $[0^\circ_4/90^\circ_4]_{3s}$ specimen was 6.35 mm thick and the $[+30^\circ_2/-30^\circ_2/90^\circ_4]_{3s}$ specimen was 6.18 mm thick. The tensile loads were applied by a screw-driven universal testing machine. The fringe patterns shown in these figures are of high quality and allow very detailed analyses. Each fringe pattern contains an abundance of data on geometric scales down to a fraction of a ply thickness. Figure 5.4.3 shows enlargements of portions of three U_θ fringe pattern segments shown in the previous figures. These fringe patterns from regions of severe deformation clearly show the potential for detailed analysis on a local level.

Manual reduction of the fringe patterns to strain has been accomplished for numerous angular locations. These analyses focused on reducing the data along the centre line of each fringe pattern segment. Manual analyses employed either finite-increment or fringe vector (tangent) methods to reduce fringe information to strain data [3]. A few angular locations were examined using digital phase shifting

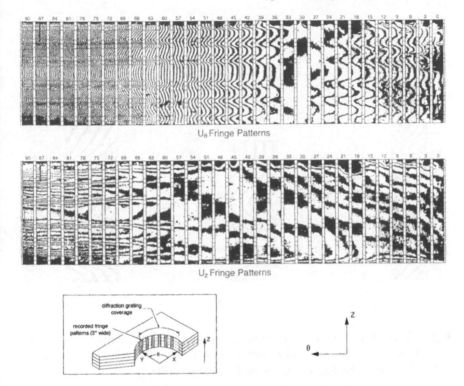

U_θ Fringe Patterns

U_z Fringe Patterns

Figure 5.4.2. Mosaic of fringe patterns from the $[+30_2^\circ/ - 30_2^\circ/90_4^\circ]_{3s}$ laminate loaded at 50 MPa. Each fringe pattern is 3° wide. The horizontal lines between fringe patterns are the approximate ply boundaries.

of the type detailed in Lassahn *et al* [4]. Phase-shifting displacement information was reduced to strain data using numerical differentiation. The confidence in all strain results is high. The experimental procedures used to obtain the fringe patterns shown in figures 5.4.1–5.4.3 and the reasons for the high confidence in these results are detailed in the next section.

5.4.3 Experimental procedure

The procedure used in this research for replicating a grating onto the cylindrical surface of an open hole in a specimen is a multi-step process. A grating was first replicated to a high-carbon steel shim in a process similar to that described in [1]. Then, a grating was replicated onto a steel sector, which had a radius equal to that of the open hole in the specimen. This replication was accomplished by bending the steel shim, with its grating, around the steel sector and allowing the liquid replication material between them to cure. The sector, with its grating, was attached to a flat steel plate with a grating replicated onto its face. This assembly

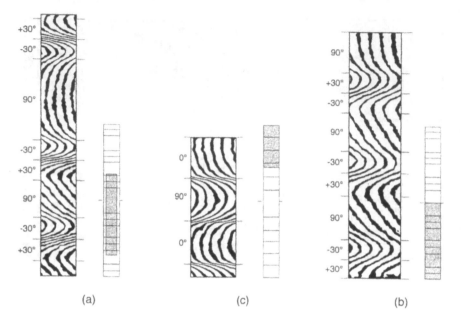

Figure 5.4.3. Enlargements of U_θ fringe patterns from (a) the $[+30^\circ_2/-30^\circ_2/90^\circ_4]_{3s}$ laminate at 75°, (b) the $[0^\circ_4/90^\circ_4]$ laminate at 75° and the $[+30^\circ_2/-30^\circ_2/90^\circ_4]_{3s}$ laminate at 66°. (Shading in the diagrams represents the enlarged regions.)

was then used to replicate a grating onto the specimen as shown in figure 5.4.4. The resulting specimen grating covered a square region around the hole on the face of the specimen and an approximately 110° arc on the cylindrical surface of the hole.

This procedure and the following procedures were developed to eliminate an uncertainty associated with the θ-direction displacements. This uncertainty is caused by unavoidable small variations in the thickness of the original grating replicated onto the steel shim. When the shim is bent into an arc, the strain at the grating surface is proportional to the distance from the neutral axis of bending. The strain would, therefore, be variable if the grating thickness is variable. This variability in the strain at the surface of the shim grating is transferred to the specimen grating in this replication process.

To record deformation information from the specimen grating, a procedure was used that was similar to the procedures used by Mckelvie and Walker [5] and by Boeman [1]. Instead of observing fringe patterns directly from the deformed specimen grating, a replica of the deformed grating was made and fringe patterns were recorded from these replica gratings. A clean steel plate and sector combination was used to make replicas of the specimen grating in a procedure that is essentially the reverse of that illustrated if figure 5.4.4.

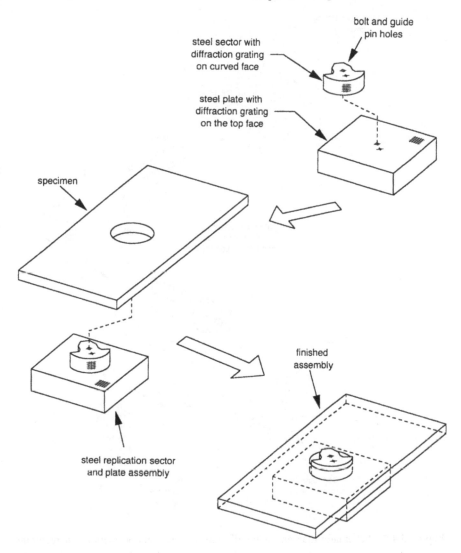

Figure 5.4.4. Schematic diagram of the process for replicating a grating onto (or from) a tensile specimen with a hole.

Fringe patterns were obtained from the curved sector gratings by placing the gratings into a interferometer, as shown in figure 5.4.5. The light diffracting from the sector gratings diverged at a large angle, and only the central portion of the output light was imaged by the fields lens. Therefore, a series of discrete fringe patterns was photographically recorded systematically at angular locations 3° apart. Thus, fringe patterns in narrow rectangular strips, each cropped to be 3° wide, were produced. These strips extended from the top face to the bottom face

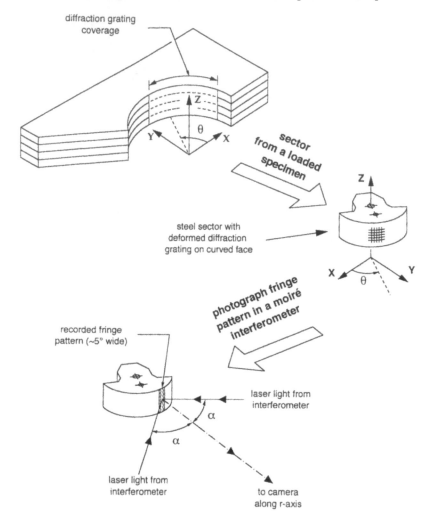

Figure 5.4.5. Schematic diagram of the discrete angular section method for recording fringe patterns from a curved sector grating. The sector is indexed by 3° for each successive fringe pattern photograph.

of the laminate. Assembly of the discrete fringe patterns into mosaics produced the piece-wise continuous images shown in figures 5.4.1 and 5.4.2.

The main advantage of recording fringe patterns in this matter from the curved sector gratings is that it completely eliminates the uncertainties associated with the grating thickness on the steel shims. The moiré interferometer can be tuned to produce a null image at a particular angular location, using the original sector that was used to replicate the gratings onto the specimen. When the null image is obtained, any circumferential strain induced by variations in the original shim grating thickness are automatically corrected at that angular location. The curved sector grating replicated from the deformed specimen is then inserted into the interferometer and the fringe image is recorded for the specified angular location. In the data reduction, the region considered was typically 1°–2° of arc, which minimized the effects of specimen curvature.

Moiré interferometric measurements from replicas of deformed gratings have proved to be extremely advantageous. Replicas can be formed from a specimen in a testing machine or in the field without the need for vibration isolation or specialized interferometers. Fringe information can then be recorded at the leisure of the researcher in the laboratory. Different carriers of rotation and/or extension can be applied to the fringe patterns to enhance the available information. Phase-shifting analysis, which can be particularly susceptible to vibration problems, can also be used on the deformed replicas. Although it was used here on curved surfaces, deformed grating replication is effective on flat surfaces, too.

5.4.4 Results

Detailed analysis of fringe patterns was applied for a number of angular locations. Figures 5.4.6–5.4.8 show results in regions of high strain or regions of special interest. Results of detailed fringe analyses of the U_θ fringe patterns for ε_θ at 75° for the $[+30_2^\circ/ - 30_2^\circ/90_4^\circ]_{3s}$ specimen are presented in figure 5.4.6. Within this figure and most subsequent similar figures, the vertical dotted lines represent the idealized ply boundaries. The filled triangles represent the approximate ply boundaries of the actual specimens. The numbers across the top of these charts represent the fibre angle of the particular ply. The finite-increment method was used to calculate the ε_θ component of strain. Figure 5.4.6 shows the ε_θ variation in relation to the ply structure of the $[+30_2^\circ/ - 30_2^\circ/90_4^\circ]_{3s}$ specimen. Local maxima occur either in the $-30°$ plies or at the interfaces between the $-30°$ and $90°$ plies. Maxima tend to occur within the $+30°$ plies.

Distributions of $\gamma_{\theta z}$ from the top surface of the laminate to the mid-plane at 56°, 58°, 60°, 62° and 64° are shown in figure 5.4.7. These data were obtained by numerically differentiating phase-shifted displacement data obtained at 2° intervals. Figure 5.4.7 shows how rapidly strain variations can change within a small angular area. The peaks located at the $+30°/ - 30°$ interfaces rapidly decay as the angular location changes from 64° to 56°. The behaviour at the $-30°/90°$ interfaces changes from highly negative to positive as the angular location changes from 64° to 56°. The behaviour of $\gamma_{\theta z}$ in the region of the $90°/ + 30°$ interfaces, however, remains relatively constant. These changes in the vicinity of the 60° angular location are visible in the U_θ fringe mosaic presented in figure 5.4.2. Examination of the mid-ply region for angular locations greater

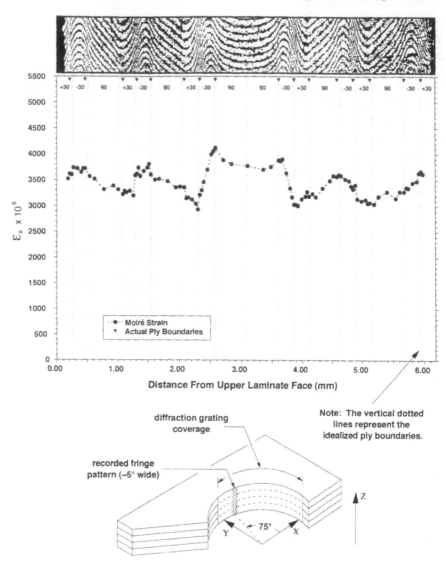

Figure 5.4.6. Circumferential (ε_θ) interlaminar strains at 75° for the $[+30_2^\circ/-30_2^\circ/90_4^\circ]_{3s}$ laminate.

than 60° shows that the central fringes are cupped toward the right. For angular locations less than 60°, these central fringes appear cupped toward the left.

Results of the fringe analyses of the U_z fringe patterns for ε_z at 84° are presented in figure 5.4.8. The fringe pattern differs from that of figure 5.4.2 because a carrier of rotation was added. The carrier fringes, however, do not influence the strains extracted from the pattern. ε_z was calculated from the fringe

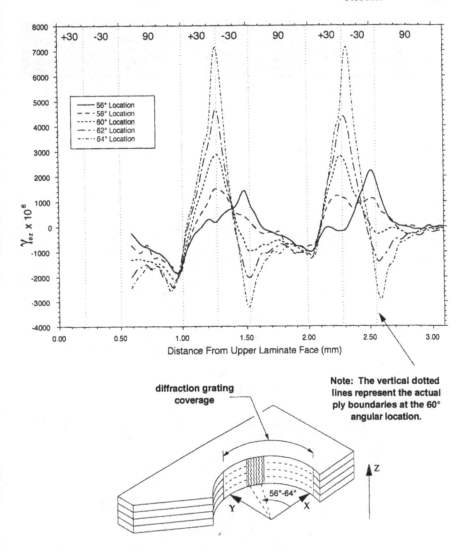

Figure 5.4.7. Interlaminar shear strains ($\gamma_{\theta z}$) at 56°, 58°, 60° and 64° from the top face to the mid-plane of the $[+30^{\circ}_{2}/-30^{\circ}_{2}/90^{\circ}_{4}]_{3s}$ laminate.

pattern using the tangent method. Sharp compressive minima are located near the interfaces between the +30° and −30° degree plies. Broad tensile maxima occur within the 90° ply groupings. A very important large-scale trend is clearly visible in figure 5.4.8. The values of ε_z from the plies near the mid-plane tend to be more compressive than in equivalent plies near the laminate faces. Results such as these have not been documented previously in the literature. This result is important in the sense that the strain behaviour in a given ply is not dependent solely on its

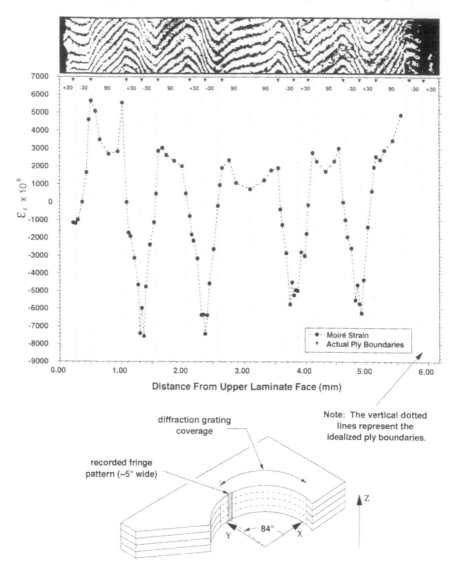

Figure 5.4.8.   Interlaminate transverse strains (ε_z) variation at 84° for the $[+30^\circ_2/-30^\circ_2/90^\circ_4]_{3s}$ laminate. (Note: the U_z fringe pattern contains a carrier of rotation.)

immediate neigbours. Instead, it depends also on the position of that ply within the laminate.

Figures 5.4.9–5.4.11 show the ε_{theta} component of strain at 66°, the $\gamma_{\theta z}$ component of strain at 75° and the ε_z component of strain at 75° for the $[+30^\circ_2/-30^\circ_2/90^\circ_4]_{3s}$ specimen, compared with predictions from a relatively new modelling

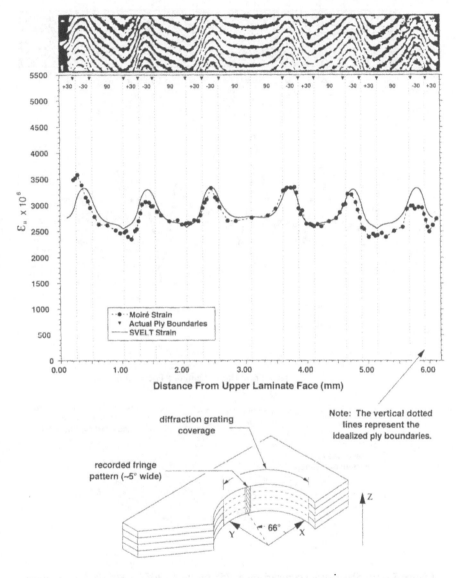

Figure 5.4.9. Circumferential strain comparison at 66° for the $[+30^\circ_2/-30^\circ_2/90^\circ_4]_{3s}$ laminate.

method based on spline approximations of displacement and interlaminar stress components [6]. The results from the experiment and the numerical model are extremely close for the ε_θ and $\gamma_{\theta z}$ components of strain. Similar verification is found at all angular locations examined. The strain trends and magnitudes measured by moiré interferometry are accurately predicted by the spline model.

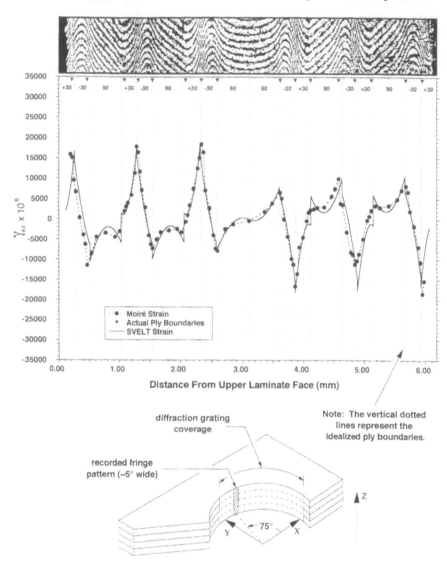

Figure 5.4.10. Shear strain comparison at 75° for the $[+30_2^\circ/-30_2^\circ/90_4^\circ]_{3s}$ laminate.

The comparison between the spline model and moiré for the ε_z component of strain is not as close. The strain trends measured by the moiré experiments are closely matched by the spline prediction. However, the strains predicted by the spline model are consistently lower in magnitude than the moiré measurements. Refinements of material properties in the numerical model might be required. The match between moiré observations and the predictions made by the new spline

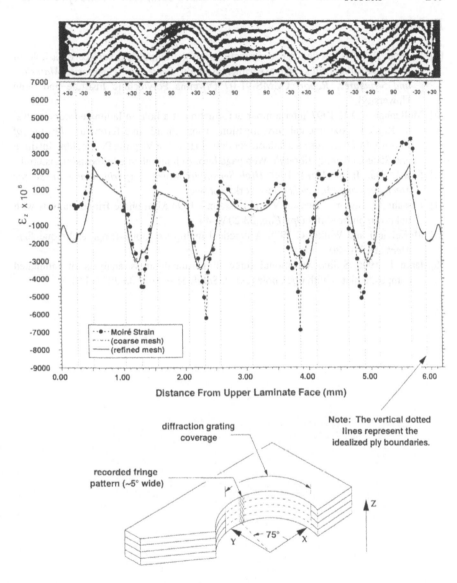

Figure 5.4.11. Transverse strain comparison at 75° for the $[+30_2^\circ/-30_2^\circ/90_4^\circ]_{3s}$ laminate using two different mesh densities to obtain the numerical data. (Note: the U_z fringe pattern contains a carrier of rotation.)

modelling technique are exceptional and important in the sense that previously published analytical predictions of the free-edge behaviour in composites have been inconsistent, even to the point of predicting the opposite sign for certain strain components.

References

[1] Boeman R 1991 Interlaminar deformations on the cylindrical surface of a hole in laminated composites: an experimental study *Center for Composite Materials and Structures, Report* CCMS-91-07 (Virginia Polytechnic Institute and State University)

[2] Mollenhauer D H 1997 Interlaminar deformation at a hole in laminated composites: a detailed experimental investigation using moiré interferometry *Doctor of Philosophy Dissertation* available for download at the Virginia Polytechnic Institute and State University library's Web page http://scholar.lib.vt.edu/thesis/thesis.html

[3] Post D, Han B and Ifju P 1994 *High Sensitivity Moiré: Experimental Analysis for Mechanics and Materials* (New York: Springer)

[4] Lassahn G, Taylor P, Deason V and Lassahn J 1994 Multiphase fringe analysis with unknown phase shifts *Opt. Eng.* **33** 2039–44

[5] McKelvie J and Walker C 1978 A practical multiplied moiré-fringe technique *Exp. Mech.* **18** 316–20

[6] Iarve E 1996 Spline variational three dimensional stress analysis of laminated composite plates with open holes *Int. J. Solids Structures* **33** 2095–118

Chapter 5.5

Thick laminated composites in compression: free-edge effects on a ply-by-ply basis

Yifan Guo
Motorola Semiconductor Products Sector, Tempe, USA

5.5.1 Introduction

Moiré interferometry is applied to study the micromechanical behaviour and smeared engineering properties of thick graphite/epoxy laminated composites. The free-edge effect, an extremely important topic in the mechanics of composites, is investigated. Laminated fibre composites behave differently near the surface, compared to their behaviour in the interior. The reason is that fibres near the surface are discontinued and less constrained by surrounding material, so they can display more readily in response to applied loads. The result is that failures often originate at the free edges. To study the free-edge effect, surface deformations have to be resolved on a ply-by-ply basis. Moiré interferometry, which features very high displacement sensitivity and spatial resolution, is an ideal tool for this investigation.

In this study, specimens were cut from thick-walled cylinders of graphite/epoxy composites. In-plane and interlaminar compression tests were conducted, with the compressive loads applied parallel and perpendicular to the plies, respectively. Note that the coordinate system maintains x perpendicular to the plies. The dimensions and stacking sequences are given in figure 5.5.1 and table 5.5.1. Quasi-isotropic laminates and cross-ply laminates were studied but only the quasi-isotropic specimens are discussed here. The full report is given in [1].

Moiré interferometry with a virtual reference grating frequency of $f = 2400$ lines/mm was used to determine the ply-by-ply deformation on one free surface of each specimen, Face A. The contour interval was $1/f = 0.417 \ \mu m$

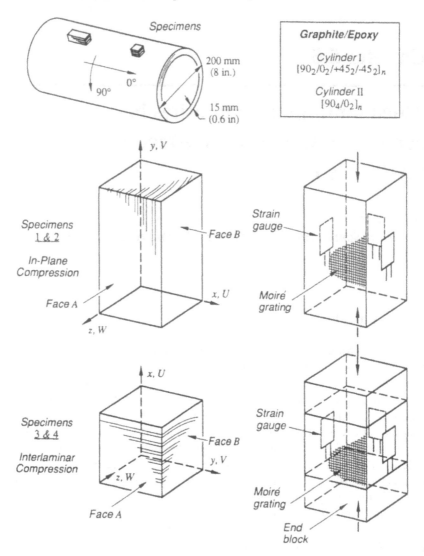

Figure 5.5.1. Specimens cut from thick-walled cylinders.

displacement per fringe order. Electrical resistance strain gages were used to measure average strain and to monitor the uniformity of the compressive loading.

5.5.2   In-plane compression

Experimental results are shown in figures 5.5.2 and 5.5.3 for the quasi-isotropic specimens. The fringe patterns represent the central region indicated by the

Table 5.5.1. Specimen dimensions.

Stacking sequence [a]	Dimensions (mm)	Loading
$[90_2/0_2 + 45_2/ - 45_2]_n$	$25 \times 13 \times 13$	In-plane
$[90_2/0_2 + 45_2/ - 45_2]_n$	$15 \times 13 \times 13$	Interlaminar

[a] Material system: IM6/2258; ply thickness: 0.19 mm.

broken box. They reveal dramatic ply-by-ply variations of displacements and strains. The symbols below the graphs designate the fibre direction in each ply: they represent, from left to right, the $90°$, $0°$, $+45°$ and $-45°$ fibre direction, respectively.

The compressive strain, $\varepsilon_y = (1/f)\,(\partial N_y/\partial y)$, is obtained from the fringe pattern in figure 5.5.2. Here, the vertical distance between adjacent fringes is essentially the same everywhere in the field, which means that ε_y is nearly the same throughout the field. The average strain was determined by measuring the distance Δy between a number of fringes (e.g. 20 fringes) and using the finite increment corresponding to Δy. Thus,

$$\varepsilon_y = \frac{1}{f}\left(\frac{\Delta N_y}{\Delta y}\right) = \frac{1}{2400}\left(\frac{-20}{3.08}\right) = -0.0027$$

where Δy is the measured distance (in millimetres) divided by the magnification of the image. The result is a uniform compressive strain of 2700×10^{-6}: it is plotted as a broken line in figure 5.5.3.

The gradient of N_y exhibits severe variations in the x direction. These relate to variations in the shear strain, determined by $(1/f)(\partial N_x/\partial y + \partial N_y/\partial x)$. The shear component $\partial N_y/\partial x$ is extracted from this pattern by the technique described in [1], which is used to enhance the accuracy of the data reduction.

In order to complete the calculation of shear strains, the term $\partial N_x/\partial y$ is needed. It is obtained from the N_x field (with no carrier) shown in figure 5.5.3(a). The change in fringe order in the y direction is everywhere zero, or essentially zero, so $\partial N_x/\partial y = 0$. Only the data from figure 5.5.2 were needed to calculate the interlaminar shear strains.

Severe strains occur near the interface between $+45°$ and $-45°$ plies, with a magnitude about five times greater than the applied compressive strain ε_y. Shears in the $90°$ plies are essentially constant through the ply thickness, as recognized by the nearly constant slope of N_y fringes across the ply. Note that nominally equal plies exhibit somewhat different strain levels. These differences reflect small ply-by-ply variations in the material, which is characteristic of composite laminates. The normal strain ε_x is determined from the N_x field. Carrier fringes of rotation [2] are used to avoid ambiguity in the fringe orders and to enhance the ease and accuracy of the data reduction. In the absence of load-induced

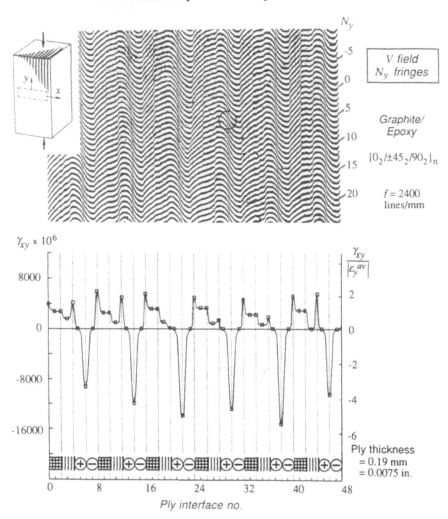

Figure 5.5.2. The N_y moiré fringe pattern in the region of the dashed box for a laminated composite in compression. Fringe orders N_y are assigned with an arbitrary zero. The graph shows the corresponding interlaminar shear strains acting along a horizontal line. Load = 22.2 kN.

fringes, the carrier of rotation is comprised of uniformly spaced fringes parallel to the x axis; therefore, $\partial N_x / \partial x = 0$ and the carrier fringes to not affect the calculated strains. The carrier fringes transform the load-induced fringes to those of figure 5.5.3(b). Again, details of the data extraction are given in [1]. For the $\pm 45°$ plies, ε_x is large and essentially constant across the thickness of the ply, consistent with the nearly constant slope of the corresponding fringes. An

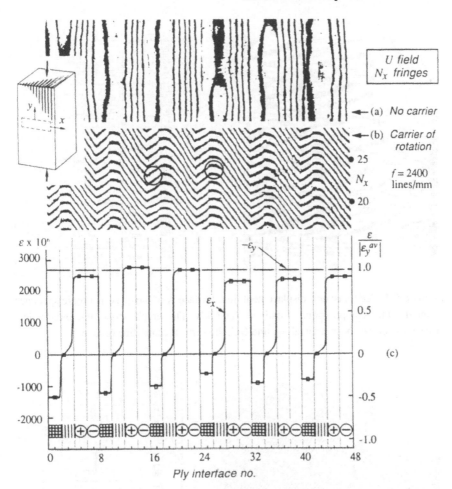

Figure 5.5.3. The N_x moiré fringe pattern for the experiment of figure 5.5.2. Carrier fringes of rotation were applied to evaluate ε_x. The graph shows the transverse normal strains on a ply-by-ply basis.

extremely abrupt change from tensile to compressive ε_x is seen at the interfaces between $-45°$ and $90°$ plies. The strain is nearly zero in $0°$ plies.

5.5.3 Interlaminar compression

Figures 5.5.4 and 5.5.5 show displacement fields for interlaminar compression. The extremely interesting V field (transverse displacement field) shows strong shear strains in repeating cycles. For analysis, the N_y fringe orders were numbered, starting with an arbitrarily selected zero. The direction of the

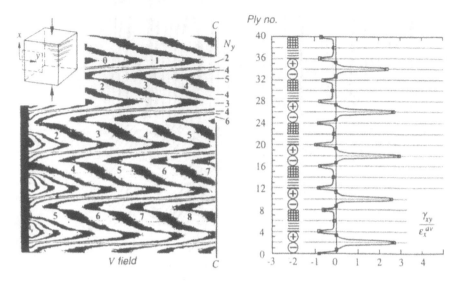

Figure 5.5.4. The transverse displacement field for face A of the quasi-isotropic specimen in interlaminar compression. Load = 6.67 kN. The graph shows the interlaminar distribution of the shear strain along line CC, the vertical centre line of face A.

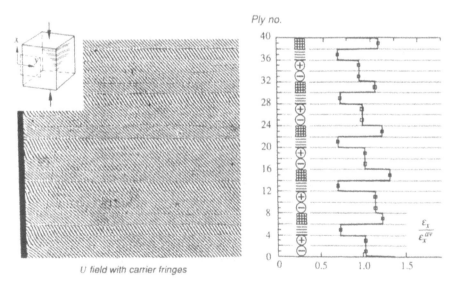

Figure 5.5.5. The vertical displacement field with carrier fringes of rotation, corresponding to figure 5.5.4. The graph shows interlaminar compressive strains ε_x on a ply-by-ply basis.

increasing fringe orders were assigned consistent with the knowledge that transverse strains were tensile, corresponding to a positive Poisson ratio.

The shear strains along the vertical centre line, CC, were determined by the following technique. First a graph of N_y *versus* x was constructed along CC: sufficient data were available using the centres of both the integral and half-order fringes. Note that x should be plotted in specimen dimensions, i.e. the fringe pattern dimensions divided by the magnification. Then, the slope of that curve was measured at various points to determine $\partial N_y/\partial x$. Prior to the introduction of carrier fringes in the U field (figure 5.5.5), the load-induced fringes were essentially horizontal, which means $\partial N_x/\partial y$ was negligible. Therefore, $\partial N_y/\partial x$. was multiplied by $1/f$ to obtain γ_{xy}.

The interlaminar shear strains at $\pm45°$ interfaces are large compared to the applied compressive strains. These occur across the whole width of the face, except near the specimen corners where the peak shears are slightly smaller. These shear strains cannot occur in the interior of the body. Instead, they are free-edge effects, acting in a narrow volume of material near the free surfaces. They can be an important factor leading to material delamination.

The fringe pattern, figure 5.5.4, shows that the transverse normal strains, ε_y, are small. They are essentially constant across the entire face of the specimen except near the corners where the fringe gradients and, consequently, the ε_y strains, are a little larger.

The compressive strains, ε_x, were determined from figure 5.5.5. The average value is

$$\varepsilon_x^{av} = \frac{1}{f}\left(\frac{\Delta N_x}{\Delta x}\right) = -0.0034$$

where Δx is a multiple of the thickness of the eight-ply sequence and ΔN is the change of fringe order in that distance. In this case, the average compressive strain is essentially uniform across the width of the face, i.e. $\Delta N_x/\Delta x$ is constant but varies vertically from ply to ply. The slope of fringes in figure 5.5.5 changes abruptly at the interfaces, corresponding to abrupt changes of compressive strains ε_x. Again the surface strains vary on a ply-by-ply basis, with extremely steep normal strain gradients at interfaces between non-parallel plies.

The alert experimentalist will notice another interesting feature in the V field (figure 5.5.4). Fringe orders are constantly lower near the top of the field than near the bottom. This means the top was displaced to the left. Lines that were vertical before loading lean slightly to the left after loading. The loading fixture did not apply a pure compressive displacement to the specimen but, instead, the displacement had a horizontal or shear component, too. The global horizontal displacement is seen as about -5 fringes in 32 plies. Thus, the corresponding global shear component is:

$$\frac{\Delta V}{\Delta x} = \frac{-5(1/2400)\ \text{mm}}{32(0.19)\ \text{mm}} = -0.00034.$$

The accidental global shear strain of this magnitude was applied together with the much larger compressive strain. In the absence of this global shear, i.e. under

Table 5.5.2. Quasi-isotropic $[90_2/0_2 + 45_2/ - 45_2]_n$ smeared material properties.

Properties [a]	In-plane	Interlaminar
E	57.0×10^9 Pa	11.7×10^9 Pa
ν_{faceA}	0.430	0.078
ν_{faceB}	0.239	0.064

[a] Up to stress level of 138 MPa for in-plane compression and 83 MPa for interlaminar compression

idealized pure compressive loading, the shear strain curve of figure 5.5.4 would be shifted to the right by approximately 0.1 which is the global shear strain value after it is normalized by the average normal strain $\varepsilon_x^{av} = -0.0034$. The peak strains would be about 4% higher and shears in the $0°$ and $90°$ plies would be essentially zero.

5.5.4   Laminate properties

Mechanical properties of the laminates were also calculated from the fringe patterns. They are the average or smeared values of the response of individual plies. Young's modules (E) and Poisson's ratio (ν) were determined by

$$E = \frac{\sigma^{av}}{\varepsilon_1^{av}} = \frac{P}{A\varepsilon_1^{av}} \qquad \nu = \frac{\varepsilon_2^{av}}{\varepsilon_1^{av}}$$

where P is the applied compressive force, A is the cross-sectional area of the specimen perpendicular to the load direction, superscript av signifies average value and subscripts 1 and 2 indicate measurements parallel (1) and perpendicular (2) to the load direction. The compressive properties of the laminates are listed in table 5.5.2.

For determination of E, the ε_1^{av} was the average strain measured on the four faces, where strain gauge data were used for the three faces and moiré data for the fourth face. Poisson's ratio was determined from the moiré data, where the strains were average values from face A, the face with the moiré grating. Young's modules and Poisson's ratio were constant up to the maximum stress level of the tests (table 5.5.1), i.e. the stress–strain properties were linear within the loading ranges.

5.5.5   Conclusions

The mechanical behaviour of thick graphite/epoxy laminated composites were characterized using moiré interferometry. Specimens cut from thick-walled cylinders were tested in in-plane compression and interlaminar compression. Interlaminar strains were determined on free surfaces on a ply-by-ply basis. Carrier fringes were used to enhance the analysis. High normal strains and very high interlaminar shear strains were measured, documenting severe free-edge effects. Average or smeared values of Young's modules and Poisson's ratio were determined for these laminates for in-plane and interlaminar compression.

References

[1] Guo Y, Post D and Han B T 1992 Thick composites in compression: an experimental study of micromechanical behaviour and smeared engineering properties *J. Composite Mater.* **26** 1930–44
[2] Guo Y, Post D and Czarnek R 1989 The magic of carrier patterns in moiré interferometry *Exp. Mech.* **29** 169–73

Chapter 5.6

Loading effects in composite beam specimens

Peter G Ifju
University of Florida, USA

The Iosipescu shear test is one of the methods used most to obtain the shear stress–strain response and shear strength of fibre reinforced composites [1, 2]. Technically, in order to determine shear strength, the entire test section of the specimen is required to be in a state of pure and uniform shear stress. This is not possible with flat coupons, such as the Iosipescu specimen, because the shear stress at the free edges must be zero. Therefore, the shear stress cannot be uniform and high shear stress gradients are present. With such shear stress gradients, equilibrium can only exist if normal stress gradients are also present. Hence, the state of stress throughout the entire test section can be neither pure nor uniform. Although it is not possible to create the ideal state of stress in flat coupons, one can create an optimal state of stress that is predominantly uniform shear throughout most of the test section. This can be accomplished by adjusting the notch geometry (angle, notch radius and depth) [3] and the method of load introduction. Since the stress distribution is also strongly dependent on material orthotropy, optimization is required for each material tested.

There has been considerable attention devoted to the optimization of notch geometry as a function of material orthotropy. The effect of load introduction on the stress distribution, however, has mostly been overlooked. Since there is a St Venant effect associated with load distribution and proximity to the test section, stress concentrations produced by contact forces adjacent to the notch can spill into the notch region and alter the antisymmetric behaviour of the specimen. In order to obtain insight into how load distribution and proximity affect the stress distributions, a series of tests was performed using moiré interferometry [4] on unnotched Iosipescu-type specimens of various material orthotropies. This research was motivated by an analytic study conducted by J M Whitney [5]

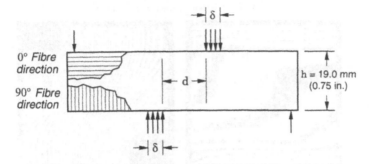

Figure 5.6.1. Specimen geometry and loading conditions.

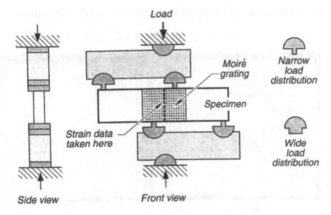

Figure 5.6.2. Loading fixtures used to investigate effects of load distribution width and spacing.

on specimens of the same geometry (figure 5.6.1) and material (AS4/3501-6 graphite/epoxy).

To duplicate the model parameters of the analytic study, a set of specialized loading fixtures were designed as illustrated in figure 5.6.2. A parametric study was performed to determine the stress distributions in the test section by adjusting the width δ of the load distribution and the spacing d of the loads. Three cases were examined: one with narrow load distributions spaced close together ($\delta/h = 0.2, d/h = 0.5$), one with narrow load distributions spaced far apart ($\delta/h = 0.2, d/h = 1$) and one with wide load distributions spaced close together ($\delta/h = 0.6, d/h = 0.5$). The geometry of the loading fixtures, shown in figure 5.6.2, were designed with the intention of performing the experimental version of the analytic study. Loading points were machined so that they could pivot to insure that the distributions were as uniform as possible. Three composite specimen types were tested: 0°, 90°, 0/90° graphite/epoxy. An aluminium specimen was also tested.

0° *Fibre direction*, δ/h = 0.6, d/h = 0.5

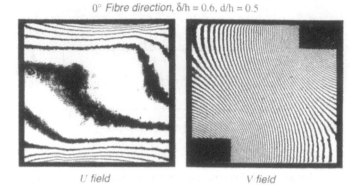

U field   V field

Figure 5.6.3. Moiré interferometry fringe patterns for a 90° fibre direction specimen.

90° *Fibre direction*, δ/h = 0.2, d/h = 1.0

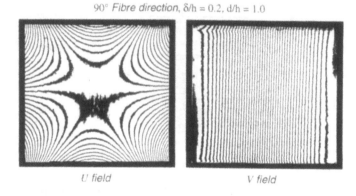

U field   V field

Figure 5.6.4. Moiré interferometry fringe patterns for a 90° fibre direction specimen.

0/90 *Cross-ply*, δ/h = 0.2, d/h = 0.5

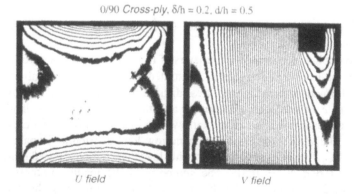

U field   V field

Figure 5.6.5. Moiré interferometry fringe patterns for a 90° fibre direction specimen.

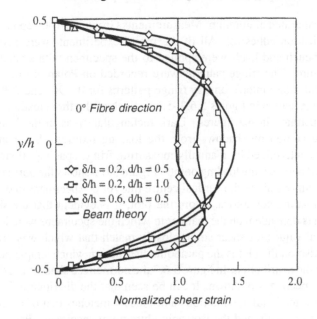

Figure 5.6.6. Shear strain distributions for the 0° specimen.

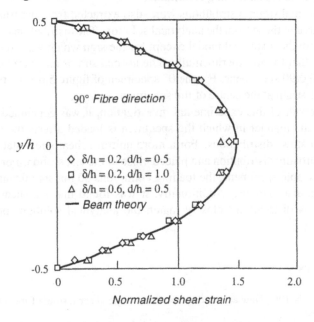

Figure 5.6.7. Shear strain distributions for the 90° specimen.

The specimen gratings for moiré interferometry were replicated from 1200 lines/mm crossed-line epoxy grating moulds which were coated with two layers of evaporated aluminium. Measurements Group PC-10C (epoxy) was used as the replication adhesive. All of the moiré experiments were performed on an optical bench and loads were applied to the specimen with a small vice-like loading fixture. The fringe patterns were recorded on Polaroid film. Examples of the typical moiré interferometry fringe patterns for 0°, 90° and 0.90 cross-ply specimens are shown in figures 5.6.3–5.6.5. Each of the three loading conditions are also illustrated in the figures. Dark rectangular areas in the V-field fringe patterns are caused by shadows from the loading fixtures. Shear and normal strains were determined by manually measuring fringe spacings along a vertical line centered between the loading points. Figure 5.6.6 shows the normalized shear stress distributions along the vertical centre line for 0° specimens under the three difference loading conditions. From the plot, it is evident that the shear strain distribution is dependent on the manner in which the specimens were loaded. The experimental values for shear strain closely match that which was predicted by Whitney. Also on the plot is the parabolic distribution which simple beam theory predicts for the distribution. In figure 5.6.7 the normalized shear stress distribution for the 90° specimens is shown. It can be seen that the distribution is not highly dependent on the loading conditions and closely matches that of a simple beam theory. The cross-ply and the isotropic aluminium specimens displayed only a modest dependence on loading conditions.

The normal strain distributions were also extracted from the fringe patterns and compared to those from the analytical solution. The distributions were similar to those from the analytical model except for the sign which was predicted to be opposite to that found experimentally. The normal strains are extremely small in the 90° and 0/90 specimens. For the 0° specimen of figure 5.6.3, ε_y rises to 45% of the shear strain at the centre of the specimen.

As a result of this experimental investigation, it was concluded that for 0° specimens the manner in which the specimen is loaded affects the shear stress and normal stress distributions. For a more uniform shear stress state, the loads should be broadly distribution and placed close to the test section. For 90°, cross-ply and isotropic specimens, the loading conditions do not significantly alter the shear stress distributions. Additionally, it was found that the shear stress and normal stress distributions closely match the analytical solutions predicted by J M Whitney.

References

[1] Iosipescu N 1967 New accurate method for single shear testing of metals *J. Mater.* **2** 537–66
[2] Adams D F and Walrath D E 1983 The Iosipescu test as applied to composite materials *Exp. Mech.* **23** 105–10
[3] Walrath D E and Adams D F 1983 Analysis of the stress state in an Iosipescu shear test specimen *Report* UWME-DR-301-102-1, Department of Mechanical Engineering, University of Wyoming Laramie, WY, NASA Grant NAG-1-272

[4] Post D, Han T and Ifju P G 1994 *High Sensitivity Moiré: Experimental Analysis for Mechanics and Materials, Mechanical Engineering Series* (New York: Springer)

[5] Whitney J M 1996 St Venant effect in the Iosipescu specimen *Proc. Am. Soc. Composites* (Dayton, OH: ASC) pp 845–52

CHAPTER 6

ELASTIC–PLASTIC HOLE EXPANSION

Chapter 6.1

Elastic–plastic hole expansion: an experimental and theoretical investigation

C A Walker, Jim McKelvie and Jim Hyzer
University of Strathclyde, Glasgow, UK

6.1.1 Introduction

Fastener holes are often expended to create a zone of compressive residual stress around the bore. Incipient cracks will see this compressive stress and their growth rate will be inhibited. This study looked at three typical geometries—a quasi-infinite plate, a thick cylinder and a simple lug—to define the nature of the residual stress field in each case.

6.1.2 Residual stress measurement

In his original discussion of the measurement of residual stresses, Mathar [1] drilled a hole into the specimen surface and measured the resulting deformation between two points diametrically opposite one another. He was able to derive equations, which related the measured deformations to the state of residual stress pre-existing in the surface before the drilling took place. The basis of this present series of studies is similar but Mathar's mechanical extensometer was replaced by a grating on the surface with deformations in this grating being monitored by an optical interferometer. The principle, though, of measuring the surface deformation remains essentially the same as in Mathar's initial work.

6.1.3 Moiré interferometry

The theory of the optical system and the grating application have been discussed elsewhere [2, 3]. In practice, deformation contours at an interval of 1.05×10^{-3} mm may be recorded across a 50 mm diameter field of view. This was adequate to encompass the whole of the areas of interest. It is a feature of the technique used that a short exposure time (e.g. 1/500 s) is adequate and the patterns of deformation are, therefore, substantially unaffected by vibration. Thus, no particular steps to eliminate environmental effects were required for the successful acquisition of interferometric data. As a result, it was possible to remove the specimens and replace them after hole expansion, hole drilling and dissection and to view the fringes after only a few moments of adjustment of the specimen position. This is a major advantage of this system of moiré interferometry.

6.1.4   Hole-drilling method

The residual stresses created by the cold expansion of a hole in a quasi-infinite plate were measured by the hole-drilling method, where the stress-relieving holes were cut with a 3 mm diameter milling cutter. It has been found [2] that, with care in the matters of tool sharpness and of limiting the pressure of the tool feed, that this can give results comparable with those obtained from abrasion or from high-speed drilling. Each hole was drilled to a depth of 3 mm.

In the initial stages of this work, the moiré contours were regarded as providing data similar to the rosette gauges normally used. The strains, however, are quite small, and rapidly varying and it was realized that a superior technique would result from a reversion to Mathar's original proposition [3, 4]. Bearing in mind that moiré fringes are contours of constant displacement, a count of the fringes, from one side of the hole to the other, gives a direct measurement of the displacement arising from the hole drilling.

In order to extract the pre-existing state of residual stresses, the deformations were measured in three directions. It has been shown that the accuracy achieved does not depend upon knowledge of the hole's position [2]. Holes were drilled at an array of points covering the grating area: by this means the whole strain field was quantified.

6.1.5   Dissection method

For the case of the thick cylinder and simple lug, the hole drilling method was less desirable since stress information right up to the edge of the bore was required; therefore, the dissection method was preferred. For this method the specimens were mounted on a rotation stage and cut radially with a 0.8 mm (1/32 in) slitting saw, at 45° intervals. Care was taken to make all cuts to a depth of 1.0 mm, and then increase all of their depths by 1.0 mm increments, until the specimens were cut halfway through their 20 mm thickness. This was done so that the relieving of residual stresses by one cut would not create plastic deformation in another section of the specimen.

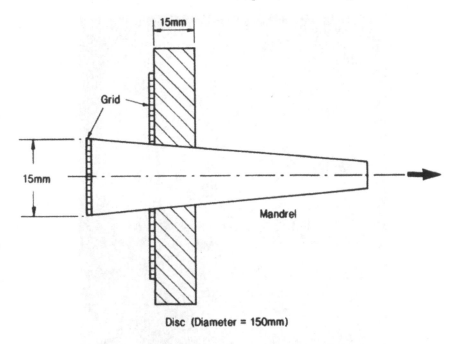

Figure 6.1.1. Expansion of a hole in a quasi-infinite plate.

To determine residual stress, two sets of moiré fringe patterns were recorded. The first recorded the total residual displacement after the cold expansion of the hole, from which the total strain (elastic plus plastic) was determined. The second was recorded after the dissection of the specimen and, therefore, relaxation of residual stresses and contained only the plastic component of strain. The elastic strain was determined as the difference between these two records and the residual stress was calculated through Hooke's law.

6.1.6   Experimental investigation

This took place in three phases: first of all a hole was expanded in a quasi-infinite plate; a thick cylinder was evaluated next, before the lug geometry was begun.

6.1.6.1   Expansion of a hole in a quasi-infinite plate

The specimen used was a plate of pure aluminium (figure 6.1.1) with a central hole. The dimensions were chosen so that the far-field elastic strains induced by hole expansion had fallen to a low level within the circumference of the specimen, simulating the behaviour of a truly infinite plate. A tapered steel mandrel was pulled into this hole. The dimensions had been calculated such that, when the

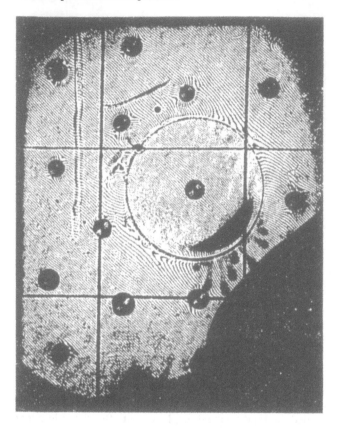

Figure 6.1.2. Interferogram of residual deformation field around the holes drilled in the quasi-infinite plate. This is the x-component, with a degree of rotational and horizontal mismatch. The fiducial lines are 20 mm apart. The circle denotes the end of the oversize mandrel.

end of the mandrel was flush with the surface, the interference between hole and mandrel was 0.5% of the original diameter of the hole.

A grating was applied, prior to expansion, over one face of the aluminium plate, and to the end of the mandrel. Since it may be presumed that the steel mandrel is in a state of uniform elastic compression, a measurement of the strain in the mandrel provides a measurement of the pressure at the mandrel/plate interface. This pressure may then be used as input for the theoretical prediction of the elastic–plastic strain distribution for comparison [5, 6]. A radial pattern of holes was drilled through the grating into the plate and a count made of the fringe differences across three diameters, as outlined previously, enabling the residual strain to be calculated at each point (figure 6.1.2).

A comparison, then, may be made between the residual strain, as a function of radius, measured experimentally and predicted from theory with the interfacial

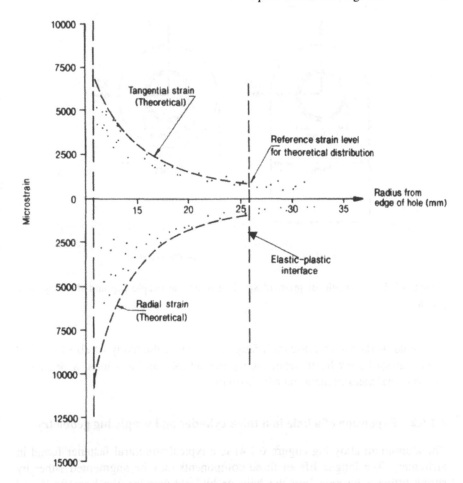

Figure 6.1.3. Comparison of experimentally measured and theoretically predicted strains in a quasi-infinite plate for an expansion of 0.5% of the hole diameter. The two sets of data were made to coincide at the elastic–plastic interface.

pressure as input (figure 6.1.3). In fact, the agreement is poor, due to the inadequacy of the thick cylinder theory, which does not take account of the real material behaviour in the plastic zone. Using the interfacial pressure as input gives an inadequate prediction of the residual strains. However, if the measured residual strain, on the elastic–plastic boundary, is used as an input condition so that the measured strain on the elastic–plastic boundary is taken as a fixed point for the theoretical distribution, then the theoretical strain curves agree well with the experimental data, forming an upper bound for the measured data. These findings are relevant to many situations in which fatigue life estimations are to be made in similar configurations—e.g. fastener holes. These fastener holes are

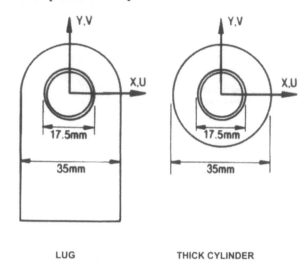

Figure 6.1.4. Schematic diagram of specimens for the simple lug and thick cylinder geometries.

often cold worked to increase their fatigue life. Since the likely levels of residual stress cannot be predicted from purely theoretical considerations, a degree of experimental measurement must be invoked.

6.1.6.2 Expansion of a hole in a thick cylinder and simple lug geometry

The aluminium alloy lug (figure 6.1.4) is a typical structural fastener found in airframes. The fatigue life of these components may be augmented either by shrink-fitting a bushing into the hole or by cold-working the bore itself and probably bushing also. This study sets out to examine this residual stress field and to compare it with a thick cylinder from an experimental and theoretical viewpoint.

In both lug and thick cylinder specimens, the bores were expanded by an oversized mandrel through an interposed bushing (figure 6.1.5). A consideration of the detailed expansion process shows it to be complex, with both axial and radial effects dominating at different times during the passage of the mandrel. The net effect, however, is of a plastic zone surrounded by an elastic zone, with a regime of residual compressive circumferential stress at the bore, inhibiting the initiation of fatigue cracks. This process is schematically shown in figure 6.1.6, where a typical element close to the bore is shown to undergo initial expansion; then, after the passage of the mandrel, the effect of the residual elastic stress field is to force this element back down into a state of residual compression. In this experiment, the mandrel diameter was 1.5% larger than the internal diameter of the bores.

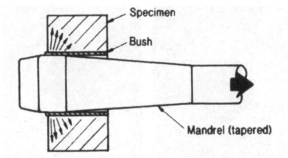

Figure 6.1.5. Cold expansion processes using an oversize tapered mandrel and a lubricating bushing.

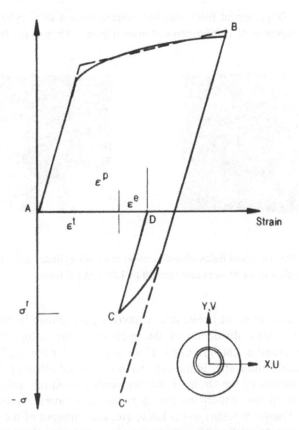

Figure 6.1.6. Elastic–plastic stress/strain path for an element close to the bore undergoing initial tension followed by compression due to the elastic recoil from elements further from the bore. The element finishes at C in a state of residual compression $-\sigma'$.

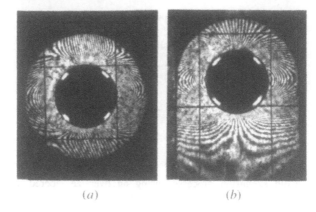

(a) (b)

Figure 6.1.7. Displacement fields after bore expansion in a thick cylinder and a lug. Horizontal component of deformation—contour interval 1.05 μm per fringe (black to black).

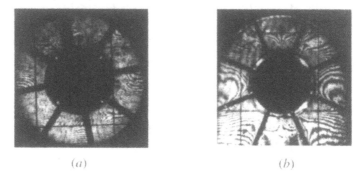

(a) (b)

Figure 6.1.8. Displacement fields after dissection in a thick cylinder and a lug. Horizontal component of displacement—contour interval of 1.05 μm per fringe.

Typical displacement fields, after expansion, are shown in figures 6.1.7(a) and 6.1.7(b). After dissection of the specimens, the deformation fields in figures 6.1.8(a) and 6.1.8(b) resulted. If we now consider the available data, the state of residual strains after initial expansion is the total effect of the elastic and plastic components of the strain. If the segments are entirely relieved of stress by the dissection, the pattern remaining is the total pattern, minus the elastic component. Figures 6.1.8(a) and 6.1.8(b) are, thus, images of the plastic strain field V-component existing after expansion: by subtracting the two patterns, the degree of elastic recovery during dissection at any point may be assessed. It is considered that, to a good approximation, with the dimensions pertaining on this instance, there is a total relief of the residual stresses. A striking feature is the almost perfect diametral symmetry of the plastic displacements in the case of the lug, despite some rotations of the individual segments .

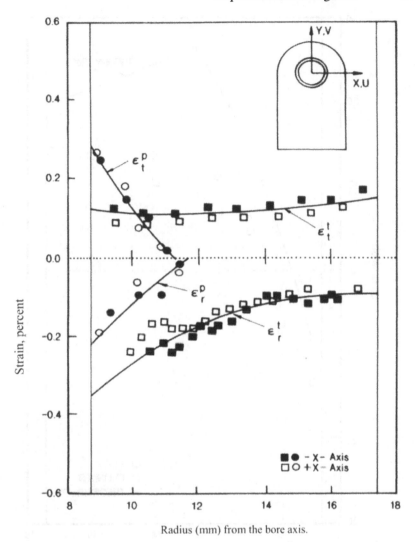

Figure 6.1.9. Residual strain distribution (x-axis) in a cold-worked lug.

Figure 6.1.9 shows a plot of the total strain in the lug geometry and the plastic component of total strain, as a function of the radius perpendicular to the line of symmetry. Both radial and tangential components are shown. Also, the calculation of the elastic recovery component of strain is illustrated. It will be seen from the plot of elastic recovery that elements close to the bore were indeed in compression after the expansion.

For comparison, the strain levels predicted theoretically [7] are shown, using the 1.5% bore expansion as input data.

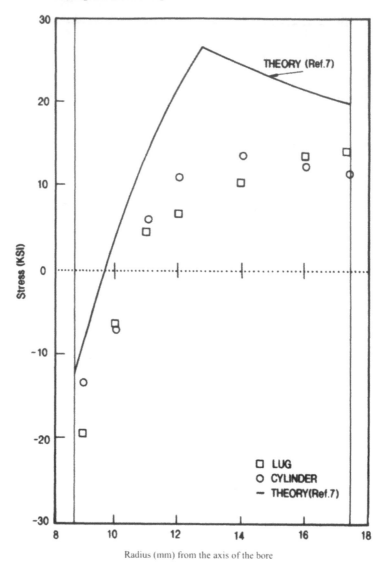

Figure 6.1.10. Residual stress in the lug geometry after expansion—comparison with thick cylinder and theoretical prediction.

Finally, using the elastic recovery data, the original state of residual stress, after mandrel expansion, may be calculated. These figures are shown in figure 6.1.10, where, for comparison, the curve predicted by theory [7] is indicated (the stress–strain behaviour of the alloy being known). While the general form of the curves is similar, the theory is not a good predictor of stress levels. This is scarcely to be expected, given the complexity of the process and the simplicity of the theoretical model.

6.1.7   Conclusion

Moiré interferometry has been used to investigate the complexities of the expansion of holes in infinite plate, thick cylinder, and lug geometries. Residual stress patterns were revealed either by a hole-drilling process, or by dissection of the specimens. In each case the results were compared with theoretical predictions, using material properties, and either bore pressure or bore dilation as input data. The level of agreement between experiment and theory was variable, depending upon the information required, and the limitations of thick-cylinder theory have been demonstrated. One benefit deriving from the use of moiré interferometry was the graphic way in which differences in strain distribution may be evidenced, and the short gauge-length over which the measurements may be made.

References

[1] Mathar J 1932 *Ermittlungen von Eigenspannugen durch Messung von Bohrlochver-formungen Archiv fur das Eisenhuttenwessen (Gruppe E, Berichte des Werkstoffanschusses No 201)* 6 pp 277–81

[2] Walker C A 1994 A history of moiré interfcrometery *Exp. Mech.* 34 280–305

[3] MacDonach A, McKelvie J, MacKenzie P M and Walker C A 1983 Improved moiré interferometry and applications in fracture mechanics, residual stress and damaged composites *Exp. Tech.* 2 20–24

[4] MacDonach A 1983 The development and application of a high sensitivity interferometer for whole-field stress analysis *PhD Thesis* University of Strathclyde

[5] Hoppe V 1986 Elastic plastic stress fields of cylindrical symmetry: an analysis with moiré interferometry *MSc Thesis* University of Strathclyde

[6] Ford H and Alexander J M 1977 *Advanced Mechanics of Materials* (Chichester: Ellis Horwood)

[7] Steele M C 1952 Partially plastic thick-walled cylinder theory *Trans. ASME J. Appl. Mech.* 19 133–40

Chapter 6.2

Roller expansion of tubes in a tube-plate

C A Walker and Jim McKelvie
University of Strathclyde, Glasgow, UK

6.2.1 Introduction

A tube-plate is a thick disc pierced with an array of holes to accept heat exchanger tubes. The tubes are sealed into the plate by expansion, by welding or by a combination of both techniques. The oldest, and most convenient, method is roller expansion, which uses three angled rollers that are rotated in the tube by a tapered mandrel. As they rotate, the rollers create zones of elastic–plastic expansion in the tube and in the surrounding area of the tube-plate. In essence, the tubes are expanded to be larger than the holes in the tube plate and, hence, are gripped by the tube-plate to create the desired sealing action. Ideally, one wants the tube plate to expand elastically, with no permanent deformation, since it is the elastic accommodation of the tube plate that is the source of the sealing force. The process is reliable and long lasting. The problem with all expansion processes, however, is that the interfacial force between the tube and tube plate, after only a small amount of expansion quickly, reaches a peak, and then falls away. Further expansion does not achieve a better seal—in fact the situation gets worse, since the plastic zone of expansion propagates into the tube plate. When adjacent or even remote holes are now rolled, any tube with a less-than-ideal seal may be found to be leaking, even when the seal had previously been tested and found to be sound. The problem is caused by the complexity of the disturbance caused to the elastic accommodation by over-enthusiastic rolling, so that even remote rolling may reduce the interfacial pressure between a tube and the tube plate to the point where the join will leak. Attempts to recover the situation by further rolling have mostly been found to be less than a brilliant success, since without a detailed understanding of the overall situation, it is unclear which tube to roll and by how much. It has to be born in mind that what is desired at the end of the day is a whole tube-sheet of non-leaking joints, not simply individual sealed pipes.

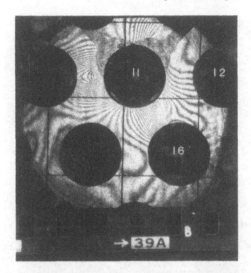

Figure 6.2.1. Live recording of tube rolling. Hole number 11 is being rolled. The shadows above the hole are of the rolling tool nosepiece, which is projecting towards to viewer. Note that lobes of deformation to the left of hole 11 and at the '5 o'clock' position, that follow the tool around as it rotates (the third lobe is hidden by the shadow of the rolling tool).

These features were observed many years ago and analysed theoretically [1, 2], but no experimental programme had been carried out prior to this study to assess the form of the elastic–plastic zones of expansion created by the rolling. This study looked at the rolling process by visualizing the expansion process in real time using moiré interferometry.

6.2.2 Experimental procedure

A model tube plate with an array of 21 holes was fixed in the field of view of an interferometer. A moiré grating was formed on the surface of the tube plate. This process was perhaps the crucial part of the programme, since ideally the grating should run to the very edge of each hole in a state of perfection and adhere to the tube-plate. Initially, the tube plate was placed on a flat glass plate and silicone rubber plugs were cast into the holes. The glass plate was removed and the grating cast over the whole area, including the holes, which were filled flush with the surface by the plugs. By cutting carefully round the edge of each hole to release the plug, and removing it, the holes were cleared.

Short lengths of tube (~2 cm longer than the thickness of the plate) were inserted into the tube plate and a rolling tool inserted from the back, i.e. on the side remote from the interferometer.

Figure 6.2.2. Montage of deformation fields around an array of hole taken after the completion of rolling of hole 11 but part way through the rolling of hole 9. The strain levels around the periphery of hole 11 lie in the range 0.5% to −1%. The fiducial lines are 20 mm apart. The moiré fringe sensitivity is 0.5 μm per fringe.

The experiment began with the tube just beginning to bite on the inside of the tube-plate bore and continued until the tube-rolling torque reached a preset limit. Interferograms of the deformation field around the tube being rolled were recorded at a rate of 4 frames s^{-1} for a total of 2 min, using a high-capacity film magazine. A typical frame shown in figure 6.2.1, where the first of the holes in the sequence is at an early stage of being rolled. It will be seen that a lobe of deformation is following each roller around as it rotates. This lobe will largely vanish when rolling ceases—i.e. the deformation is elastic—but it can be seen that such deformations, transferred even remotely, may disturb the sealing of other tubes.

At a later stage of the rolling—(see figure 6.2.2, which is a montage of several deformation fields)—it can be seen how the effects of rolling one tube may be transferred to the nearest and next-nearest neighbours, even at this early stage in the rolling of the whole tube-plate. Hole number 11 has been rolled to completion. It is evident that the residual stress field extends well beyond the immediate hole where the rolling has been carried out. In fact, the deformation around hole 11 is into the plastic regime. One may understand how interaction with further amounts of rolling from adjacent holes may result in a reduction in the interfacial pressure and, in the worst case, a leak developing.

In essence, the result of this study was a quantitative reiteration of the conclusions of the 1940 studies, namely that in a large tube-plate, rolling should be carried out delicately, with the larger picture in view. The tube-plate cannot be considered to be simply an assemblage of discrete tubes surrounded by an elastic deformation zone.

References

[1] Goodier J N and Schoessow G J 1943 The holding power and hydraulic tightness of expanded tube joints: analysis of the stress and deformation *Trans. ASME* **65** 489–96

[2] Grimson E D and Lee G H 1943 Experimental investigation of tube expanding *Trans. ASME* **65** 506–22

CHAPTER 7

HYBRID MECHANICS

Chapter 7.1

Some applications of high sensitivity moiré interferometry

C Ruiz
Oxford University, UK

7.1.1 Introduction

The application of moiré techniques to the determination of the state of strain over a wholw field is particularly attractive when compared to point-by-point measurements such as resistance strain gauges. The two applications described here illustrate the power of the technique.

7.1.2 Interferometric study of dovetail joints

The fatigue strength of dovetail joints between blades and discs in axial compressors has been shown to depend on the parameter

$$k = \sigma \tau \delta$$

where σ is the stress acting on a plane normal to the contact surface, τ is the shear stress between the two surfaces in contact and δ is the relative slip [1, 2]. Using a 1200 lines/mm cross grating provides a nominal sensitivity of 0.42 μm/fringe, quite adequate for the amount of slip found between blade and disc. The technique developed by Post for the production and replication of these interference gratings [3] has been used to obtain the parameter k and the whole field displacement map [4]. The optical arrangement was very simple: light from a ruby laser was first expanded through a pinhole and collimated by a parabolic mirror as shown in figure 7.1.1. The collimated light (A), deflected by a flat mirror, illuminated the test piece at the appropriate angle to produce a first-order diffracted beam normal to the test piece (A (+1)). Part of the collimated light (B) was further reflected by a second mirror (b) to produce the second diffracted beam (B (−1)). The resulting

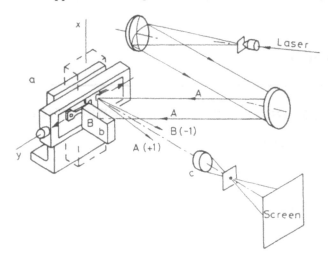

Figure 7.1.1. Experimental set-up for the study of a dovetail joint.

image was projected on a screen. To obtain displacements in the horizontal and in the vertical directions, the rig used to load the test piece was mounted on a rotating table (a) as shown in figure 7.1.1. Typical fringe patterns are shown in figure 7.1.2. The high fringe density is sufficient to obtain the strains from the derivatives of the displacement without any need to increase the sensitivity by, for example, phase shifting. Along the flank of the dovetail, where most of the slip occurs, the diffraction grating has peeled off causing a loss of definition. Although it is possible to work out the displacements by extrapolation from the adjoining area, this process introduces unknown errors. One way out is to produce the diffraction grating on the surface of the test coupon by a photoresist method. This has been done very successfully for the study of crack-tip deformation in fatigue [5].

7.1.3   Application to the study of cracks

The four-beam optical arrangement shown in figure 7.1.3 has superseded the original system used for the study of the dovetail. The collimated beam illuminates the specimen and the three mirrors M_1, M_2, M_3. Half the incident beam B_1 impinges directly on the specimen surface while the other half B_2 impinges on M_3 and then reflects to the specimen. In figure 7.1.3(*b*), beams A_1 and A_2 illuminate the grating lines normal to the y-axis after being deflected by M_1 and M_2 to produce the V displacement fringes. The four beams A_1, A_2, B_1 and B_2 illuminate the specimen simultaneously. Shutters F_x, F_y are used to occlude A_1 and A_2 or B_1 and B_2 so as to permit viewing of the U or V displacement fringes only. The interferogram is collected by a charged coupled device (CCD) camera whose lens serves the dual purpose of decollimating

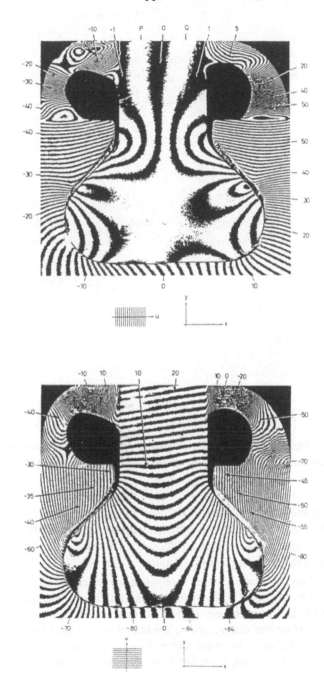

Figure 7.1.2. Typical fringe patterns for displacements in the horizontal and vertical directions.

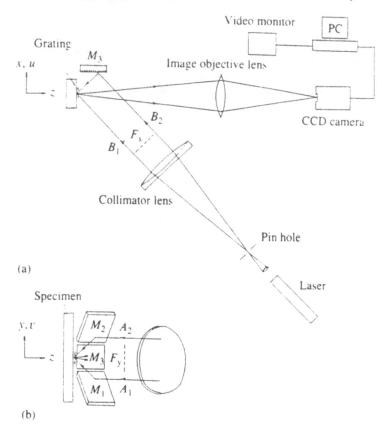

Figure 7.1.3. Optical arrangement for simultaneous illumination in the horizontal and vertical directions. B_2 is produced by mirror M_3; A_1 and A_2 by mirrors M_1 and M_2.

and focusing the image on the CCD sensor which consists of a scanned array $(512 \times 512$ pixels) of light-sensitive diodes with processing hardware to convert the scanned output to an analogue video signal format. The signals are then recorded and analysed using a desk computer.

When studying small features such as sharp notches and crack-tips, it is advantageous to use some form of fringe multiplication procedure to increase the density of data over the whole field in order to obtain an accurate value of the strains from the displacements measured in the interferogram. In the case of cracks, the stress intensity factor K may be obtained directly from the displacements. In particular, according to linear elastic fracture mechanics,

$$K_{\mathrm{I}} = \lim_{r \to 0} \left(\frac{uE}{4} \sqrt{\frac{2\pi}{r}} \right)$$

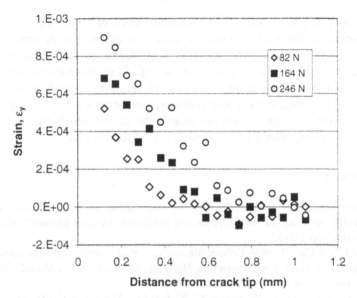

Figure 7.1.4. Variation of strain normal to crack centre line ahead of the crack-tip in a carbon fibre reinforced laminated beam with a central part-through thickness crack.

where u is the crack opening displacement at a distance r from the crack-tip [7]. In practice, crack-tip plasticity invalidates this equation for small values of r and far-field stresses have the same effect remote from the crack-tip. As a result the experimental value of K_I determined from this equation is of doubtful accuracy. It is then preferable to use a hybrid experimental–numerical technique in which the displacements are measured from the interferogram, the stresses are found by finite elements and the two are combined to obtain the strain energy release rate from Irwin's crack closure integral,

$$G_I = \lim_{r \to 0} \frac{1}{2\Delta} \int_0^\Delta \sigma_y(\Delta - r, 0)u(r, \pi)\, dr$$

where G_I is the strain energy release rate, σ_y is the normal stress at a distance $(\Delta - r)$ ahead of the crack-tip along the x-axis (crack centre line) and $u(r, \pi)$ is the crack opening at a distance and from the crack-tip. Details of the technique are found in [7].

The measurement of K_I in composites presents a real challenge. While moiré interferometry, applied to metallic specimens, has been shown to provide reliable and accurate results, the same cannot be said for composite materials. Figure 7.1.4 [8] shows the strain distribution ε_y ahead of the crack-tip in a carbon fibre reinforced epoxy laminate beam consisting of 21 layers with a crack extending over four layers under three load levelsa, 82, 164 and 246 N. The scatter of the results makes it impossible to draw smooth strain-position curves from which to evaluate the limiting values of either K_S or G_I. The layer thickness, which is 0.125 mm, coupled with the oscillations introduced by residual stresses are believed to be responsible for the scatter. High-sensitivity moiré interferometry is, in this case, such a fine and sensitive instrument that, by picking up disturbances at the microstructural level shows up the limitations of the continuum solid mechanics approach followed in theoretical fracture mechanics. The experimental technique always reveals the reality which may not always agree with the simplified theoretical models used for analysis.

References

[1] Boddington P H B and Ruiz C 1983 A biaxial fatigue test for dovetail joints *Proc. ASME Int. Conf. on Advances in Life Prediction Methods (Albany)* (New York: ASME) pp 277–83

[2] Ruiz C and Chen K 1986 Fatigue life assessment of turbine blade-disc joints *Proc. Int. Conf. Fatigue of Engineering Materials and Structures, I.Mech.E (Sheffield)* (London: IMechE)

[3] Post D 1983 Moiré interferometry at VP1 & U *Exp. Mechanics* **23** 203–10

[4] Ruiz C, Post D and Czarnek R 1985 Moiré interferometric study of dovetail joints *J. Appl. Mech.* **52** 109–14

[5] Fellows L J, Nowell D and Hills D A 1997 Comparison of crack closure measurements using moiré interferometry and results from a Dugdale-type crack closure model *Adv. Fract. Research, Proc. 9th Conf. Fracture (Sydney)* (London: Pergamon) pp 2551–8

[6] Poon C Y, Kujawinska M and Ruiz C 1993 Spatial-carrier phase shifting method of fringe analysis for moiré interferometry *J. Strain Anal.* **28** 79–88

[7] Poon C Y and Ruiz C 1994 Hybrid experimental–numerical approach for determining strain energy release rates *Theor. Appl. Fracture Mach.* **20** 123–31

[8] Ruiz C, Hallett S R and Gungor S 1998 Application of fracture mechanics criteria to laminated CFRP beams under impact *Proc. ECCM 8* (Cambridge: Woodhead) vol 1, pp 118–24

Chapter 7.2

Hybridizing moiré with analytical and numerical techniques: differentiating moiré measured displacements to obtain strains

R E Rowlands
University of Wisconsin, USA

7.2.1   Introduction

Engineering components frequently contain geometric discontinuities that give rise to highly localized stresses. Such geometric discontinuities range from holes or side notches to cracks. While solutions to many such cases are available for isotropic materials, the ability to obtain analytical solutions for such situations in orthotropic composites is significantly more challenging. Stresses now depend on material properties and perforated composites need not fail at the location of maximum stress. Furthermore, traditional finite-element analyses (FEA) around holes or notches require many small elements and appreciable computer storage capacity and run-time. Numerical FEA reliability is aggravated if the external shape and/or loading is complicated or not well defined. The nature of the orthotropy (directions of material axes) can influence where on the boundary of a cutout the maximum stress occurs, thereby necessitating even more elements. Classical moiré analyses of such problems can be hampered by difficulties in determining reliable experimental information on the boundary of a discontinuity. Since moiré interferometry typically measures displacements, the recorded data must be differentiated to get strains or velocities.

Recognizing this, this section discusses the advantages of hybridizing moiré-measured displacements with numerical and theoretical concepts to evaluate reliable strains and stresses in isotropic and orthotropic composite members under static or impact loading.

7.2.2 Analyses

7.2.2.1 General comments

In solid mechanics, one often seeks spatial or temporal derivatives of the displacements. Moiré-measured displacements typically contain some scatter and such scatter aggravates obtaining reliable derivatives. It can, therefore, be advantageous to smooth the measured data prior to differentiating. However, one wants to be careful not to remove relevant information during a smoothing process. Splines, regression analysis and finite-element concepts can be advantageous for processing and differentiating moiré data.

7.2.2.2 Splines and regression analysis

One-dimensional splines are effective when the situation under consideration involves differentiating a single measured displacement, particularly if slopes are only needed in one, consistent direction. 'Smoothing splines' enable measured data to be smoothed as well as differentiated. While one-dimensional splines can be used to differentiate multiple displacements, i.e. both in-plane displacement components of a plane-stress problem, the situation is now more labour-intensive and cross-derivatives (say shear strains) are often less reliable. In [1–3] one-dimensional 'smoothing' cubic splines are used to smooth and differentiate moiré data. Differentiating moiré fringes by traditional methods is most reliable when the differentiation direction is fairly normal to the fringes or, at least, when it cuts across many fringes. Nevertheless, the chain rule of differential calculus can be combined with one-dimensional splines to compute accurate derivatives in directions of relatively low fringe density [3].

One-dimensional splines are suitable for processing moiré data in local regions. However, more full-field approaches are convenient if one desires to determine, from moiré measured U- and V-displacements, the strains and stresses throughout substantial regions of a structure or its entirety. Two-dimensional splines or regression analysis can then be used to evaluate the three in-plane strains/stresses full-field from the moiré measured horizontal and vertical displacements such as those of figure 7.2.1 [4].

7.2.2.3 Finite-element concepts

Several ideas based on various finite-element concepts are effective for processing experimental information, including converting moiré-measured displacements into strains [5–12]. Based on prior research by Segelman *et al* [5], Feng and colleagues [6–10] developed and applied a very effective scheme for smoothing and differentiating both in-plane measured displacements. The technique (called SFER, for smoothing finite-element representation) results in a governing equation identical in form to that for displacement-based FEA but the (quasi) stiffness matrix is independent of material properties and the (quasi) load vector

Figure 7.2.1. Two-dimensional deformation field ($u + v$) around a hole.

involves moiré-measured displacements at discrete locations rather than loads. The method is applicable to two and three dimensions. Applications of the method to composites and fracture mechanics are included in [7–10], whereas [11] extends the technique to transient problems. Figure 7.2.2 demonstrates the use of SFER to evaluate the shear strain around the edge of a hole in a perforated composite from moiré measured displacements.

Motivated by an earlier paper on hybrid FEA [13], Rhee *et al* [12] published a hybrid moiré numerical and analytical method for evaluating accurate strains and stresses on the edge of geometric discontinuities from limited moiré-measured displacements well away form the discontinuity figure 7.2.3. This figure illustrates the ability to determine the tangential stress on the boundary of a hole accurately by this approach. The experimentally determined tangential stress distribution in figure 7.2.3 is based on moiré-measured displacements which originated no closer than $2.4R$ from the hole edge, where R is the hole radius. Satisfying the traction-free conditions exactly on the hole boundary of figure 7.2.3 by analytic continuation greatly contributes to the reliability of the moiré results based on such distant measured data.

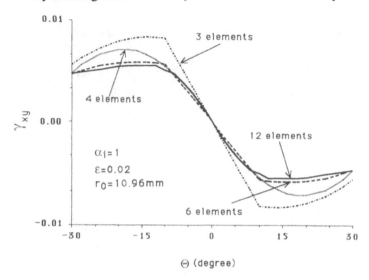

Figure 7.2.2. Shear strain around the edge of a hole in a composite plate. (Moiré processed by SFER.)

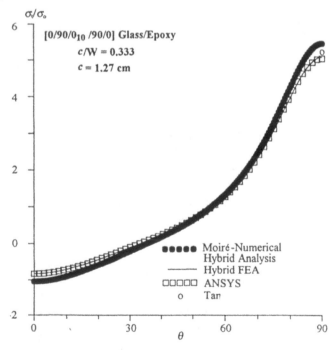

Figure 7.2.3. Tangential stress on boundary of hole in orthotropic tensile composite strip.

7.2.3 Summary

The effectiveness of employing numerical and analytical concepts to help process and differentiate moiré-measured displacement data has been demonstrated.

References

[1] Rowlands R E, Liber T, Daniel I M and Rose P G 1973 Higher-order numerical differentiation of experimental information *Exp. Mech.* **13** 105–12

[2] Sciammarella C A and Rowlands R E 1974 Numerical and analog techniques to retrieve and process information *Proc 5th Int. Conf. on Experimental Stress Analysis (Udine)* (Bethel, CT: SEM)

[3] Rowlands R E, Jesen J A and Winters K D 1978 Differentiating along arbitrary orientations *Exp. Mech.* **18** 81–6

[4] Rowlands R E, Winters K D and Jensen J A 1978 Full-field numerical differentiation *J. Strain Anal.* **13** 177–83

[5] Segelman D J, Woyak D B and Rowlands R E 1979 Smooth spline-like finite-element differentiation of full-field experimental data over arbitrary geometry *Exp. Mech.* **19** 1–12

[6] Feng Z and Rowlands R E 1987 Continuous full-field representation and differentiation of three-dimensional experimental vector data *Computers and Structures* **56** 979–90

[7] Feng Z, Sanford R J and Rowlands R E 1991 Determining stress intensity factors from smoothing finite-element representation of photomechanical data *Eng. Fracture Mech.* **40** 593–601

[8] Feng Z, Rowlands R E and Sanford R J 1991 Stress intensity determination by an experimental-numerical hybrid technique *J. Strain Anal.* **26** 243–51

[9] Feng Z and Rowlands R E 1991 Smoothing finite-element and experimental hybrid technique for stress analysing composites *Computers and Structures* **39** 631–9

[10] He S, Feng Z and Rowlands R E 1997 Fracture process zone analysis of concrete using moiré interferometry *Exp. Mech.* **37** 367–73

[11] Yang Y Y 1994 A finite-elements hybrid method for representing and differentiating transient photomechanical data *PhD Thesis* University of Wisconsin, Madison, Wisconsin

[12] Rhee J, He S and Rowlands R E 1996 Hybrid moiré-numerical stress analysis around cutouts in loaded composites *Exp. Mech.* **36** 279–87

[13] Gerhardt T D 1984 A hybrid/finite element approach for stress analysis of notched anisotropic materials *ASME J. Appl. Mech.* **51** 804–10

CHAPTER 8

RESIDUAL STRESSES

Chapter 8.1

Automated moirè interferometry for residual stress determination in engineering objects

Małgorzata Kujawińska and Leszek Sałbut
Warsaw University of Technology, Poland

8.1.1 Introduction

Manufacturing processes, i.e. the action of load and thermal stresses, create residual stresses in engineering elements [1]. These stresses acting in conjunction with those produced by live loads can threaten the safety of an engineering structure by creating and propagating fatigue cracks. The difficulty in measuring them non-destructively, the unpredictability of their magnitude, sense and direction, their adverse ability to combine with stress corrosion and environmental and fatigue situations and the difficulty in removing them can render them extremely troublesome. Residual stresses should be evaluated under operating conditions to assess their effect on the intended service capability of the component. This problem has been investigated for many years in the search for an inexpensive and reliable method of determining these stresses.

In order to carry out an experimental examination of residual stresses, two questions have to be answered: (1) *how to determine* residual stresses from a mechanical point of view, i.e. how to reach inside a 3D body; and (2) *how to measure* these stresses from an experimental point of view, i.e. which experimental technique to apply. Non-destructive techniques that can reach and measure phenomena inside a 3D body include x-ray spectroscopy, neutronography and ultrasonography. These have the capability, to an extent, of penetrating the interior of an investigated body and measuring the stress state. However, the results obtained can differ depending on the material of the element and often they are only qualitative.

Another way of determining residual stresses is to rely on comparative measurements of the displacement/strain before and after stress relief. The measurements are performed by strain gauges, photoelastic coatings and moiré or interferometric techniques. Stresses may be relieved by hole drilling (trepanning), sectioning a body into small pieces or by annealing.

Here, automated moiré (grating) interferometry (GI) has been selected as an experimental technique for determining the strain on the surface of a sample. This technique, in conjunction with numerical analysis based on a theoretical stress model, enables the original 3D residual stresses to be determined.

This chapter describes the methodology and instrumentation for measuring two types of engineering elements, namely a railway rail and a laser beam weldment. Although these elements are very different, they both suffer from high values of localized residual stresses caused by manufacturing and thermal processes.

8.1.2 Objects of measurements

8.1.2.1 Railway rail

To verify analytical and experimental methods for residual stress analysis in railway rails, a controlled laboratory loading process is necessary. This can be realized by rolling a specially designed wheel along the top of the testing rail (see figure 8.1.1(a)) with a known contact force. A new UIC 60 rail (made in Huta Katowice steel works) was rolled on an EMS-60 testing machine [2] in the Central Research Institute of the Polish State Railways, Warsaw. The rolling process was performed with the load 150 kN, at a frequency of 1 Hz and the number of cycles applied was 5×10^5.

Usually, the so called Batelle technique [3, 4] is used for 3D residual stress determination in rails. This is based on the general concept of destructive stress relief and consists of two main sectioning steps:

- Yasojima–Machii (YM) slicing (for δ_{xx}, δ_{yy}), which is done by cutting a thin transverse slice from a rail slab and dicing it;
- Meier sectioning (for δ_{zz}), which is done by cutting a rail into longitudinal rods (4 mm $\times$ 4 mm $\times$ 500 mm).

Here, the in-plane displacement fields in the YM slices were determined, for comparison purposes, by three methods:

1. By strain gauges placed on the sections of rail in which the stress had been relieved by cutting. The cutting was performed with great care (slow speed, heavy cooling) in order to avoid the introduction of additional stresses.
2. By moiré interferometry while the in-plane displacement was being measured at the centre of the same sections of rail slice as mentioned earlier.

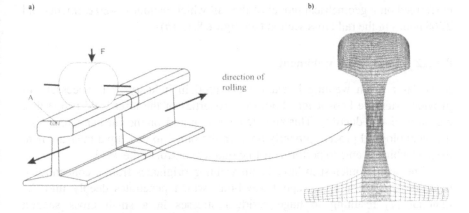

Figure 8.1.1. Schematic diagram of the rolling process (*a*) and a geometrical model of the rail slice (*b*). A is the area considered for the analysis of wheel/rail contact region.

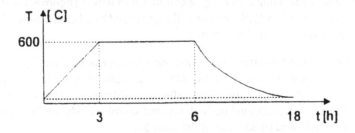

Figure 8.1.2. Temperature change *versus* time during the annealing process of the YM slice.

3. By moiré interferometry while the residual stress in the samples was being released by an annealing process, which was performed according to the temperature scheme shown in figure 8.1.2.

For the moiré interferometry, the YM slices underwent an additional preparation stage: a cross-type diffraction grating with frequency 1200 lines/mm was attached to the upper part of the slice. When the stress was released by sectioning, the aluminium master grating was transferred through epoxy. However, stress relief by the annealing process required the application of a high-temperature-resistant grating. For this, the process developed at the National Physical Laboratory (UK) was adopted [5].

An analytical model for predicting residual stress in rails was developed by Orkisz *et al* [6] and it relies on the notion of a shakedown state established by the highest load to which the rail has been subjected. This model has been simplified by Magiera [4] for the in-plane strain and stress components in the rail slice, which can be compared with our experimental results. Calculations were performed on a geometrical model of the rail which includes 2480 elements and 2768 nodes in the rail cross section (see figure 8.1.1(*b*)).

8.1.2.2 Laser beam weldment

Since laser beam welding is extensively used in industry, it is necessary to provide extensive knowledge about the properties of the laser weld properties, e.g. strength and ductility. This knowledge is required, on the one hand, to design the technological process properly and, on the other, to develop a measurement method which is able to perform on-line quality control of the weld.

The special interest in laser beam welding originates from a concentrated heat source generating a vapour key-hole, which penetrates deeply into the material [7] resulting in huge residual stresses in a small cross section (figure 8.1.3). Two types of welded specimen were considered:

1. a ferritic steel, overlap-joint welded specimen, FS, produced by a beam power of 4 kW using a welding speed of 1.8 m min^{-1} (figure 8.1.4(*a*));
2. a structural steel welded specimen, SS, produced by a beam power of 10 kW, using welding speed of 0.8 m min^{-1}.

The problem of residual stress determination was approached here by relying on comparative measurements of the in-plane displacement/strain before and after the stress was released. Cross-type reflection diffraction gratings (frequency 1200 line/mm) were replicated on both welded samples as shown in figure 8.1.4. Stresses were released by sectioning the samples:

• For FS, this was done by cutting through the gratings perpendicular to the weld direction. Due to the limited size of the specimen grating (45 mm × 45 mm), the displacements were measured separately at the areas K and L located symmetrically at both sides of the weld at a distance 2–3 mm from the weld centre and at the area P which covered the weld zone (figure 8.1.4(*a*).
• For SS, the sample was cut along the weld (figure 8.1.4(*b*)).

8.1.3   Experimental set-up and procedure

The residual stresses in the rails and welded samples have to be determined in a system which allows the large areas of interest to be analysed. The one-channel moiré interferometer presented in figure 8.1.5 is designed to test components with a measuring area of up to 100 mm × 100 mm [8]. In the system two mutually coherent light beams symmetrically illuminate the specimen grating. The first diffraction orders of the two illuminating beams interfere and produce a fringe pattern with information about the in-plane displacement in the direction perpendicular to the grating lines. Residual stresses cause the deformation state

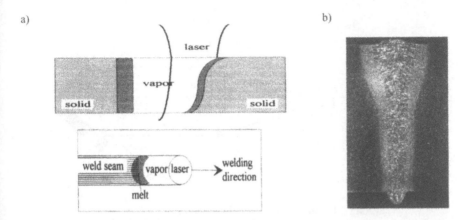

Figure 8.1.3. The scheme for the laser beam welding process (*a*) and the laser weld geometry (*b*).

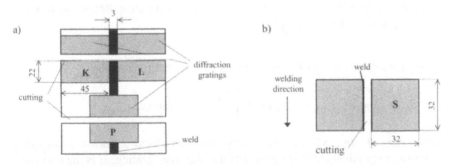

Figure 8.1.4. The welded specimens under study: ferritic steel, FS (*a*) and structural steel, SS (*b*). K, L, P, S are the areas with gratings.

which, when released by cutting or annealing, causes, in turn, the opposite deformation of the sample under test. The displacement measurement depends on this opposite deformation.

In the system presented here, one mirror head and the rotary mounts of the specimen and CCD camera enable the U and V displacements to be measured. The rotary mounts provide high accuracy (a fraction of a pixel), so that the U and V displacement maps are correlated for further strain calculations. The specimen is also repositioned very accurately (a fraction of a fringe), so that any systematic errors connected with the specimen grating and aberration of the system can be eliminated by simple subtraction of the displacement maps before and after stress release.

The interferograms obtained in the system represent contour maps of the in-plane displacements (alternatively $u(x, y)$ and $v(x, y)$) with a half specimen

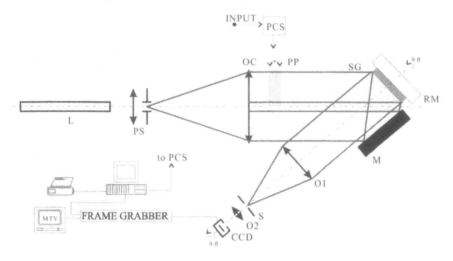

Figure 8.1.5. Diagram of the one-channel moiré interferometry system: L, He–Ne laser; PS, pinhole system; OC, collimator objective; SG, specimen grating; RM, M, mirrors; PP, parallel glass plate; PCS, phase-shifting control.

grating period sensitivity (417 nm per fringe) and are expressed by

$$I_i(x, y) = a(x, y) + b(x, y) \cos \left[\frac{4\pi}{d} (u(x, y) \text{or} v(x, y)) + \delta_i \right] \qquad (8.1.1)$$

where $a(x, y)$, $b(x, y)$ are the background and local contrast modulation functions, respectively, d is the period of the specimen grating, δ_i is the ith phase shift of the interferogram, $\delta_i = i2\pi/N, i = 1, \ldots, N$.

The interferograms are captured by CCD camera using imaging objectives 01 and 02 and the video signals are digitized using a frame grabber. The image is analysed by an automatic fringe pattern analyser which employs a temporal phase-shifting method [9]. The phase shifts are introduced automatically by rotating the parallel glass plate PP. In order to minimize the errors connected with possible phase-shifter misalignment and some nonlinearities in the detector, the five-intensity self-calibrating algorithm, with the phase shift $\delta_i = i\pi/2$ [10] was applied and the phase modulo (2π) was calculated as follows.

$$\phi(x, y) = \text{arc tg} \frac{2(I_2 - I_4)}{2I_3 - I_1 - I_5} \qquad (8.1.2)$$

where I_i is the intensity in (x, y) coordinates of the interferogram with the ith phase shift.

The phase was than unwrapped using the minimum spanning tree algorithm and scaled for displacement values (in nm). Later, according to requirements, the strains $\varepsilon_x, \varepsilon_y, \gamma_{xy}$ and material constants such as Poisson's ratio were calculated.

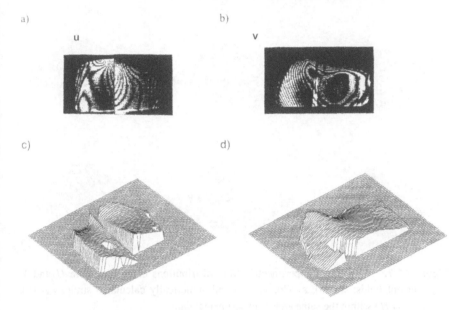

Figure 8.1.6. Interferograms representing in-plane u_1 (*a*) and v_1 (*b*) displacement fields in the YM slice of the rail after the residual stress has been released by annealing and 3D plots of the calibrated displacement fields $u = u_1 - u_0$ (*c*) and $v = v_1 - v_0$ (*d*).

The measurements were straightforward for samples with the cross-grating transferred by epoxy, as it was not necessary to capture the initial interferograms for calculating and subtracting the so called 'zero displacement' fields. The high-temperature gratings suffer from significant distortion (a few fringes in the field of view) and, therefore, the following measurement process was carried out:

1. The sample was mounted on the rotary mount RM and adjusted.
2. The interferograms for the U direction were registered and after rotating the mount and CCD camera by 90°, for the V direction.
3. The u_0 and v_0 in-plane displacement fields were calculated,
4. The sample was annealed.
5. The sample was mounted on the rotary mount in the same position as in step 1. This operation is possible due to the three-point support system used.
6. The u_1 and v_1 displacement fields are registered and calculated.
7. The displacement values $u = u_1 - u_0$ and $v = v_1 - v_0$ are calculated. In this step the initial distortion of the specimen grating and the interferometer imperfections are removed.
8. The strains ε_x, ε_y and γ_{xy} are calculated.

Figure 8.1.7. 3D plots of experimental strain distributions calculated from U and V displacement fields: ε_x (a), ε_y (b), γ_{xy} (c) and numerically calculated strains ε_{xt} (d), ε_{yt} (e), γ_{xyt} (f) within the same area as experimental data.

8.1.4   Experimental results and discussion

8.1.4.1   Residual strain determination for railway rails

Full-field information about the residual displacement and strains in rails was obtained by analysing the annealed slice of rail. The interferograms obtained after stress release are shown in figure 8.1.6 together with the displacement fields (u, v) which, according to the methodology described in section 8.1.3, were obtained by subtraction of the final and initial displacement fields ($u = u_1 - u_0$, $v = v_1 - v_0$). The experimental strain fields calculated from U and V displacements are shown in figures 8.1.7(a)–8.1.7(c) and they are compared with their numerical predictions (figure 8.1.7(d)–8.1.7(f)) calculated according to analytical the model described by Magiera and Orkisz [4].

For the comparison, the interferogram obtained for the U displacement map at the rail slice with the stress relieved by cutting is presented in figure 8.1.8. The interferogram consists of several separated fringe patterns and the component fringe patterns may also have severe disturbances at the edges due to the cutting process. Here the modified software enables, by a semi-manual procedure, a mask to be put on the distorted areas. Further processing was performed on modified interferograms and average strain values ε_x and ε_y were calculated for each component interferogram.

The theoretical and experimental ε_x strain distributions determined by strain gauges (YM) and moiré interferometry, while the stress was being released by cutting (IM-C) and by annealing (IM-W), in the cross sections B–B and C–C

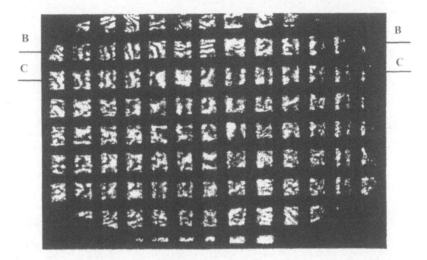

Figure 8.1.8. The rail slice with the stress released by cutting together with the interferograms representing the U displacement fields.

(as marked in figure 8.1.8) are shown in figure 8.1.9. The strain distributions for THEOR and IM-W for the cross section (C–C) were very similar; however, the experimental values were lower. This may arise from the fact that the stress was not completely released by annealing. The pair of strain distributions in C–C obtained for the YM and IM-C methods in the samples in which the stress was released by cutting are similar in character and value; however, they differ from THEOR and IM-W. This proves that the physical nature of the stress release influences the values of determined stress.

The cross section B–B, which is closer to the high plasticity region produced by the contact area rail/wheel, indicates significant differences between the strain determined theoretically and experimentally. The theoretical predictions in this area may be incorrect due to model deficiencies within the plastic region and incorrect local material constant values applied during calculations. Therefore, experimental data especially in this region are genuinely necessary.

Introducing imaging optics with a higher magnification into the system presented in figure 8.1.5 enables the area A (indicated in figure 8.1.1) to be analysed. The interferogram, U displacement and ε_x strain within the region close to the contact area rail/wheel are shown in figure 8.1.10. These experimental data allow the rail behaviour in the contact area to be modelled properly.

8.1.4.2 Residual stress determination in laser beam welds

As indicated in section 8.1.2, two types of welded specimen with different ways of relieving the residual stresses were considered.

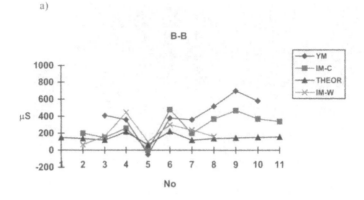

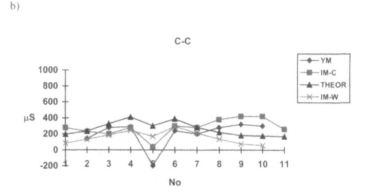

Figure 8.1.9. The strains ε_x distributions obtained due to stress ease by cutting and by strain gauges (YM) and grating interferometry (IM-C) measurements, by annealing and grating interferometry measurements (IM-W) compared with numerical calculations (THEOR). The results in cross-section B–B (*a*) and C–C (*b*) as shown in figure 8.1.8.

The measurements of the ferritic steel specimen with K, L, P gratings enabled the determination of two displacement/strain fields [11]:

- those at the weld and heat-affected zone (area P—figure 8.1.11(*a*)–8.1.11(*d*) and figure 8.1.12(*a*)–8.1.12(*c*)); and
- those outside the weld area in order to see how far from the weld strains can be recognized (areas K, L–figure 8.1.11(*e*)–8.1.11(*h*) and figure 8.1.12(*d*)–8.1.12(*f*)).

Due to the limited size of the specimen grating, the tests were performed separately; however, they allowed us to combine the results as shown in figure 8.1.12(*g*)–8.1.12(*i*). The displacement distributions obtained for areas K

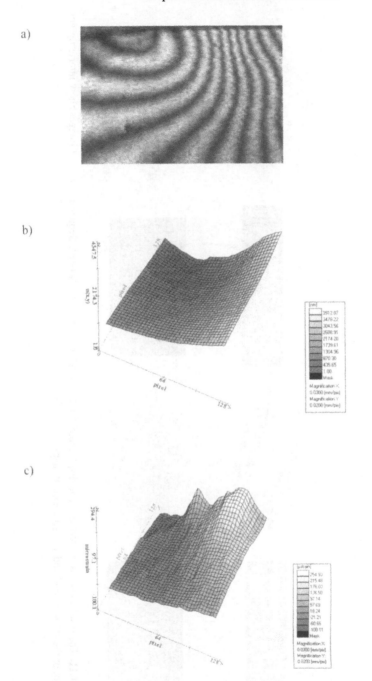

Figure 8.1.10. The analysis of wheel/rail contact region A: the interferogram representing *U* displacement (*a*), 3D plot of *U* displacement (*b*) and 3D plot of ε_x strain (*c*).

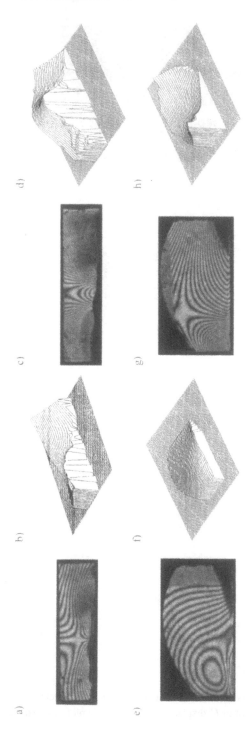

Figure 8.1.11. The residual stress related in-plane displacement in the area P (*a*)–(*d*) and L (*e*)–(*h*). Sequential pairs of images show interferogram and 3D plot of the respective U and V displacement maps.

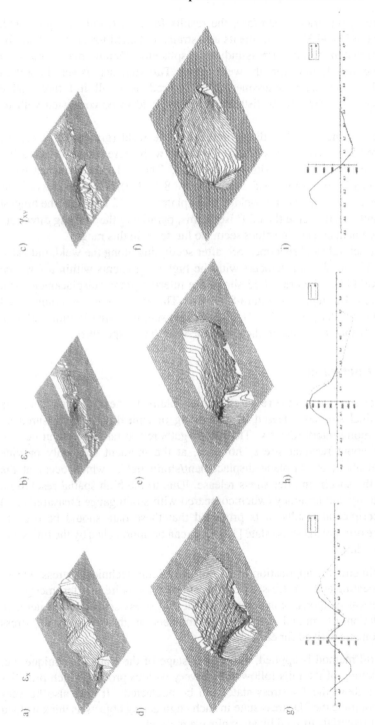

Figure 8.1.12. The residual stress related strain maps ε_x, ε_y and γ_{xy} respectively in the area P (*a*)–(*c*) and L (*d*)–(*f*) and their central cross-sections for (*g*) ε_x, (*h*) ε_y and (*i*) γ_{xy} strains when combined for the P and L areas.

and L were symmetrical; therefore, the results for area L only are presented. Analyzing the U and V displacements and strains obtained for area L, it can be clearly seen that the residual-stress-induced displacement/strains may be detected at a distance over 40 mm from the weld centre. This statement is very important when on-line weld quality assessment is considered, as it will, in future, enable measurements at a significant distance from the weld to be correlated with its quality.

Analyzing the U and V displacements in the weld (grating P), it can be seen (figure 8.1.11(*b*)) that u is zero at the centre with decreasing and increasing values to the left (−3.5 μm) sides of the weld. The V displacement reaches a maximum (18.3 μm) at the weld centre (figure 8.1.11(*d*)). This result confirms the experience from laser beam welded thick plates ($t = 12$ mm), i.e. the biggest displacements occur inside the 600° isotherms, parallel to the welding direction. The displacement and strain values seem, so far, to be in this range.

The structural steel specimen SS, after sectioning along the weld, indicated significant U and V displacements with the highest gradients within a few mm from the weld [12]. Figure 8.1.13 shows the interferograms, displacements and strains obtained in region S (figure 8.1.4(*b*)). The *P–V* values of strains equal $\varepsilon_x = 3550 \times 10^{-6}$, $\varepsilon_y = 3950 \times 10^{-6}$ in the measurement area 30 mm × 30 mm (figure 8.1.7) and are similar to those obtained for the FS specimen.

8.1.5   Conclusions

This chapter summarizes the research results obtained by experimental analysis of the residual stresses released by sectioning or annealing and measured by automated moiré interferometry. The results gathered so far support our current knowledge about residual stress; however, at the moment they only provide information about the in-plane displacement/strain fields, which occur at the surface of the specimen after stress release. Due to the high spatial resolution, high sensitivity and accuracy (when compared with strain gauge measurements) of the experimental results, it is proposed that these data should be used to compute the original 3D stress state [13]. This can be approached by the following hybrid procedure:

- Experiment: the application of an oblique slicing technique, stress release by annealing and 2D stress state determination by moiré interferometry;
- Numerical methods: computation of the 3D stress state through iterative calculation of mutual accumulative impacts of the out-of-plane stress components derived for each slice.

If this hybrid method is applied, the second stage of the Batelle technique, i.e. Meier sectioning of the rails followed by a very tedious process which provides information about the δ_{zz} stress state, can be neglected. It may also be very helpful to compute the 3D stress state in such engineering objects as thick welded elements or materials formed by an explosive method.

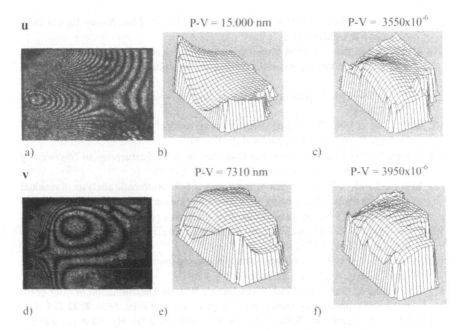

Figure 8.1.13. The results of the welded specimen SS analysis after the residual stress releasing by cutting (a,d) interferograms and (b,e) 3D plots of U and V in-plane displacements; (c,f) 3D plots of ε_x, ε_y strains respectively.

However, even without getting into the determination of more sophisticated 3D residual stresses, moiré interferometry has provided an excellent tool to extend our knowledge about local residual stresses in welded and heat-affected zones as well as in the neighbourhood of wheel/rail contact.

This extended knowledge about the displacement caused by residual stress may, in future, provide the basis for developing a method for on-line quality control of welding. This statement is especially valid as we have shown here that the residual stress extends a few centimetres away from the centre of the weld.

In general, the methodology described is capable of providing the data for simulating or experimental control of laser beam welding or railway rails, for testing as well as supporting and checking the results of numerical calculations.

References

[1] Rowlands R E 1987 Residual stress in SEM *Handbook of Experimental Mechanics* ed A S Kobayasahi (Englewood Cliffs, NJ: Prentice-Hall)
[2] Świderski Z and Wójtowicz A 1992 Plans and progress of controlled experiments on rail residual stress using the EMS-60 machine *Residual Stress in Rails* vol 1 (Dordrecht: Kluwer) pp 57–66

[3] Groom J J 1983 Determination of residual stresses in rails *Final Report* for US DOT No DOT/FRA/ORD-83/05

[4] Magiera J and Orkisz J 1994 Experimental-numerical analysis of 3D residual stress state in railroad rails by means of oblique slicing technique *Proc. SPIE* **2342** 314–25

[5] Forno C 1994 High temperature resistant grating for moirè interferometry *Proc. SPIE* **2342** 238–43

[6] Orkisz J *et al* 1990 Discrete analysis of actual residual stress resulting from cyclic loadings *Computers and Structures* **35** 397–412

[7] Gordon R 1995 *Residual Stress and Distortion in Welded Structure–An Overview of Current US Research Initiatives* IIW-Doc. XV-878-95

[8] Kujawińska M, Sałbut L, Olszak A and Forno C 1994 Automatic analysis of residual stresses in rails using grating interferometry *Recent Advances in Experimental Mechanics* ed S Gomez *et al* (Rotterdam: Balkema) pp 699–704

[9] Kujawińska M 1993 The architecture of a multipurpose fringe pattern analysis system *Opt. Lasers Eng.* **19** 261–8

[10] Schwider J *et al* 1993 Digital wave-front measuring interferometry: some systematic errors sources *Appl. Opt.* **22** 3421–32

[11] Kujawińska M, Sałbut L, Weise S and Jüptner W 1996 Determination of laser beam weldment properties by grating interferometry method *Proc. SPIE* **2782** 224–32

[12] Holstein D, Jüptner W, Kujawińska M and Sałbut L 2001 Hybrid experimental-numerical concept of residual stress analysis in laser weldments *Exp. Mech.* **41** 343–50

[13] Kujawińska M 1996 Experimental-numerical analysis of 3D residual stress state in engineering objects *Akademie Verlag Series in Optical Metrology* **3** 151–8

Chapter 8.2

Determination of thermal stresses near a bimaterial interface

Judy D Wood
Virginia Polytechnic Institute, USA

In every case where dissimilar materials are bonded together and undergo a subsequent change in dimension (from a change of temperature, loading or moisture content), stresses will develop at the interface due to the mismatch of material properties. Numerical solutions predict very high stress gradients near the interface [1], but experimental measurements are needed to supplement these studies. Moiré interferometry was chosen to study this problem because full-field capabilities were required to measure the high strain gradients near the interface. In this study, the complex state of elastic strains and corresponding stresses in the region of a bimaterial interface were evaluated using moiré interferometry, where the loading was a uniform temperature change.

8.2.1   Experimental procedure

The specimen configuration that was evaluated here is illustrated in figure 8.2.1. More details of the experimental procedure are given in [2–4]. The steel and brass plates were joined by a very thin, high-temperature, silver-solder film along the mating surfaces. The specimen was then machined on all surfaces to its final size. The specimen gratings were replicated at elevated temperature (157 °C) and evaluated at room temperature (24 °C). Thus, the experimental analysis measured the deformations caused by a change of temperature of −133 °C. The small gratings near the traction-free corners were used to measure the coefficients of thermal expansion (contraction) of each material. The large grating that spanned the two materials deformed as a result of two effects: (a) the free thermal contraction of the steel and brass; and (b) the state of stress caused by the mutual constraint along the joint interface. A special technique [2] was used for the

287

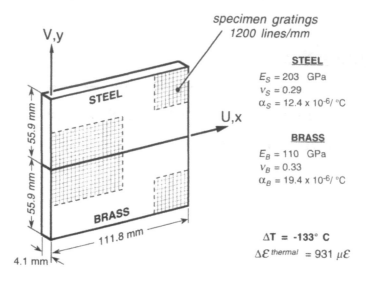

Figure 8.2.1. Bimetal specimen and specimen gratings.

thermal stress analysis, where the small gratings were used to separate the two effects.

The elevated-temperature procedure is illustrated in figure 8.2.2. The process was different than the standard replication technique [5] by three factors: (a) the mould was a cross-line diffraction grating on an ultra-low expansion (ULE) glass substrate; (b) the replication was carried out at elevated temperature in an oven; and (c) the reflective metallic film was applied after replication. At the elevated temperature, after the epoxy solidified and cured, the mould was separated. At that temperature the lines of the specimen grating were uniformly spaced. As they cooled to room temperature (24 °C), the specimen and overlying grating deformed. The thermal expansion of the mould (on the ULE substrate) was negligible compared to that of the metals, so the mould had the same frequency at room temperature as at elevated temperature. The mould was also used to tune the moiré interferometer to a null field, which corresponded to the elevated temperature frequency of the specimen grating. Thus the moiré pattern observed in the interferometer revealed the absolute change of displacements incurred in cooling from elevated room temperature.

8.2.2 Extracting the coefficients of thermal expansion

The total deformation is the result of the stress-induced strain and the thermal strain. The stress-induced strain, alone, is revealed when the constant uniform thermal strain is subtracted from the total deformation,

$$\varepsilon^{\text{stress}} = \varepsilon^{\text{total}} - \varepsilon^{\text{thermal}}. \tag{8.2.1}$$

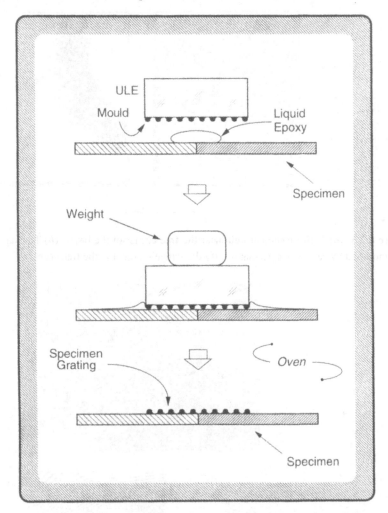

Figure 8.2.2. Procedure to produce the specimen grating at elevated temperature. ULE glass is an ULE substrate (by Corning Glass Works).

This equation applies to the steel and brass, separately.

First the interferometer was tuned to produce null U and V fields, using the ULE mould in place of the specimen in the interferometer. This provided initial conditions corresponding to the specimen grating frequency at the elevated temperature (157 °C). Then, the specimen was mounted on a moveable stage so that each of the three gratings could be centred in the moiré interferometer and angular adjustments could be made. In the free corners, the stress-induced strain is zero and the strain resulting from the thermal change can be measured, directly, in each material.

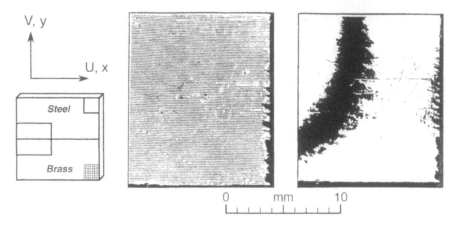

Figure 8.2.3. (*a*) *V*-displacement field near the free corner of the brass. (*b*) The same *V* field modified by carrier fringes that nullify the fringe gradient at the free corner.

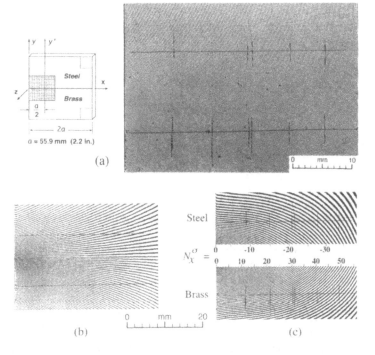

Figure 8.2.4. *V*-Displacement fields in shaded region for $\Delta T = -133\,°\mathrm{C}$: (*a*) contours of the total displacements; (*b*) *V* field modified by carrier fringes of extension; and (*c*) contours of the stress-induced displacements in steel and in brass.

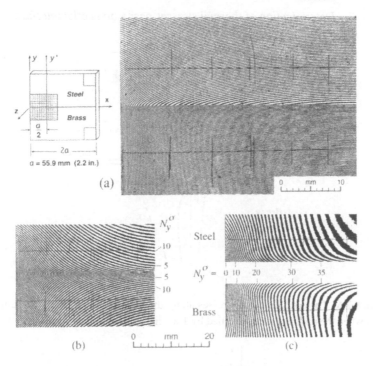

Figure 8.2.5. U-displacement fields in shaded region for $\Delta T = -133\,°C$: (a) contours of the total displacements; (b) U field modified by carrier fringes of extension; (c) contours of the stress-induced displacements in steel and brass.

8.2.2.1 Separation of thermal strains and stress-induced strains

In the course of the experiments, carrier fringes [6] were used to emphasize features of the deformation field. In addition, a special technique (whereby carrier fringes cancelled the uniform deformation fringes of free thermal contraction) was applied to produce patterns that represent the stress-induced displacements alone [2]. An example of this is shown in figure 8.2.3, where the fringe gradient is (a) was cancelled by carrier fringes.

8.2.3   Results

The direct results of moiré interferometry are shown in figures 8.2.4 and 8.2.5. Figure 8.2.4(a) is the total V-displacement field. It is dominated by the thermal contraction of extension to emphasize the abrupt changes near the interface. Figure 8.2.4(c) is the result of subtracting the thermal strains from each material, using carrier fringes. Thus, figure 8.2.4(c) is the stress-induced deformation pattern. Similar to figure 8.2.4, figure 8.2.5(a) is the total U-displacement field

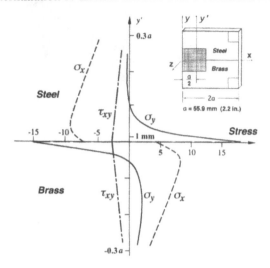

Figure 8.2.6. The distribution of stresses along the y′ axis. The units on the stress scale are psi when the numbers are multiplied by 100, and MPa when multiplied by 6.9.

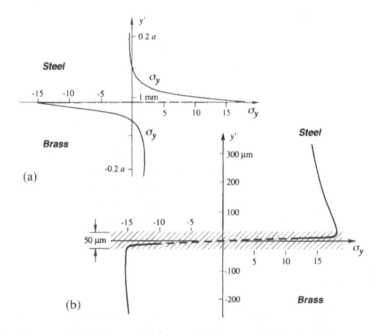

Figure 8.2.7. (*a*) The σ_y stresses along y′. (*b*) Expanded vertical scale showing the σ_y peaks near the interface and a severe stress gradient between the peaks. The numbers denote stresses in psi when multiplied by 1000, and MPa when multiplied by 6.9.

resulting from the change of temperature and figure 8.2.5(c) is the stress-induced deformation pattern for each material.

The stress-induced strains were determined from the experimental data and the corresponding stresses, σ_x, σ_y and τ_{xy}, were evaluated using the known material properties. The stresses along a line crossing the interface are given in figure 8.2.6. Extraordinarily large stress gradients were detected near the interface. They were investigated in detail, in a supplemental study using microscopic moiré interferometry [7–9]. The supplemental study provided deformation measurements much closer to the interface. It showed that the maximum stress in the steel and the minimum stress in the brass occurred away from the interface, resulting in a severe stress gradient but a continuity of stress across the interface. The results are shown in figure 8.2.7. The experimental analysis and the results apply to the free surface of the specimen.

References

[1] Chan W T and C W Nelson 1979 Thermal stresses in bonded joints *IBM J. Res. Development* **23** 179–88

[2] Post D and Wood J D 1989 Determination of thermal strains by moiré interferometry *Exp. Mech.* **29** 318–22

[3] Post D, Wood J D, Han B, Parks V J and Gerstle F P Jr 1994 Thermal stresses in a bimaterial joint: an experimental analysis *ASME J. Appl. Mech.* **61**

[4] Wood J D 1992 Determination of thermal strains in the neighbourhood of a bimaterial interface *PhD Dissertation* Virginia Polytechnic Institute and State University, Blacksburg, Virginia

[5] Post D, Han B and Ifju P 1994 *High Sensitivity Moiré: Experimental Analysis for Mechanics and Materials* (New York: Springer) pp 293–312

[6] Guo Y, Post D and Czarnek R 1989 The magic of carrier fringes in moiré interferometry *Exp. Mech.* **29** 169–73

[7] Wood J D and Han B 1992 Micromechanical study of thermal strains near the interface of a bimaterial joint by microscopic moiré interferometry *Proc 7th Int. Congress on Experimental Mechanics (Las Vegas, NE)* (Bethel, CT: SEM)

[8] Han B and Post D 1992 Immersion interferometer for microscopic moiré interferometry *Exp. Mech.* **32** 38–41

[9] Han B 1992 Higher sensitivity moiré interferometry for micromechanics studies *Opt. Eng.* **31** 1517–26

Chapter 8.3

Interior strains by a hybrid 3d photoelasticity/moiré interferometry technique

C W Smith
Virginia Polytechnic Institute and State University, USA

The complex problem of measuring displacement and strain fields inside a solid body can be solved by combining stress freezing photoelasticity with moiré interferometry measurements. The hybrid method utilizes a three-dimensional photoelastic model, loaded and processed by a stress-freezing cycle. The model is then dissected and sliced. High-frequency moiré gratings are applied to the (previously) internal planes and the slices are annealed to relieve the deformation. Analysis by a moiré interferometer reveals full-field maps of the U and V displacements released by the annealing step.

8.3.1   Frozen stress photoelasticity

When a two-dimensional transparent photoelastic model is placed in a circularly polarized monochromatic light field, dark and bright fringes will appear which are proportional to the applied load. These fringes are called stress fringes or isochromatics and they reveal the magnitudes of the maximum in-plane shear stresses throughout the body.

Some transparent materials exhibit mechanical and optical diphase characteristics when raised above a certain temperature, called the critical temperature (T_c). these materials are used for three-dimensional photoelastic models. The material, while still elastic, will exhibit a fringe sensitivity of about 20 times the value obtained at room temperature and its modulus of elasticity will be reduced to about one-six-hundredth of its room temperature value. By raising the model temperature above T_c, loading and then cooling slowly to room temperature, the stress fringes developed at the elevated temperature will be retained when the material is returned to room temperature. Since the material is so much more sensitive to fringe generation above T_c than at room temperature, fringe reduction upon unloading at room temperature is negligible. The three-dimensional model can then be sliced without disturbing the 'frozen in' fringe pattern and each slice can be analysed as a two-dimensional model but one which contains the three-dimensional effects.

8.3.2 The hybrid method

Using a replicating plastic for which the glass transition temperature is below the model T_c a moiré grating is replicated on the surface of each slice. Then with each slice placed on a smooth surface (sprinkled with a non-sticking talcum powder to nullify friction), the slices are subjected to a second stress-freezing cycle to anneal out the strain field imposed by the initial stress freezing. The resulting moiré patterns represent the inverse of the imposed internal deformation fields. The moiré interferometer used to determine the in-plane displacement fields is adjusted by using the mould previously employed to replicate the specimen grating, so that the virtual reference grating frequency projected by the interferometer represents the specimen in zero load condition.

The hybrid method has been applied effectively to studies of linear elastic fracture mechanics, where both photoelastic and moiré results were utilized: examples are reported in [1] and [2]. The hybrid method can be applied in other fields, too, to measure deformations at interior regions of three-dimensional bodies.

References

[1] Epstein J S 1983 On the variation of the first classical eigenvalue of fracture mechanics in three dimensional transitory stress fields *PhD Thesis* Virginia Polytechnic Institute and State University, Blacksburgh, VA

[2] Smith C W, Post D and Nicoletto G 1983 Experimental stress intensity distributions in three dimensional cracked body problems *Exp. Mech.* **23** 378–82

CHAPTER 9

HIGH TEMPERATURE EFFECTS AND THERMAL STRAINS

Chapter 9.1

Moiré interferometry using grating photography and optical processing

Gary Cloud
Michigan State University, USA

9.1.1 Case study: whole-field strain analysis to 1370 °C

This case study focuses on the design, set-up and testing of a procedure to measure surface strain distributions in turbine superalloys at high temperatures using a variant of the moiré method (e.g. [1, 2]).

In this particular project, strains were to be measured at temperatures ranging from room temperature to 'as high as possible' in flat tensile specimens with and without notches. In the end, the measurements were performed up to 1370 °C, roughly the melting temperature of mild steel. Biaxial strains and whole-field mapping were required. The expected strain magnitudes ranged from 0–2%, although larger values were possible.

This problem and its solution illustrates the principle of choosing and adapting a technique to fit the problem rather than modifying the problem to fit available or favourite techniques, as is often done.

9.1.2 Problem characteristics

This problem fits into a broad class of experimental mechanics problems for which whole-field strain measurement is needed but conventional optical techniques, including speckle methods, geometric moiré, moiré interferometry and holography, are not suitable.

The breadth of the class is illustrated by other problems that have been analysed using similar techniques, including

- plastic deformation of metals such as by coldworking,

- three-dimensional analysis of strain,
- measurement of crack-opening displacement, crack length and strains around a crack-tip at various temperatures,
- thermal strain gradients at weldments and
- strain mapping in flexible composites and polymers.

These problems share certain characteristics that dictate the choice of experimental approach, including the following ones.

- Whole-field analysis is needed, so that resistance strain gages and similar methods cannot suffice.
- The sensitivity required is beyond the capability of geometric moiré.
- The strain range is beyond the limitations of moiré interferometry.
- The environment is hostile or, at least, difficult, so that remote sensing and observation are necessary.
- The testing might occur in discrete irreversible steps.
- Out-of-plane deformation might take place.
- In the three-dimensional problem, observations must be taken inside the specimen, perhaps at several different strata.
- The observations must be *in situ* or at the testing machine, where interferometric methods, have not typically performed well because of vibrations, noise and dirt.

9.1.3 Moiré interferometry with optical Fourier processing

The project requirements described earlier dictated the adaptation of an intermediate-sensitivity moiré technique that involves high-resolution photography of specimen gratings. The replica photographs are subsequently superimposed with chosen submaster gratings and spatially filtered in an optical Fourier processor to create moiré fringe patterns with appropriate sensitivity and pitch mismatch. The fringes are then digitized for the construction of displacement and strain maps by digital computer.

Space limitations preclude offering a complete description of the methodology and examples of results. Interested readers are referred to the book by the author [3, Chs 10–12] or to the several papers on the subject (e.g. [4–11]).

There are several attractive aspects of the moiré technique that incorporates spatial filtering. The fundamental idea is to take advantage of the sensitivity multiplication and noise reduction offered by optical Fourier processing of moiré grating photographs that have been recorded for various states of a specimen.

Particular characteristics and advantages of the approach include the following ones:

- The sensitivity of the method can be controlled *after* the experimental data are recorded, an unusual advantage in any experiment.

- The sensitivity and range are between those of geometric and interferometric moiré.
- The method is very flexible in that any two specimen states can be compared easily, a useful feature when irreversible processes over a broad range are involved.
- The original data are permanently recorded for leisurely study later.
- Certain common errors are automatically eliminated.
- Fringe visibility is usually much improved over what is obtained by any method of direct, photographic or optical superimposition.
- The method is useful in difficult environments requiring remote observation and tolerance to vibration and dirt.
- The method is easy to set up and use.

9.1.4   Outline of the technique

Here the steps in intermediate-sensitivity moiré using Fourier optical processing are listed.   Since the steps extend from specimen preparation through data reduction, the list appears rather lengthy.   Some of the steps are, of course, common to other procedures, especially geometric moiré. Recognize that many variations in the procedure are possible and the method is easily adapted to resources at hand.

Step 1:   Master gratings, typically of 1000–2000 lines/inch (lpi) (39.4–79.8 lines/mm) are obtained and reduced photographically to create a set of working submasters of various grating spatial frequencies including the fundamental frequency.

Step 2: Gratings of the fundamental frequency are applied to the specimen. Often, photoresist and contact printing are used but other methods have been developed to fit the test conditions.

Step 3: Fiducial marks and identity labels are applied to the specimen surface.

Step 4: The specimen is placed in the loading system and the grating photographed at low magnification, say between 1:1 and 2:1. Slotted apertures tuned to harmonics of the grating fundamental frequency can be used at this stage to gain sensitivity multiplication [12].

Step 5: The specimen is loaded to the first load.

Step 6: The grating, now deformed with the specimen, is photographed again, using the same set-up.

Step 7: The photographic plate of the undeformed grid is superimposed with a submaster grating having a spatial frequency of one to three (sometimes higher or lower, depending on the sensitivity desired) multiplies by the frequency of the photographed specimen grating plus or minus a small frequency mismatch.

Step 8: The assembly of specimen grating photograph and submaster grating is placed in a coherent optical processor and adjusted to produce the correct baseline (zero strain) fringe pattern at the processor output. This fringe pattern is photographed.

Step 9: Steps 7 and 8 are repeated with the deformed grating photoplate.

Step 10: Steps 7 and 8 are repeated as desired with other submaster gratings to produce fringe patterns with different pitch mismatch and, in some cases, different sensitivity multiplication factors.

Step 11: Steps 5–10 are repeated for each specimen load state that is of interest in the experiment.

Step 12: The fringe patterns are enlarged and printed.

Step 13: The prints of the fringe patterns are sorted and coded for identification during the data analysis procedure.

Step 14: The locations of the fringes along the desired axes are determined and recorded as arrays of numbers showing fringe order and location from a specified fiducial mark: this is best done on a digitizing tablet if available.

Step 15: The arrays of fringe data are converted to plots, and differentiations yield strain; computerized reduction is appropriate.

An important characteristic of this procedure is that it operates in a differential mode, which, as a rule, is best in any experiment. To understand this remark, notice that fringe data are collected for the specimen in each load state, including a baseline data set at zero or beginning load. A requirement is that nothing is changed between creation of the data sets. The data sets are then always processed in pairs often, but not always, including the baseline data. There are several advantages to this procedure, these being typical of differential measurements. For one, the effects of any residual initial fringe patterns, whether accidental or deliberately induced as by pitch mismatch, are automatically eliminated. There is no false initial strain. Another distinct advantage is that any two of the specimen states may be easily compared, since the grating at each state is permanently stored on glass photoplates. Such a capability is very useful when the specimen is loaded in several irreversible stages or where the total range is very large, as in studies of plastic deformation or fracture.

The following sections contain some technical details. These must be kept brief given space limits and potential users of the technique should review the literature cited earlier. Also, the details mentioned here are specific to high-temperature moiré analysis.

9.1.5   Gratings

A master grating of good quality is assumed to be in hand. Submaster gratings may be made by contact printing or by using high-resolution photography. It is desirable to have submasters of several different pitches to facilitate the use of pitch mismatch during optical filtering.

The replication of gratings on the specimen is accomplished using techniques that are chosen to fit the problem at hand. Often, photoresists, stencils and vacuum deposition can be used to create a grating of appropriate pitch and contrast and of materials that can tolerate the test environment. For most testing of this type, two-way gratings (a grid as opposed to a grill) are used in order that biaxial strains can be measured. Three-way gratings have also been employed to obtain the complete strain state via the strain rosette equations over the entire field (see [3, ch 8] and [13]).

Extensive trials were required to develop gratings for the high-temperature project. Two related approaches that use ceramic coatings were found best. A refractory paint (Pyromark 2500 from Tempil Division, Big Three Industries) that withstands extreme temperatures served at once to protect the specimen surface from oxidation, to bond a metallic grating to the specimen and to provide contrast for photography. The grating is a nickel or stainless steel mesh of the sort used in the printing industry and it is available in spatial frequencies from a few hundred lines/inch up to about 3000 lines/inch (Buckbee-Mears, St. Paul, MN, USA).

In the first approach, the ceramic paint is sprayed onto the specimen to a thickness of 3–4 μm. Immediately a piece of the nickel mesh is pressed into the paint using a Teflon block that is not wetted by the paint. It is important that the mesh is not allowed to sink into a layer of paint that is too thick. Also, if the paint is too thick, it will crack and flake at the high temperatures. Getting the correct paint thickness and inserting the mesh are the only tricky parts of the whole process and some experience born of trial and error is useful. The paint is allowed to dry and then it is cured and, if desired, vitrified at increasing temperatures. The metallic mesh is left in place and, since it remains highly reflective, it provides excellent photographic contrast with the diffuse white paint background.

The second approach to specimen grating preparation follows the same steps except that the nickel grating is pulled away from the paint after the paint has partially dried. In this case, the grating is created by the blocks of paint that are left behind. This type of grating does not reinforce the specimen and it is highly photographable under oblique lighting because it is a relief grating.

Figures 9.1.1 and 9.1.2 are photomicrographs of a nickel mesh–ceramic substrate grid before and after testing at high temperature in air. The grating is damaged by the severe environment but it is still serviceable. Figure 9.1.3 shows a ceramic grid after testing to extremely strains at room temperature. Note the degree of deformation that is possible even though the ceramic is somewhat brittle. In fact, under high-strain conditions, normal and shear strains can be estimated by measurements of local mesh dimensions in the specimen or in a grating photograph with a measuring microscope

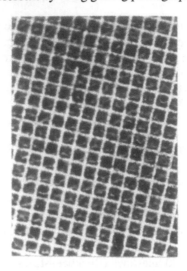

Figure 9.1.1. Photomicrograph of 40 lines/mm ceramic grid with an embedded nickel mesh before testing.

Figure 9.1.2. Photomicrograph of 40 lines/mm ceramic grid with embedded nickel mesh after cycling to various temperatures up to 1370 °C.

9.1.6 Grating photography

Photography of the specimen grating during the test is the most important and critical part of the process. High-resolution techniques of the sort developed for the printed circuit industry are called for.

Figure 9.1.3. Photomicrograph of 40 lines/mm ceramic grid (nickel mesh removed) after testing to 30% strain at room temperature.

In the study at hand, a 4 × 5 view camera with its frame stiffened with aluminum plates was mounted on a sturdy stand that could be brought close to the testing machine, as shown in figure 9.1.4. Among others, a Zeiss Planar lens optimized for finite conjugates was used at wide aperture ($f5.6-f11$). A fine ground glass made from the same photoplates as those to be used in the photography was mounted in a plate holder from which the separator was removed: this measure ensures that the focused image and emulsion will be in the same plane. A microscope is mounted on a cross slide attached to the camera frame and adjusted so its motion is exactly perpendicular to the optical axis. Magnification and focus are then adjusted in turn. The microscope makes it possible to get an absolute sharpest focus of the specimen grating over the whole plane of the photoplate. Typically, focus is performed in white light and the photography is undertaken with flash-lamps mounted off-axis to enhance grating contrast and to reduce the effects of vibration and convection currents. High-resolution photoplates should be used: blue-sensitive holographic plates have served well. A blue or green filter reduces contamination of the photographs by radiation from the specimen as temperatures increase.

At this stage, some sensitivity multiplication can be realized by the use of slotted apertures in the camera lens so that the grating frequency is multiplied [12]. That is, one can force a 1000 lines/inch grating of high contrast (approximating a bar-and-space grating) to be photographed as, say, a 3000 lines/inch grating.

Figure 9.1.4. High-resolution camera, testing machine and induction heating unit for cyclic testing at high temperature.

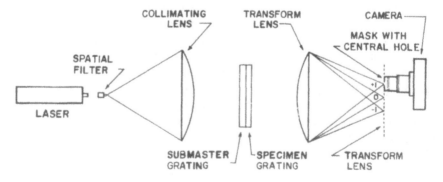

Figure 9.1.5. Schematic diagram of a typical optical spatial filtering system for obtaining moiré fringe patterns from grating photographs.

9.1.7   Optical Fourier processing

If the grating replica photoplates are superimposed on one another or if one is superimposed on a submaster grating, it is unlikely that moiré fringes will be seen or, at least, the fringes will not be useful owing to optical noise and poor contrast.

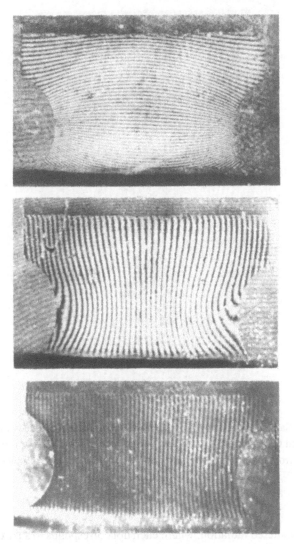

Figure 9.1.6. Typical moiré fringe patterns for various test conditions: (top) for longitudinal strain under load at 870 °C; (middle) for transverse strain under load at 870 °C and (bottom) for transverse strain under no load at 1370 °C.

Rather, the two photoplates or submaster-plus-specimen replica are clamped together and placed into an optical Fourier analyser that allows spatial filtering to be performed, such as that shown schematically in figure 9.1.5. Space precludes a treatment of Fourier optical theory here but the process is actually quite simple [3, chs 10–11]. One is creating a diffraction pattern wherein distance corresponds to frequency in the input plane. The pattern will be an array of bright dots or patches

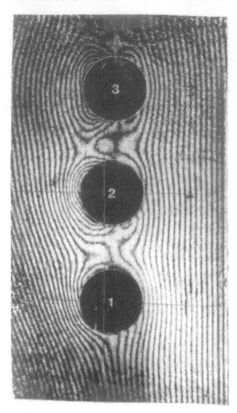

Figure 9.1.7. Sample moiré fringe pattern obtained by these techniques for measurement of strain caused by coldworking of fastener holes in aluminum alloy at room temperature. This result is from a study of the effect of cold-expanding a row of holes that lie near a plate edge.

in a field of scattered radiation, where each dot corresponds to a particular grating spatial frequency and orientation. The light in one of the bright patches is selected for use in creating an image (that is, the inverse Fourier transform) by another lens or a camera (or the eye) and all the other bright dots are blocked. The result is a reconstructed image made with the energy corresponding to only one grating direction and pitch. Noise and the effects of the cross-grating are eliminated and fringe visibility is greatly improved.

At this stage, additional sensitivity multiplication can be realized by choosing from the diffraction pattern a bright patch that corresponds to a higher diffraction order [3, 7, 12]. One is effectively using the higher harmonics of the gratings that actually exist in the photoplate when the grating lines are recorded as something approximating bar-and-space characteristics.

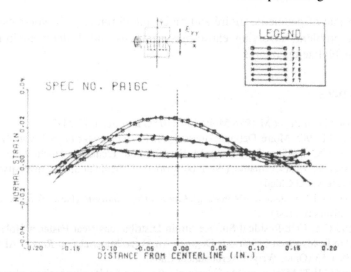

Figure 9.1.8. Sample strain plots for two tensile load levels for stainless steel at 870 °C. Note: The data reduction software plots 100 data points in the span: the plotted points shown are for curve identification.

Figure 9.1.6 shows some typical moiré fringe photograph for a simple notched specimen at two temperatures, with two of the samples being at load. Figure 9.1.7 is a pattern obtained during a study of the coldworking of fastener holes in aluminum alloy.

9.1.8 Data processing

The strains are extracted from the moiré fringe photographs by differentiating the fringe *versus* distance relationship and multiplying by the grating pitch. For extensive studies, this is best done by computer. Figure 9.1.8 shows two strain plots from the high-temperature study. These were obtained by digitizing the fringe orders as a function of position and then using a computer program especially written for the purpose [4]. It uses cubic-spline curve fitting with a variable fitting parameter and subsequent finite-difference differentiation. Nowadays, the recording of moiré fringe patterns and subsequent calculation of strains can be performed by video-computer systems, with a considerable increase in speed.

9.1.9 Closure

This method, which employs high-resolution grating photography and optical spatial filtering, is easy to implement and is very effective for whole-field strain measurements in a broad class of important problems. Its sensitivity tends to fall between that of geometric moiré and moiré interferometry. It works especially well for problems involving elasto-plastic strains and difficult environments, including high temperatures.

References

[1] Cloud G and Bayer M 1988 Moiré to 1370 °C *Exp. Tech.* **12** 24–7

[2] Bayer M 1985 Moiré Dehnungsmessungen im Temperatur-bereich von 870 °C bis 1370 °C Studienarbeit. Rheinish-Westfalische Techniche Hochschule Aachen, Aachen, West Germany: (Work performed at Michigan State University; contact Professor G Cloud)

[3] Cloud G 1995 *Optical Methods of Engineering Analysis* (New York: Cambridge University Press)

[4] Cloud G L 1978 Residual Surface Strain Distributions Near Fastener Holes Which Are Coldworked to Various Degrees Air Force Material *Lab Report* AFML-TR-78-1-53 (Ohio: Wright Aeronautical Labs)

[5] Cloud G 1979 Measurement of elasto-plastic strain fields using high-resolution moiré photography, coherent optical processing and digital data reduction *Proc. 8th Congress of International Measurement Confederation (Moscow)* **S7** 13–26

[6] Cloud G 1980a Measurement of strain near coldworked holes *Exp. Mech.* **29** 9–16

[7] Cloud G 1980b Simple optical processing of moiré grating photographs *Exp. Mech.* **20** 265–72

[8] Cloud G, Radke R and Peiffer J 1979 Moiré gratings for high temperatures and long times *Exp. Mech.* **19** 19N–21N

[9] Cloud G and Sulaimana R 1981 An experimental study of large compressive loads upon strain fields and the interaction between surface strain fields created by coldworking fastener holes *Air Force Technical Report* AFWAL-TR-80-4206 (Ohio: Wright Aeronautical Labs)

[10] Cloud G and Paleebut S 1984 Surface and internal strain fields near coldworked holes obtained by a multiple embedded-grid moiré method *Eng. Fracture Mech.* **19** 375–81

[11] Cloud G and Paleebut S 1992 Surface and interior strain fields measured by multiple embedded grid moiré and strain gages *Exp. Mech.* **32** 273–81

[12] Cloud G L 1976 Slotted apertures for multiplying grating frequencies and sharpening fringe patterns in moiré photography *Opt. Eng.* **15** 578–82

[13] Cloud G, Cesarz W and Leeak J 1979 A true whole-field strain rosette *Proc. 8th Congress of Int. Measurement Confederation (Moscow)* **S7** 27

Chapter 9.2

Heat-resistant gratings for moiré interferometry

Colin Forno
Visiting Professor, City University, London, UK

9.2.1 Introduction

Moiré photography [1, 2] is a powerful technique for examining the full-field deformation behaviour of both large civil engineering structures, such as buildings and bridges and small components. It is based on applying a regular pattern to the surface and taking photographs with a specially modified camera whilst the structure undergoes a deformation process. One particular feature of the technique is a method for preparing a pattern that can withstand heating. A suspension in alcohol of titanium dioxide pigment is sprayed through a very thin metal mesh lightly attached to the component's surface. When the mesh is removed, what remains is an array of dots of pure pigment which retains its white opacity at temperatures in excess of 1300 °C. This allows the study of thermal and mechanical behaviour to be carried out in oxidizing environments [2, 3] and the approach is especially useful where the levels of anticipated strain are relatively high, for example at the weld interface of two materials with different expansion coefficients.

In order to resolve low-level strains, consistent with studies of creep or residual stress release, a more sensitive technique is required. With a basic displacement resolution of 0.42 nm, moiré interferometry offers an adequate sensitivity. However, the gratings which traditionally form the surface pattern are cast on to the surface in either epoxy-resin resin or silicone rubber and will only survive an operating temperature below about 200 °C. For higher temperatures Ifju and Pool [4] have reported a 'zero thickness' grating involving an elaborate and expensive process of ion etching that is not amenable to some industrial applications.

A straightforward photoresist/thin-film grating process has been developed which will withstand temperatures of at least 1100 °C. It can be applied to metals and ceramics and has been used for examining the behaviour of welded joints following heat cycling and in the measurement of residual stress in rails, in the form of residual strain release when the component is annealed [5]. Although so far our trials have been confined to examinations in cool conditions, it is possible to envisage applications to measurements made at high temperature.

9.2.2 Basic grating procedure

The photoresist/metal film process involves producing a resist grating on the surface that forms a mould for a subsequent thin layer of heat-resistant metal. When heated, the resist is destroyed but the corrugated metal film retains the form of the original grating and, being in contact with the component, follows the mechanical deformations.

There are comparatively few stages in the grating preparation:

(1) The component is mechanically worked to produce a polished finish. This need not be particularly flat but problems can occur with the field coverage of the viewing system if it is too curved.
(2) A layer of photoresist is applied. To achieve a uniformly thin layer the method of spinning is recommended, but good results can be obtained by fine spraying using an airbrush.
(3) The resist layer is then exposed to an interference pattern at 1200 lines/mm derived by two-beam interference from a blue wavelength laser.
(4) For a complete displacement analysis, the component is re-exposed after rotating it through 90°, so forming a latent image of a crossed grating.
(5) A final, heavy exposure is given through a mask bearing a transparent crossed-line array of approximately 10 lines/mm.
(6) After the resist is developed, it is coated with a heat-resistant metal by vacuum evaporation or electroless deposition.

Except for the fifth stage where the coarse 10 lines/mm grid is applied, the procedure for preparing metalized gratings is conventional. This last treatment is, however, important. Even a spun resist coating on anything other than an optical flat will invariably result in a layer of non-uniform thickness, with the final grating structure extending in some regions down to the component's surface while elsewhere supported on a continuous layer of resist. In this condition, when the grating is coated with metal and the component is subsequently heated, the resist beneath the film will begin to melt. Those areas supported on a continuous layer will tend to flow in an uncontrollable fashion, causing the metallic film to break up.

To preserve the continuity of the film, with its embossed grating structure, the resist is 'cut' with the coarse grid through to the specimen surface. In this

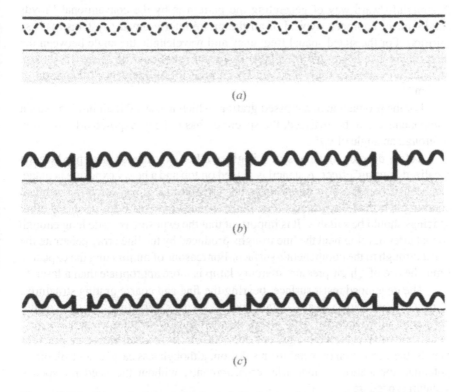

Figure 9.2.1. Stages in the preparation of the heat-resistant grating: (*a*) exposing the virtual grating, (*b*) exposing the coarse grating, developing and coating with heat-resistant metal and (*c*) following heating.

way, the evaporated metal is locked onto the component, so preventing localized movement or break-up of the grating. Figure 9.2.1 illustrates the processes involved.

9.2.3 Detailed preparation

The surface of the component is first polished. This is preferably carried out with as gentle a mechanical working as possible and under well-lubricated conditions in order to minimize work-induced stresses. Next, it is coated with a layer of photoresist [6]. If the component is circular, the method of applying the resist and then spinning is recommended. Alternatively, good results can be achieved by fine spraying using an airbrush. So as to sensitize and harden the resist, it is baked at 60 °C for 30 min. It is then exposed to a 1200 lines/mm pattern.

A suitable pattern can be derived by interference between two beams from a blue wavelength laser, such as a helium–cadmium laser operating at 426 nm.

A straightforward way of generating the pattern is by the conventional Lloyds single-mirror approach (typical of all the moiré interferometry systems that use mirrors). For the stated spatial frequency and wavelength, the angle between the direct incidence beam and the plane of the mirror should be set to 15.4°. As a guide, an exposure time of 1 min is required for an incident illumination of $2\ \mathrm{W\ m^{-1}}$.

For the production of a crossed grating, which is needed if all in-plane strain components are to be analysed, the specimen has to be re-exposed after rotating it through an angle of 90°.

Before developing the resist, it is given a last exposure so as to produce the localized 'cutting' effect. A stencil is placed on top and a heavy exposure is made. The stencil consists of a fine transparent crossed-line array which has a much coarser pitch than the grating—one with 10 lines/mm has been used but other spacings should be suitable. It is important that the exposure is made long enough so that after development the fine troughs produced by the line array penetrate the resist through to the component's surface. For reasons of minimizing the exposure time, the use of a high-pressure mercury lamp is more appropriate than a laser.

The developed resist surface, bearing the fine and coarse grating structures, is next coated with a heat-resistant metal, such as Nichrome, or the cobalt–chromium alloy Stellite. This is best carried out by vacuum evaporation to a thickness which is preferably greater than 100 nm. For nickel and some precious metals, the deposition of metal from solution, although less easy to control, offers potential application in industrial environments, without the need for special vacuum procedures.

It has been found that this two-stage grating approach is tolerant to large variations in the thickness of the resist layer. In addition, if the component is initially taken slowly through the decomposition temperature of the resist, it is better able to withstand thermal shock.

The stated procedure is intended for thermal studies in air where the material of the component does not exhibit surface degradation through oxidation. Stainless steel and ceramics are two such materials that can be heated in the oxidizing atmosphere of a conventional furnace. For ferritic and other oxidation-prone materials, there is the optimum choice of heating in a vacuum or inert gas. An alternative but less reliable method is first to protect the untreated component with a resistant layer. Nichrome can be used on ferritic steels at temperature up to 800 °C or a thick coating of the same material as that used on top of the resist grating.

9.2.4 Polishing-induced stresses

There are inevitable stresses introduced on and below the surface, even when polishing is carried out carefully. A question can be posed as to the effects of this polish-induced stressed layer on the measurement of the bulk stresses. It is difficult to assess quantitatively because, in the removal of these topmost layers by etching, the surface is roughened to a degree that prevents the application of a good quality grating.

Tests using x-ray diffraction on our carefully polished specimens, which have then been incrementally etched, show that although such stresses may be high they are confined in a surface layer only a few micrometres deep [8]. It is reasonable to assume then that providing the component is much thicker than the stressed layer—in the example described here approximately 1000 times thicker—then the layer will dissipate its stresses throughout the bulk of the material during heating. As a result, whereas the release of bulk residual stress present in the component will modify the grating frequency and so be detected, the release of the polish-induced stresses will not provide a significantly measurable contribution to the overall deformation.

Verification that the layer has a negligible effect was carried out on a carefully annealed sample of carbon steel approximately 15 mm thick. This was polished and coated with a heat-resistant grating. A comparison of the moiré interferograms of the specimen in its initial condition and after subjecting it to a second annealing cycle appeared very similar. This appears to confirm that the initial low-level bulk stresses have not been influenced by the annealing of the thin stressed layer introduced by polishing.

9.2.5 Optical system aberrations

Unless expensive high-quality optical elements are employed in the recording system the grating will exhibit variations. Such effects are of little consequence providing that before heating the grating is examined in a moiré interferometer (which will have its own aberration characteristics) so as to record the initial condition. Any initial errors will be displayed as fringes. When the component is deformed through mechanical stress or heating, a different fringe pattern will be generated which, after subtraction of the initial pattern, can be related to actual surface displacements and by computation to local strains.

9.2.6 Annealing studies

The high-temperature grating process has been applied to several studies, including a through-section sample of steel railway track. Residual stresses were present close to the top surface as a result of plastic deformation caused by the regular passage of trains.

Advice from material experts indicated that heating the specimen to an annealing temperature would cause a redistribution and relaxation of residual stresses. The suggested regime was to hold it at 600 °C for 30 min, followed by slow cooling, over 12 hr in order to prevent the re-introduction of stresses

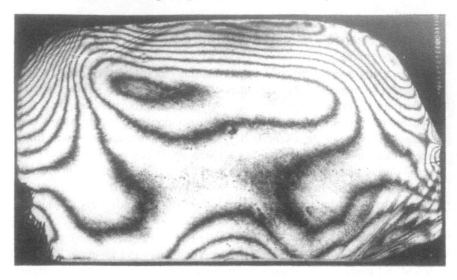

Figure 9.2.2. The y-displacement map following the release of internal stresses by annealing.

generated by internal thermal gradients. The annealing process was carried out inside a vacuum furnace.

Figure 9.2.2 shows the interference fringe pattern relating to displacements in the y-direction. There is a concentration of fringes along the top face which is consistent with a release of stresses induced by wheel loading. Along the right-hand side, there is a region of finely spaced fringes that are associated with compression by the wheel flange striking the side of the rail. Differentiation of the displacement map reveals the distribution of strain (figure 9.2.3), in this case as an intensity-coded map. In view of the calculated high levels of related strain, it is possible that the upper left-hand corner is a potential site for the initiation of failure.

9.2.7   High-temperature studies

At the time of developing this heat-resistant grating, there was no facility available for experimenting with specimens held at elevated temperatures. Until the recent discovery by McKelvie [8], it was assumed that with an open furnace the refractive disturbance in the heated airspace between the interferometer and the component would destroy the stability of the wavefronts of the interrogating laser beams. The consequent effects, in the form of dynamic turbulence of the moiré fringes, would be unacceptable. McKelvie shows that if the beams are aligned at a low angle, either side of the grating surface, the effects of thermal disturbances are greatly reduced. There are no additional problems in preparing

Figure 9.2.3. The intensity-coded map showing the *y*-strain distribution.

higher spatial frequency gratings for low-angle illumination and the possibility of making measurements on structures that are heated in air is attractive.

References

[1] Burch J M and Forno C 1982 High resolution moiré photography *Opt. Eng.* **21** 602–14
[2] Forno C 1988 Deformation measurement using high resolution moiré photography *Opt. Lasers Eng.* **8** 189–212
[3] Webster J M *et al* 1981 Strain determination in a transition joint at high temperature *Exp. Mech.* **21** 195–200
[4] Ifju P and Post D 1991 Zero thickness specimen gratings for moiré interferometry *Exp. Tech.* **15** 45–7
[5] Holstein D, Salbut M, Kujawinska W and Jüptner W 2001 Hybrid experimental-numerical concept of residual stress analysis in laser weldments *Exp. Mech.* **41** 343–50
[6] Shipley photoresist type: 1400-17
[7] Middleton J C 1993 British Steel, Swindon Laboratories, Rotherham Private communication
[8] McKelvie J 1997 Thermal-insensitive moiré interferometry *Opt. Lett.* **22** 1–3

CHAPTER 10

MICROMECHANICS AND CONTACT PROBLEMS

Chapter 10.1

Grid method for damage analysis of ductile materials

Jin Fang
Peking University, Beijing, China

10.1.1 Introduction

Fracture failure of many ductile materials takes the form of accumulated damage from the growth and coalescence of microviods. The voids often grow due to plastic straining of the surrounding material where flow localization takes place. When the ligaments between the neighbouring voids become sufficiently thin, void coalescence occurs by successive propagation of cracks. In recent decades, some constitutive models have been developed for void-containing materials. Based on a rigid-plastic solution for a spherically symmetric deformation applied around a spherical inclusion, Gurson [1] presented a theory to predict the plastic flow in voided ductile media. Since then some modifications have been made to this model, for example that by Tvergaard and Needleman [2] who introduced more parameters into the approximate yield conditions for a matrix with periodically arranged voids. For void growth and coalescence in materials under dynamic loading, Curran *et al* [3] have presented a detailed review.

Experimental investigations have been essential to the study of the behaviour of void growth and coalescence in ductile materials and to the development of failure models or constitutive relations of the materials containing microvoids. In recent years, using a caustic method and scanning electron micrography, Theocaris [4] has examined the variation in thickness of the ligament between pairs of circular holes with a diameter of 2 mm in tensile polycarbonate specimens. He studied the deformation produced by the plastic mode transition with the alignment and orientation of two holes with respect to the direction of the load [5]. By observing the flow localization within the ligaments between neighbouring holes, Geltmacher *et al* [6] studied the coalescence of holes with a diameter of 1.6 mm that are 'randomly' distributed in sheet specimens under uniaxial and equal-biaxial tension.

In this chapter, a grid method is presented to study the growth and coalescence of the voids in a mesoscopic scale (about 15–30 μm in void diameter) rather than large-scale holes. Gratings with a density of 150 line/mm are used in both the quasi-static and impact tests to store the plastic deformation of the material around voids. Digital image processing is used to obtain the plastic strain field in the ligament between the neighbouring voids so that a quantitative analysis can be made to understand the behaviour of the flow localization and the failure mechanism of void-linking in ductile materials.

10.1.2   Gratings and experiments

10.1.2.1   Specimens and zero-thickness gratings

Sheet specimens are manufactured from copper foils of thickness 0.2 mm. To observe the material damage, cylindrical voids with a diameter of about 30 μm are 'drilled' by a laser beam focusing on the centre part of the long test sheets, in which the width (6 mm) is much larger than void size so that the effect of the specimen boundary can be fully eliminated. After the voids have been arranged by drilling in different orientations and spacing, the copper sheets are annealed from 400°C with the temperature decreased slowly so as to eliminate the residual stress around the voids and to increase the ductility of the material.

A zero-thickness grating [7] is produced on the specimen to record the plastic deformation around the voids. The surface of the copper specimens is well polished and then covered uniformly with a layer of photoresist. A glass model of a cross-line amplitude grating with frequency of 150 line/mm is attached to the specimen so that the grid shadows are left on the surface after exposure to light and development. The corrosion process is then performed directly on the metallic specimen surface to produce a specimen grating with an orthogonal array of dots and a depth less than 0.1 μm which has little influence on material deformation and damage.

10.1.2.2   Quasi-static loading and impulsive loading

To observe void growth and coalescence in the material under a quasi-static load, a uniaxial tension is produced by a loading device attached to a microscope. The ductile metal has a macroscopic behaviour of plastic work-hardening with a maximum extensibility of 25% for perfect specimens. This is reduced to 13%–15% on the introduction of artificial voids, which represents the damage produced by voids on the apparent properties of the material. During the procedure of loading, the plastic deformation of the matrix matter around the voids is recorded with the help of deformed grids following the specimen deformation.

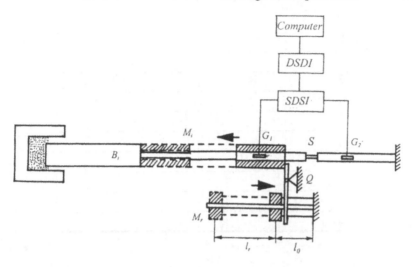

Figure 10.1.1. Schematic illustration of a split Hopkinson bar to produce an impact load on a specimen.

To study the dynamic fracture behaviour of the material under impact, the specimens are impulsively loaded using a split Hopkinson bar (SHB), as illustrated in figure 10.1.1. Two rubber strips, R_1 and R_2, are stretched for a certain elongation of l_r by a mass block M_r so as to store an elastic energy. With the help of a lever beam rotated at point Q, the sudden release of those rubber strips produces a rapid movement in an impact block M_i in its slide-rail system towards the stepped surface of an input bar B_i. Through this procedure the propagation of the stress wave generates an impulsive tension on the specimen S with various strain rates. A pair of strain gauges, G_1, are mounted symmetrically on the input bar to measure the strain ε_i of the incident wave, by a superdynamic amplifier circuit and an oscilloscope connected to a computer. Meanwhile, another pair of strain gauges, G_2, are symmetrically mounted on the output bar to obtain the strains ε_t of the transmitted wave. With a uniform stress distribution in the specimen, the strain ε_s, stress σ_s, and the strain rate $\dot{\varepsilon}_s$ of the specimen, can be obtained by

$$\varepsilon_s = -\frac{2C}{L_s} \int_0^{t_0} (\varepsilon_t - \varepsilon_i)\, dt \tag{10.1.1a}$$

$$\sigma_s = E\left(\frac{A}{A_s}\right) \varepsilon_t \tag{10.1.1b}$$

$$\dot{\varepsilon}_s = -\frac{2C}{L_s}(\varepsilon_t - \varepsilon_i) \tag{10.1.1c}$$

where E, A, and C are Young's modulus, the cross-sectional area and the elastic wave speed of the input bar and A_s and L_s are the cross-sectional area and the

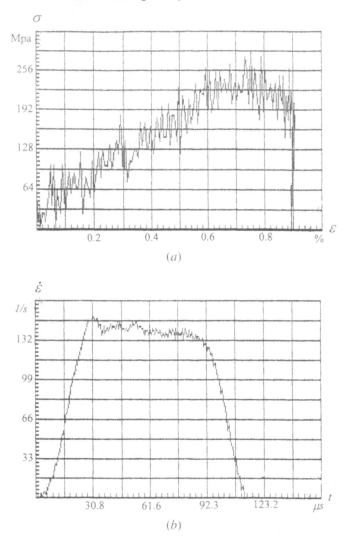

Figure 10.1.2. (*a*) A dynamic relation of impulsive stress and tensile strain of a material containing voids. (*b*) The history of the strain rate of the impulsively loaded specimen.

length of the specimen, respectively. Figure 10.1.2(*a*) shows the dynamic stress–strain relation of the specimen and figure 10.1.2(*b*) presents the history of the strain rates that has a mean level of 140 1 s^{-1}.

10.1.3 Image processing of deformed gratings

After the deformed grating has been captured by a CCD camera, either from real-time recording during the quasi-static tension or from post-recording of the specimen impulsively loaded, digital image processing is utilized to obtain the plastic strain components in the matrix material around the voids. Two main methods are introduced in the following sections.

10.1.3.1 Determination of strain by cross grid

The etched cross grating consists of tiny islands of dots that deform freely to follow the material deformation around microvoids. Thus the plastic strain components can be obtained by the comparison of the distorted array of cross points with the originally orthogonal grids.

The determination of the location of the cross points is the first step of image processing. Usually, one can make use of grey transformation with a threshold value to obtain the dot images with enhanced contrast. This technique is suitable for the case in which the intensity background is relatively uniform so that the contrast stretching can be performed to get the binary images. When the grey background of the grid images differs greatly from the whole-field, manual determination of some left dots can be carried out by magnifying the cross dot in the local region involved and calculating the geometrical centre of those dots. Since the distance between the cross grid is larger than the dimension of the dots, this kind of processing is accurate enough to determine the array points under the condition of image magnification. Figure 10.1.3 gives some dot images resulting from the deformed grating images, which shows clearly the plastic deformation around two voids which are aligned with their link-line of centres perpendicular to the apparent tensile direction.

Based on this pre-processing to eliminate the grey noise and to obtain the deformed cross points, the displacement gradients can be computed by a comparing the algorithm of the deformed grid with the initial one without any distortion. Through examining the relative position changes of the neighbouring dots, or $X_I = x'_i - x_i$, $Y_I = y'_i - y_i$ ($I, i = B, C, D, E$), as illustrated in figure 10.1.4, the displacement derivatives at centre point $A(x, y)$ are given by

$$\left.\frac{\partial u}{\partial x}\right|_A = \frac{X_D - X_B - 2P_0}{2P_0} \qquad \left.\frac{\partial u}{\partial y}\right|_A = \frac{X_C - X_B}{2P_0}$$

$$\left.\frac{\partial v}{\partial x}\right|_A = \frac{Y_D - Y_B}{2P_0} \qquad \left.\frac{\partial v}{\partial y}\right|_A = \frac{Y_E - Y_C - 2P_0}{2P_0} \qquad (10.1.2)$$

where P_0 is the pitch of the original cross grid. Therefore, the Eulerian strain components can be obtained given by

$$\varepsilon_x = \sqrt{1 + 2\frac{\partial u}{\partial x} + \left(\frac{\partial u}{\partial x}\right)^2 + \left(\frac{\partial v}{\partial y}\right)^2} - 1 \qquad (10.1.3a)$$

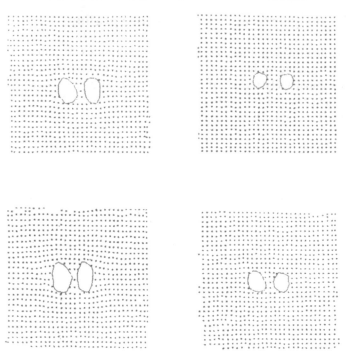

Figure 10.1.3. Some dot images of the deformed grid around enlarged voids.

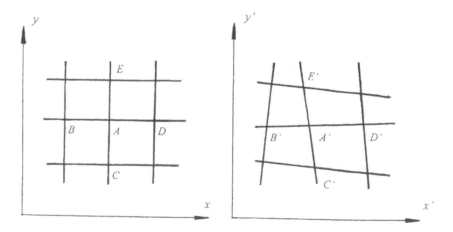

Figure 10.1.4. Determination of displacement derivatives by position changes of the cross-points of the grid.

$$\varepsilon_y = \sqrt{1 + 2\frac{\partial v}{\partial y} + \left(\frac{\partial v}{\partial y}\right)^2 + \left(\frac{\partial u}{\partial x}\right)^2} - 1 \qquad (10.1.3b)$$

$$\gamma_{xy} = \arcsin \frac{\dfrac{\partial u}{\partial y} + \dfrac{\partial v}{\partial x} + \dfrac{\partial u}{\partial x}\dfrac{\partial u}{\partial y} + \dfrac{\partial v}{\partial x}\dfrac{\partial v}{\partial y}}{(1+\varepsilon_{xx})(1+\varepsilon_{yy})} \qquad (10.1.3c)$$

Figure 10.1.5 presents the contours of those plastic strain components around two voids in a specimen subjected to a macroscopic strain of about 12% in figure 10.1.3. It is obvious that the strain ε_x in the tensile direction (figure 10.1.5(a)) is generally larger than other components and the maximum points exit in the ligament between the two voids which causes the coalescence of the voids with the initiation and propagation of a crack. Because both voids shield the localized deformation in the ligament from the lateral contraction of the material outside these two voids, the compressive strain components ε_y (figure 10.1.5(b)) in the ligament region is less than that outside the two lateral sides. The shear component γ_{xy} (figure 10.1.5(c)) is asymmetrical about the link-line of the void centers. It shows that the maximum values exist at the points in the outside area near the voids, in the orientation of 45° with respect to the loading direction, where the shear bands can be observed in the early stage of flow localization.

10.1.3.2 Displacement determination by Fourier transform

The grid method in the preceding section is a spatial domain technique based on grey-level mapping that is flexible to deal with the images with local specifications like noise of low frequency or the boundary of inner voids. A frequency domain method of Fourier transform is presented in this section that processes the grid images in the whole field, to make the solution of displacement an automatic procedure of computation [8].

To describe the technique in a general manner, an orthogonal grid is considered here with rectangular brightness distribution of the width b_x and b_y in the x and y directions, respectively. The brightness function $f(x, y)$ can be expressed as

$$f(x, y) = \sum_{m=-\infty}^{+\infty} \sum_{n=-\infty}^{+\infty} a_{m,n} \exp[j2\pi(mf_{x0}x + nf_{y0}y)] \qquad (10.1.4)$$

wherethe coefficient $a_{m,n}$ are

$$a_{m,n} = a_0^2 \frac{\sin(m\pi b_x/p_{x0})}{m\pi b_x/p_{x0}} \frac{\sin(n\pi b_y/p_{y0})}{n\pi b_y/p_{y0}} \qquad (10.1.5a)$$

and f_{x0}, f_{y0} are the initial frequencies of the grid pattern, related to the grid pitches p_{x0} and p_{y0} given by

$$f_{x0} = \frac{1}{p_{y0}} \qquad f_{y0} = \frac{1}{p_{y0}} \qquad (10.1.5b)$$

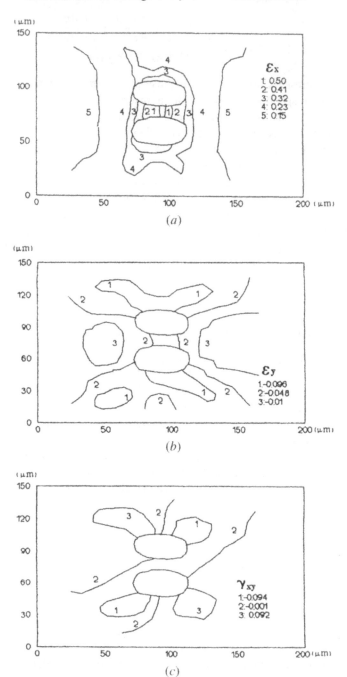

Figure 10.1.5. Schematic illustration of plastic strain contours for components of ε_x (a), ε_y (b) and γ_{xy} (c).

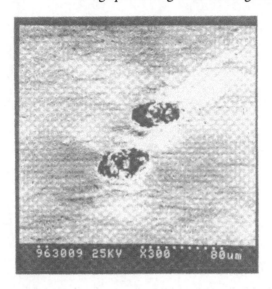

Figure 10.1.6. A grating image of distorted grid around two voids aligned at 45° inclined to the tensile direction.

respectively. In this test, the pitches in the x and y directions are identical.

Together with the plastic flow around voids, the grids deform with the material surface, as shown by an example in figure 10.1.6. It is seen in the figure that a distorted grid distributed around two voids, which are aligned at 45° with respect to the tensile direction of the specimen. The displacement components u and v, represented by a spatial description of Eulerian variables ξ and η as coordinates, can be expressed as

$$u(\xi, \eta) = \xi - x \qquad v(\xi, \eta) = \eta - y. \qquad (10.1.6)$$

Thus the brightness $g(\xi, \eta)$ of the deformed grating, corresponding to $f(x, y)$ at the initial points (x, y), can be written as

$$
\begin{aligned}
g(\xi, \eta) &= f[\xi - u(\xi, \eta), \eta - v(\xi, \eta)] \\
&= \sum_{-\infty}^{+\infty} \sum_{-\infty}^{+\infty} a_{m,n} \exp\{j2\pi[mf_{x0}(\xi - u(\xi, \eta)) + nf_{y0}(\eta - v(\xi, \eta))]\} \\
&= \sum_{-\infty}^{+\infty} \sum_{-\infty}^{+\infty} b_{m,n} \exp[j2\pi(mf_{x0}\xi + nf_{y0}\eta)] \qquad (10.1.7)
\end{aligned}
$$

where

$$b_{m,n} = a_{m,n} \exp\{-j2\pi[f_{x0}u(\xi, \eta) + f_{y0}v(\xi, \eta)]\}. \qquad (10.1.8)$$

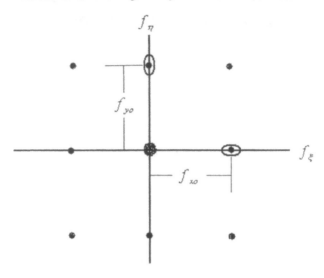

Figure 10.1.7. Illustration of the spectrum distribution of figure 10.1.6.

Making a computation of the fast Fourier transform (FFT) of equation (10.1.7), the spectrum of the deformed grating can be obtained given by

$$G(f_\xi, f_\eta) = FT\{g(\xi, \eta)\} = \sum_{-\infty}^{+\infty} \sum_{-\infty}^{+\infty} B_{m,n}(f_\xi - mf_{x0}, f_\eta - nf_{y0}). \quad (10.1.9)$$

Figure 10.1.7 shows the spectrum of figure 10.1.6, in which two pairs of the first-order impulses are distributed symmetrically about the origin of the frequency axes f_ξ and f_η, because the cross grid is, in fact, an amplitude type. To obtain the displacement components $u(\xi, \eta)$ and $v(\xi, \eta)$, one from each pair of the first-order components of spectrum is filtered by a small window, respectively. They are then moved to the origin of the frequency coordinates so as to subtract the basic frequency of the standard grid. An inverse Fourier transform can then be performed for those two first harmonics given by

$$FT^{-1}\{B_{10}\} = b_{10} = a_{10} \exp\left\{-j\frac{2\pi}{P_{y0}}u(\xi, \eta)\right\} \quad (10.1.10)$$

and

$$FT^{-1}\{B_{01}\} = b_{01} = a_{01} \exp\left\{-j\frac{2\pi}{P_{y0}}v(\xi, \eta)\right\}. \quad (10.1.11)$$

Figure 10.1.8 presents the real (10.1.8(*a*)) and imaginary parts (10.1.8(*b*)) of equation (10.1.10), which involves information about the displacement $u(\xi, \eta)$. Figure 10.1.9 shows the real (10.1.9(*a*)) and imaginary parts (10.1.9(*b*)) of equation (10.1.11), which involves information about $v(\xi, \eta)$, respectively. The

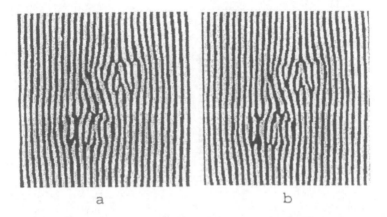

Figure 10.1.8. Diagram of the real (*a*) and imaginary part (*b*) of the frequency component at (1, 0) in the spectrum.

displacement components u and v can thus be obtained through those two-dimensional distributions:

$$u(\xi, \eta) = \frac{1}{2\pi p_{x0}} \tan^{-1} \left\{ \frac{\text{Re}\{b_{10}\}}{\text{Im}\{b_{10}\}} \right\} \qquad (10.1.12a)$$

$$v(\xi, \eta) = \frac{1}{2\pi p_{y0}} \tan^{-1} \left\{ \frac{\text{Re}\{b_{01}\}}{\text{Im}\{b_{01}\}} \right\} \qquad (10.1.12b)$$

where $\text{Re}\{\cdot\}$ and $\text{Im}\{\cdot\}$ are the real and imaginary parts of the complex variables b_{10} and b_{01}, respectively. The Eulerian strain components at any point can, therefore, be obtained by substituting them into equations (10.1.3c) as before. Figure 10.1.10 gives three contours of obtained strain components of ε_x (10.1.10(*a*)), ε_y (10.1.10(*b*)) and γ_{xy} (10.1.10(*c*)), respectively, distributed in matter around voids of figure 10.1.6 when the macroscopic strain of the tensile specimen is about 9.7%.

10.1.4 Some analyses of void damage

10.1.4.1 Interaction between the neighbouring voids

Experimental results have shown that the spatial arrangement of the voids affects not only the characteristic of flow localization of the ligaments but also the coalescence form among the neighbouring voids. An example is seen in figure 10.1.11 for a specimen containing three artificial voids which form an equilateral triangle. Under the impulsive tensile loading as mentioned in section 10.1.2.2, the voids grow individually during the first impact (figure 10.1.11(*a*)). But the two voids, which are aligned perpendicular to the tensile direction, are linked together by a crack caused by the rupture of the

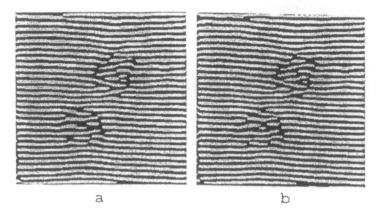

a   b

Figure 10.1.9. Diagram of the real (*a*) and the imaginary part (*b*) of the frequency component at (0, 1) in the spectrum.

ligament between them during the second impact (figure 10.1.11(*b*)). It is clear that in the first step the extent of the damage with the three voids is nearly the same with their shape changing from circular to elliptical. The impulsive tension makes the major axis increased significantly in the tensile direction. The minor axis in the perpendicular direction, however, does not change very much. The impulsive tension causes the nucleation and coalescence of the smaller voids near the equator region of the main voids, where the maximum tensile strains exist, as shown in figure 10.1.12(*a*). Even for the areas of those voids which are nearly the same under the first impact, the distance between the two voids aligned perpendicular to the tensile direction becomes shorter than that between one of them and other voids. The second impact on the specimen initiates a crack that propagates in the matter between the two voids to result in ligament failure. At this time, the macroscopic strain of the sheet specimen is about 10% and the plastic strain distribution around the voids is illustrated in figure 10.1.12. Both the pattern pictures and the strain contours show that the plastic flow occurs mainly in the material around the two linked voids, with little change in the area and shape of the other void or the strain distribution around it. This implies that the interaction between these two linking voids is much stronger than that between one of them and the other void. The dynamically plastic flow quickly develops within the intervoid ligament to absorb part of the impact energy through material damage.

The interactive behaviour of a pair of the closest neighbouring voids has the typical characteristics to represent the growth process of voids [9]. Based on the area measurement of two voids which are aligned in different directions θ with respect to the tensile direction, the relation between K_s, the ratio of the damaged void area S to the initial void area S_0, and the far-field stress σ_f under quasi-static load is shown in figure 10.1.13. For those specimens in which the

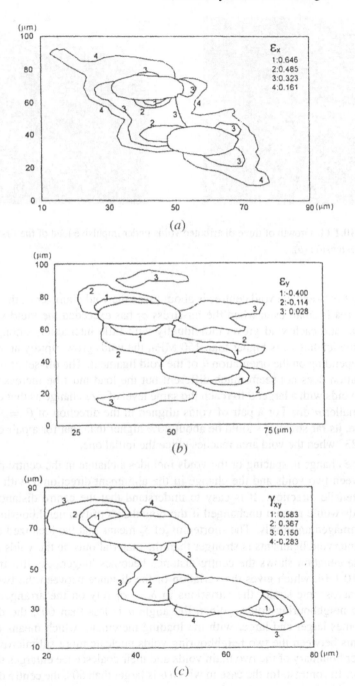

Figure 10.1.10. Plastic strain components ε_x (a), ε_y (b) and γ_{xy} (c), distributed in material around the two voids in figure 10.1.6.

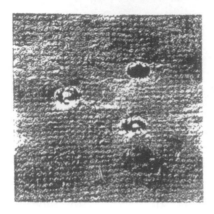 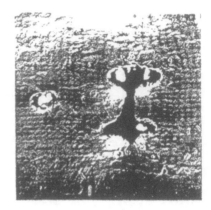

Figure 10.1.11. Growth of three distributed voids under impulsive load of the first (*a*) and second extension (*b*).

spacing l between the void centres is about twice the void diameter d, the change in the area is small even when the far-stress σ_f has exceeded the yield stress of 60 MPa, and each void grows individually with little interaction among them. When the loading σ_f is higher than 150 MPa, the voids grow rapidly at different rates depending on the orientation θ of the void ligament. The off-set of this flow localization does not seem much different but the load must be increased for a pair of voids with a larger θ to reach the same level of size change as that of voids with smaller θ do. For a pair of voids aligned in the direction of $\theta = 90°$, for instance, its far-stress σ_f should be about 20% higher than that for a pair of voids at $\theta = 23°$ when the void area reaches twice the initial one.

The change in spacing of the voids includes a change in the centre distance δ_c between two voids and the change in the alignment direction θ with respect to the tensile direction. It is easy to understand that the centre distance δ_c of the voids would remain unchanged if the boundaries of two neighbouring voids were damaged uniformly. The shortening of δ_c means that the localized damage of the intervoid ligaments is stronger than the material outside the voids and the opposite situation shows the centre distance becomes longer. As illustrated in figure 10.1.14, which gives the variation of the distance between the two voids with macroscopic strain, the variations in $\delta_c - \sigma_f$ rely on the arrangement of the two neighbouring voids. When the angle θ is less than 60°, the distance δ_c becomes larger and larger with the loading increment, which means that the ligaments between the two neighbouring voids nucleate more submicrovoids in the inner boundary of the two main voids and their coalescence enlarges the void distance. In contrast, for the case in which θ is larger than 60°, the centre distance of two voids decreases with an increase in the far-field stress, which implies that the damage to the material outside the two voids is more serious than that to the material in the inner part. A critical degree, e.g. about $\theta = 60°$ for the void

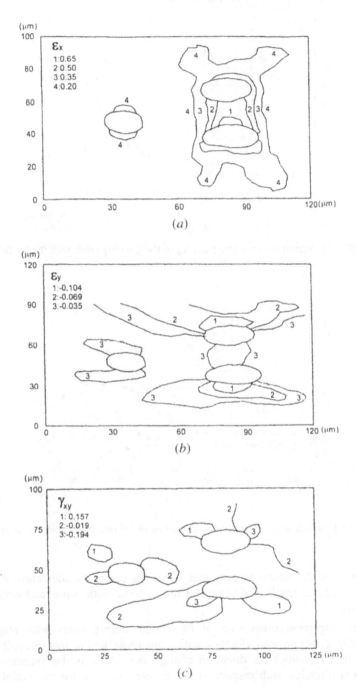

Figure 10.1.12. Plastic strain distribution around microvoids in figure 10.1.11 when the macroscopic strain of the specimen is about 11%.

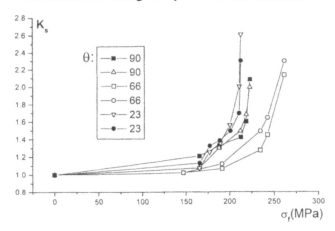

Figure 10.1.13. Variation of the area ratio K_s of the growing voids with the far-field stress σ_f.

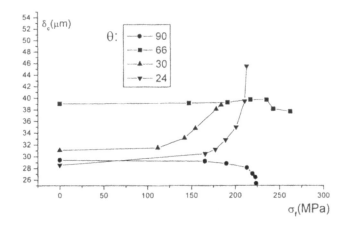

Figure 10.1.14. Variation of spacing δ_c of the centres of two voids with the far-stress σ_f.

distance $\delta_c = 2d$, seems to exist that keeps the centre distance constant. This implies the same levels of damage for the material both inside and outside the two voids.

The alignment orientation of two neighbouring voids with respect to the tensile direction also changes with anincrease in the macroscopic strain. Figure 10.1.15 shows the direction change $\Delta\theta$ which is the decrease in the alignment direction with respect to the tensile direction for the initial angles $\theta = 17°$, $23°$ and $66°$, respectively. It is clear that for two special situations, $\theta = 0°$ (i.e. two voids are aligned in the tensile direction) and $\theta = 90°$ (i.e. the link between two voids is perpendicular to the tensile direction), there is no

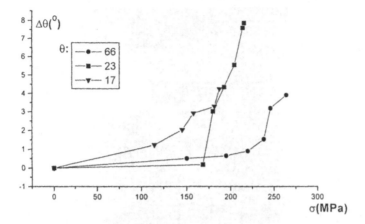

Figure 10.1.15. Orientation change $\Delta\theta$ of between two voids with the far-stress σ_f.

change in direction for $\Delta\theta$ because of the symmetry of between the loading and the arrangement of the voids. Between these two specific cases, the alignment orientation θ with respect to the tensile direction decreases depending on the arrangement of the voids and the extent of damage (nucleation of microvoids and their coalescence) along the main void boundaries. This type of direction change develops rapidly as the localization damage in the ligaments is dominated by the plastic flow of the yielded material.

10.1.4.2 Flow localization and ductile failure of ligaments

From the specimen surface, the localized shear bands can be clearly observed in the material around voids accompanying the void growth under the action of both tensile and shear stress. It is noteworthy that the plastic strain distribution obtained from grid image processing, as described in the previous sections, indicates quantitatively the shear band development. Figure 10.1.16 presents a photograph of two voids at 45° inclined to the tensile direction of the far-field, in which bright shear bands can be observed along the 45° direction at the locations near the voids. A contour illustration of the maximum shear strain $\gamma_{max} = 2[(\varepsilon_x - \varepsilon_y)^2 + \gamma_{xy}^2]^{1/2}$ is given in figure 10.1.17. This is obtained from digital processing the deformed grid pattern in figure 10.1.16. Figure 10.1.17 shows clearly that the shear bands result from the plastic flow localization caused by the maximum shear stress in the material between and outside the two voids at 45° to the tensile direction. Similar results can also be observed from the other cases of void arrangement in which the distribution of the maximum shear strain is in good agreement with that of shear band development.

Though the behaviour of the shear flow localized around voids can be interpreted by controlling the maximum shear strain, void linking or coalescence within the intervoid ligament is caused by both tensile and shear stresses in the

Figure 10.1.16. Shear bands produced in the near region of the voids at 45° inclined to the tensile direction.

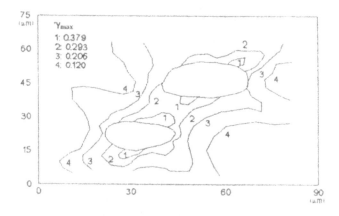

Figure 10.1.17. A contour of the maximum shear strain γ_{max} corresponding to figure 10.1.16.

ligament. An effective strain, or the von Mises equivalent strain,

$$[\varepsilon^*] = \sqrt{\tfrac{1}{2}[(\varepsilon_1 - \varepsilon_2)^2 + \varepsilon_1^2 + \varepsilon_2^2]}$$

for the case of plane strain in our tests, can be used to predict the ductile failure of the intervoid material, based on measurement of the deformed grids and data processing the plastic strain field. Figure 10.1.18 shows a contour map of $[\varepsilon^*]$ for two voids aligned perpendicular to (10.1.18(*a*)) and at 45° inclined to (10.1.18(*b*))

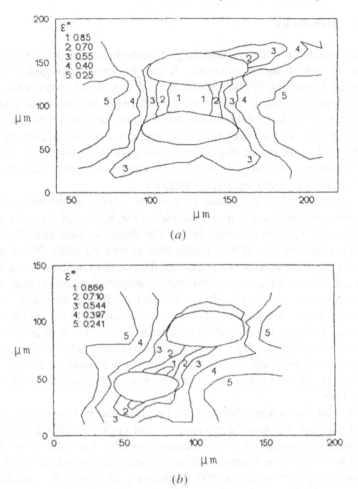

Figure 10.1.18. The von Mises equivalent strain resulting from strain components, predicting the linking of the voids aligned perpendicular to (*a*) and at 45° inclined to (*b*) the loading direction.

the tensile direction of far-field tension, respectively. It is obvious that the maximum value of the effective strain (about seven times the far-field strain ε_f) exists within the intervoid ligament to cause a plastic flow in the ligament and the nucleation of the submicrovoids along the inner boundary of the two voids, and finally to generate a crack. With the increase in flow localization, ductile fracture happens in the ligament by the propagation of a crack resulting in void linking.

10.1.5   Remarks

The experimental results show that the grid method is a useful tool in microvoid damage analysis of ductile materials. A zero-thickness cross-grid grating can be used to follow the plastic deformation of a material under either quasi-static or dynamic loading. For the image patterns of the deformed grids, both geometrical comparison of the cross dots in spatial domain processing and the Fourier transform of the whole-field from frequency domain processing can be utilized to obtain the plastic displacement and strain components. By masuring the void geometry, the process of void growth can be analysed as their interaction causes the changes in spacing and orientations.

Using this method the plastic flow in the vicinity of the microvoids can be quantitatively analysed to reveal the failure mechanism of a ductile material governed by void growth and coalescence. The obtained plastic strain field shows clearly the characteristics of flow localization around the voids. The loci of the maximum shear stress component indicate the positions where the shear bands are formed. The von Mises equivalent strains, obtained from processing the strain field, is shown to be a reasonable criterion for predicting the plastic flow that causes void linking in the ligaments between the voids. For the ductile material in our tests, a single void normally grows to be three times as large as its initial area when the specimens fail under quasi-static tension. Due to the interaction between two neighbouring voids, with an initial centre distance about twice the void diameter, the two voids coalesce each other when the void area is about twice the initial one.

The void interaction influences not only the plastic flow of the ligaments between the voids but also the growing form of each void. For a pair of voids with an initial spacing between two and three times the void diameter, the distance between the void centres becomes shorter or longer as the orientation to the tensile direction is larger or smaller than 60°, respectively. Moreover, the variation in the alignment angle of the voids with respect to the loading direction depends also on the orientation arrangement of the voids. These results imply that the extent of damage to the void boundaries near the intervoid ligament differs from that near the boundaries outside the two neighbouring voids.

Acknowledgment

The support of the NNSFC Grant No.10125211 is gratefully acknowledged.

References

[1] Gurson A L 1977 *J. Eng. Mater. Tech. Trans. ASME* **99** 2–15
[2] Tvergaard V and Needleman A 1984 *Acta Metall.* **32** 157
[3] Curran D R, Seaman L and Shockey D A 1987 *Phys. Rev.* **147** 253–388
[4] Theocaris P S 1986 *Int. J. Solids Struc.* **22** 445–66

[5] Theocaris P S 1995 *Eng. Frac. Mech.* **51** 239–64
[6] Geltmacher A B, Koss D A, Matic P and Stout M G 1996 *Acta Mater.* **6** 2201–10
[7] Post D 1987 *Moire Interferometry, Handbook on Experimental Mechanics* ed
    A S Kobayashi (Englewood Cliffs, NJ: Prentice-Hall) pp 314–87
[8] Fang J and Ni S Y 1998 *J. Mater. Sci. Lett.* **17** 223–6
[9] Fang J, Ni S Y, Yang H W, Bai S L and Dai F L 1997 *Acta Mech. Sinica* **13** 280–8

Chapter 10.2

A contact problem

Jin Fang
Peking University, Beijing, China

10.2.1 Introduction

Contact mechanics has been an important subject in engineering as it is related to force or moment transition, rolling and sliding between bodies and with many friction and lubrication problems. Analytical, experimental and numerical methods are three techniques for analysing the stress and deformation of two solid bodies into contact. Many analytical solutions, such as Hertz' theory for the contact between two elastic bodies with parabolic profiles [1], Boussinesq's solution for small area contact on semi-infinite space, etc [2], form the basis of contact mechanics. Among the experimental methods for solving contact problems, optical techniques are useful for determining contact deformation and stresses. In fact, just from Newton's experiments on interference fringes in a gap between two glass lenses, in 1882 Heinrich Hertz developed his theory for classical contact thus laying the earliest foundation of contact mechanics [3]. With the development of modern optics, other optical techniques have been applied to contact problems. For instance, a combination of the isochromatic fringes of photoelasticity, that show the difference of the principal stress, and the isopathic fringes from holographic photoelasticity, which presents the sum of principal stresses, can be used to separate the normal contact stresses by data subtraction in the near region of the contact surface [4]. The shadow optical method of caustics indicates the stress concentration or singularity by simple caustic curves and the contact force of the in-plane concentrated touch can be easily determined by measuring the maximum diameter of the caustic shadows beneath the contact location [5].

Its strong nonlinearity make the solution of contact problems complicated and difficult. Pure analytical solutions are applicable to only a few contact cases and pure experimental methods are also limited by data processing and their applications. With the rapid development of digital computers in recent decades, some numerical methods have been employed for contact analysis. Among them the finite-element and boundary-element methods are both popular and successful [6]. For practical contact analysis of engineering materials or structures, the actual contact conditions, e.g. contact sizes and profiles, interface roughness and friction, provide the basis for numerical simulations. Therefore, a combination of analytical, numerical and experimental techniques seems to be an important way of determining the stress and deformation of solid bodies in contact with each other. For example, in the case of an existing strain singularity at the end points of a finite contact zone of two in-plane plates, the end-load can be determined by measuring the caustic curves and then employing them as initial values in the recurrence formula to solve the stress distribution in the contact area [7]. With the boundary conditions derived from photoelasticity or any other stress-measuring method, the finite-element method can be used to acquire the contact stress [8]. By speckle photography of the displacement, the contact deformation of two touching bodies with rough surfaces can be determined. Also the displacement-prescribed boundary condition can be provided for the boundary-element method to obtain the contact stresses of the normal and tangential traction with little restriction on the contact profiles of two finite plane bodies [9].

In this chapter, a method for combining moiré interferometry with numerical computation is presented to show how the stress distribution of in-plane contact can be evaluated by this type of hybrid technique [10]. With high-density gratings moiré interferometry is a highly sensitive method for displacement measurement. Apart from the advantage of recording in real time, it is more suitable for plane contact problems because it gives the deformation components directly, without the necessity for component separation as in holographic interferometry or photoelasticity. Meanwhile, the high quality of the fringes near the contact zone makes it possible to measure the deformation of the body boundaries very accurately, which ensures an accurate solution for the contact stress without initial errors.

In the next section, a basic analysis of the contact between an arbitrarily distributed load on a semi-infinite plane is described. This analysis requires deformation measurement of the contact zone. Several techniques to obtain the displacement and their derivative fringe patterns are presented in section 10.2.3. With the experimental results, the hybrid solution of the contact stress is given in section 10.2.4 with examples of various contact distributions. Some conclusions are then presented in the final section.

10.2.2   The in-plane contact of two bodies

When the plane dimension of a plate is larger than the area of contact with another body, the contact stresses are highly concentrated in that contact region with little direct influence from the geometry of the plates and the supporting conditions far

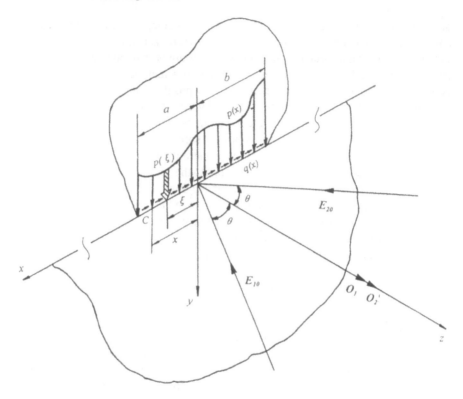

Figure 10.2.1. Schematic illustration of contact stress distribution and illumination of a dual beam.

from the contact region. This situation can be seen as the problem of a half-plane loaded by arbitrarily distributed tractions produced by the contact force. When the thickness of the plate is small and the load is applied in the plate plane, the contact stresses can be regarded as uniform over the plate thickness.

The analysis by Flamant [11] provided an elastic solution for a half-plane loaded by a concentrated normal force p at a point $(\xi, 0)$. When a shear force q is added at the same point in the contact zone, as illustrated in figure 10.2.1, in which the x axis is coincident with the boundary surface, and the y axis is directed into the plate, the stress components can be expressed as follows

$$\sigma_x = \frac{-2}{\pi} \frac{1}{(x^2 + y^2)^2} [px^2 y + qx^3] \qquad (10.2.1a)$$

$$\sigma_y = \frac{-2}{\pi} \frac{1}{(x^2 + y^2)^2} [py^3 + qxy^2] \qquad (10.2.1b)$$

$$\tau_{xy} = \frac{-2}{\pi} \frac{1}{(x^2 + y^2)^2} [pxy^2 + qx^2 y]. \qquad (10.2.1c)$$

By substituting these stress expressions into Hooke's law to yield the strains and then into the relation for strain displacement, the displacement components $u(x, 0)$ and $v(x, 0)$ along the contact boundary can be expressed as follows

$$\bar{u}(x, 0) = -\frac{1-v}{2E}p - \frac{2}{\pi E}q \ln r + C_1 \tag{10.2.2a}$$

$$\bar{v}(x, 0) = --\frac{2}{\pi E}p \ln r - \frac{1+v}{\pi E}p + \frac{1-v}{2E}q + C_2 \tag{10.2.2b}$$

where C_1 and C_2 are constants.

In a general plane contact, the normal pressure $p(x)$ and tangential traction $q(x)$ are distributed over the interface of the two plates in an arbitrary manner. Taking the tractions $p(\xi)d\xi$ and $q(\xi)d\xi$ on an elemental area $d\xi$ as the concentrated pressure force and tangential force, respectively, which are applied at $(\xi, 0)$ by a distance of $x - \xi$ from a point $C(x, 0)$, the displacement components $u(x, 0)$ and $v(x, 0)$ due to the complete load distribution can thus be obtained by integrating the element displacement expressed by equations (10.2.2a) and (10.2.2b), over the contact zone given by

$$u(x, 0) = -\frac{1-v}{2E}\left\{ \int_{-b}^{x} p(\xi)\, d\xi - \int_{x}^{a} p(\xi)\, d\xi \right\}$$
$$-\frac{2}{\pi E} \int_{-b}^{a} q(\xi) \ln |x - \xi|\, d\xi + C_1 \tag{10.2.3a}$$

$$v(x, 0) = -\frac{2}{\pi E} \int_{-b}^{a} p(\xi) \ln |x - \xi|\, d\xi$$
$$-\frac{1+v}{\pi E} \int_{-b}^{a} p(\xi)\, d\xi + \frac{1-v}{2E}\left\{ \int_{-b}^{x} q(\xi)\, d\xi - \int_{x}^{a} q(\xi)\, d\xi \right\} + C_2. \tag{10.2.3b}$$

In these equations, it is necessary to split the integration range as the step in the displacement changes. Resulting from the pressure traction $p(\xi)$, the horizontal displacement $u(x, 0)$ on the left-hand side of the point $C(x, 0)$, is opposite to that on the right-hand side. A similar situation exists for the vertical displacement $v(x, 0)$ resulting from the tangential traction $q(\xi)$ which shows upwards and downwards deformation of the contact interface, respectively, on the left- and right-hand sides of the action point.

With moiré interferometry it is possible for us to measure the displacement components $u(x, 0)$ and $v(x, 0)$ of the contact interface. By substituting them into the previous equations, we can solve the contact tractions by inversion. In practice, to eliminate the constants C_1 and C_2 involved in the displacement distribution, we can also use the forms for the displacement derivatives $\partial u/\partial x$, $\partial v/\partial x$, which represent the tangential component of strain and the slope of the deformed contact interface, respectively. By differentiating the equations (10.2.2) and (2-2b) with respect to variable x, the displacement derivatives $\partial u/\partial x$, $\partial v/\partial x$

are given by

$$\frac{\partial u(x,0)}{\partial x} = -\frac{1-\nu}{2E}p(x) - \frac{2}{\pi E}\int_{-b}^{a}\frac{q(\xi)}{x-\xi}\,d\xi \qquad (10.2.4a)$$

$$\frac{\partial v(x,0)}{\partial x} = -\frac{2}{\pi E}\int_{-b}^{a}\frac{p(\xi)}{x-\xi}\,d\xi + \frac{1-\nu}{2E}q(x). \qquad (10.2.4b)$$

Thus the determination of the contact tractions relies on the measurement of The displacement or displacement derivatives along the contact interface. Some related measuring techniques from moiré interferometry are presented in the following sections.

10.2.3 Contact deformation measured by moiré interferometry

10.2.3.1 Moiré interferometry for displacement patterns

By the superimposition principle of geometric moir interferometry for the low-density gratings, the interpretation of moiré interferometry is simple. When two coherent laser beams illuminate the specimen symmetrically about the grating normal, a 'virtual' grating is formed in the front by interference. By superimposing this virtual grating on the real grating on the specimen surface generates moiré fringes that are related to the deformation of the bodies. This type of interpretation is easily understood. It is not applicable, however, to some other cases of moiré interferometry especially to some measurements of displacement derivatives as presented in the following sections, where the two illuminating beams are incoherent with each other and no 'virtual' grating is formed on front of the specimen grating.

Mathematical processing based on light propagation of optical complex amplitudes may be an available way to describe the fringe pattern formation involved in those techniques. We present here the case of displacement measurement. As a phase-type reflecting grating with a spatial frequency f_g is illuminated by a collimated beam at an angle θ, the propagating direction φ_m of the mth order diffraction satisfies

$$p(\sin\theta + \sin\varphi_m) = m\lambda \qquad (10.2.5)$$

where $p = 1/f_g$ is the pitch of the specimen grating. When the illuminating direction θ is identical to the angle of the first diffraction, or $|\theta| = \sin^{-1}(\lambda f_g)$ by letting $\phi_1 = 0$ in equation (10.2.5), first-order diffraction will be propagated along the normal to the specimen. As the contact region with the specimen grating is illuminated by two beams (figure 10.2.1), E_{10} at $-\theta$ and E_{20} at $+\theta$, in the horizontal plane $x-O-z$ symmetrically about the grating normal, respectively, the complex amplitudes O_1 and O_2, which are the positive first-order diffraction of

E_{10} and the negative first-order diffraction of E_{20}, respectively, can be expressed as

$$O_1(x, y) = A \exp\{i[\varphi_0(x, y) + \varphi_1(x, y)]\} \qquad (10.2.6a)$$
$$O_2(x, y) = A \exp\{i[\varphi_0(x, y) + \varphi_2(x, y)]\} \qquad (10.2.6b)$$

where $\varphi_0(x, y)$ is the initial phase of the grating, $\varphi_1(x, y)$ and $\varphi_2(x, y)$ are the phase changes due to the in-plane displacement $u(x, y)$ and the out-of-plane displacement $w(x, y)$ of the contact bodies. They are given by

$$\varphi_1(x, y) = \frac{2\pi}{\lambda}\{u(x, y) \sin\theta + w(x, y)(1 + \cos\theta)\} \qquad (10.2.7a)$$

$$\varphi_2(x, y) = \frac{2\pi}{\lambda}\{-u(x, y) \sin\theta + w(x, y)(1 + \cos\theta)\} \qquad (10.2.7b)$$

The superimposing of those two wavefronts results in moiré fringes with an intensity distribution

$$I(x, y) = |O_1 + O_2|^2 = 4A^2 \cos^2\frac{\alpha(x, y)}{2} \qquad (10.2.8)$$

where the phase difference $\alpha(x, y)$ is related to the in-plane contact displacement $u(x, y)$ by

$$\alpha(x, y) = \varphi_1(x, y) - \varphi_2(x, y) = \frac{4\pi}{\lambda}u(x, y) \sin\theta. \qquad (10.2.9)$$

Similarly, another dual-beam illumination in the vertical plane of y–O–z by symmetrical arrangement produces a phase difference $\beta(x, y)$ between the two diffractions $\psi_1(x, y) - \psi_2(x, y)$, which presents the component $v(x, y)$ in the y direction:

$$\beta(x, y) = \psi_1(x, y) - \psi_2(x, y) = \frac{4\pi}{\lambda}v(x, y) \sin\theta. \qquad (10.2.10)$$

Figure 10.2.2 gives a fringe pattern for the displacement $v(x, y)$ near the boundary of a plate in contact with another plate of smaller size, showing clearly the variation in the deformation of the contact interface.

10.2.3.2 A polarized shearing method for derivative patterns

Ithas been shown in section 10.2.2 that the displacement derivatives can be used to solve the contact stress. By data interpolation, it is possible to determine the fractional fringe numbers in the displacement patterns and to obtain the strain components by finite difference calculation. Some experimental techniques to obtain the whole-field contour maps of the displacement derivatives directly have been developed in moiré interferometry. Mechanical differentiation is

Figure 10.2.2. A fringe pattern of the displacement $v(x, y)$ in the contact region.

a relatively simple method to obtain the fringes of the in-plane displacement derivatives, in which a carrier pattern is introduced by slightly changing the directions of the dual-illuminating beams and double exposure is performed by shifting the recording plate laterally in between. Optical filtering is then used to subtract the derivative fringes [12]. Optical differentiation directly shifts optical images so that real-time measurement can be realized without double exposure and recording plate shifting. In this section, a polarized shearing method for moiré interferometry that eliminates not only the disturbance of the out-of-plane displacements but also the background fringes of the in-plane displacements, to give a pure pattern of the in-plane displacement derivatives by optical filtering is presented [13].

Two laser beams with linearly orthogonal polarization are used to illuminate the specimen grating at the inclined direction $+\theta$ and $-\theta$, respectively, as illustrated in figure 10.2.3. Neither amplitude- nor phase-type gratings change the polarization property and the two first-order diffraction patterns propagate along the normal to the specimen, towards a glass wedge G with small angle β and thickness t_0 in maximum. The glass plate is placed at an angle α to the optical axis z to reflect the diffraction light by its front and rear surfaces. In the image plane (ξ, η) where the light intensity is recorded, the complex amplitudes of the wavefronts V_{1f} and V_{2f} from the front surface of the glass can be represented by Jones matrices:

$$V_{1f}(\xi, \eta) = \begin{pmatrix} e^{i\varphi_1(\xi,\eta)} \\ 0 \end{pmatrix} \tag{10.2.11a}$$

$$V_{2f}(\xi, \eta) = \begin{pmatrix} 0 \\ e^{i\varphi_2(\xi,\eta)} \end{pmatrix} \tag{10.2.11b}$$

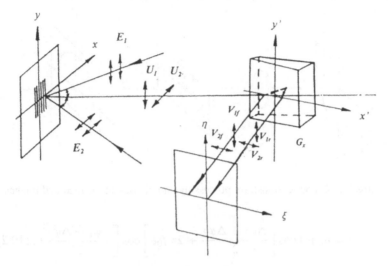

Figure 10.2.3. Schematic illustration of dual-beam illumination with linearly orthogonal polarization and shearing recording of the diffraction.

respectively. The thickness of the glass and the angle of light reflection provide a lateral shift of the wavefront from the rear surface in comparison with that from the front surface. Meanwhile, the wedge angle β also generates a linear change in the optical path with a carrier function

$$F(\xi) = (2\pi/\lambda)(\tan\beta)\xi = 2\pi f_0 \xi \qquad (10.2.12)$$

where $f_0 = \tan\beta/\lambda \approx \beta\lambda$ is the spatial frequency of the carrier fringes. Therefore, the complex amplitude V_{1r} and V_{2r} from the rear surface of the glass wedge can be written as

$$V_{1r}(\xi, \eta) = \begin{pmatrix} \exp^{i[\varphi_1(\xi,\eta)+\Delta\varphi_1(\xi,\eta)+2\pi f_0\xi]} \\ 0 \end{pmatrix} \qquad (10.2.13a)$$

$$V_{2f}(\xi, \eta) = \begin{pmatrix} 0 \\ \exp^{i[\varphi_2(\xi,\eta)+\Delta\varphi_2(\xi,\eta)+2\pi f_0\xi]} \end{pmatrix} \qquad (10.2.13b)$$

where the phase increments $\Delta\varphi_1(\xi, \eta)$ and $\Delta\varphi_2(\xi, \eta)$ caused by image-shearing are

$$\Delta\varphi_1(\xi, \eta) = \varphi_1(\xi + \Delta\xi, \eta) - \varphi_1(\xi, \eta) = [\partial\varphi_1(\xi, \eta)/\partial\xi]\Delta\xi \quad (10.2.14a)$$
$$\Delta\varphi_2(\xi, \eta) = \varphi_2(\xi + \Delta\xi, \eta) - \varphi_2(\xi, \eta) = [\partial\varphi_2(\xi, \eta)/\partial\xi]\Delta\xi. \quad (10.2.14b)$$

Therefore, the intensity distribution $I(\xi, \eta)$ on the recording plate is

$$I(\xi, \eta) = (V_{1f} + V_{1r} + V_{2f} + V_{2r})^*(V_{1f} + V_{1f} + V_{1f} + V_{1f})$$

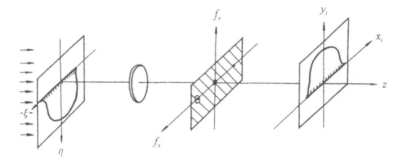

Figure 10.2.4. Fourier transform processing and filtering of the recorded images.

$$= I_0 + 4\cos\left[\frac{\Delta\varphi_1 + \Delta\varphi_2}{2} + 2\pi f_0\xi\right]\cos\left[\frac{\Delta\varphi_1 - \Delta\varphi_2}{2}\right]. \quad (10.2.15)$$

This is a fringe carrier consisting of two cosine terms with a basic frequency f_0 modulated by a deformation gradient. The first cosine term includes the sum of the phase increments which is related to the out-of-plane displacement derivatives $\partial w(\xi, \eta)/\partial\xi$ and the second involves the difference in the phase increments that indicates the in-plane derivatives, as presented by equations (10.2.7a) and (10.2.7b), respectively. Even though these gradient components are coupled to each other in this intensity distribution, the existence of the carrier frequency f_0 makes the spatial frequency of the first cosine term much higher than that of the second. Optical filtering can thus be employed to obtain the separated patterns of the in-plane derivative fringes.

By illuminating the film H, that has recorded the intensity of equation (10.2.15), in an optical Fourier transform system of figure 10.2.4, an optical spectrum appears in the focal plane (f_x, f_y) of the transform lens L, with complex amplitude distribution of

$$U(f_x, f_y) = U_o(f_x, f_y) + U_1 FT\left\{\cos\left[\frac{\Delta\varphi_1(\xi, \eta) - \Delta\varphi_2(\xi, \eta)}{2}\right]\right\}$$
$$\times \{\tfrac{1}{2}\delta(f_x - f_0, f_y)FT[\exp(j\omega\varphi)]$$
$$+ \tfrac{1}{2}\delta(f_x + f_0, f_y)FT[\exp(-j\omega\varphi)]\} \quad (10.2.16)$$

where U_0 and U_1 are the light background, $\delta(\cdot)$ is the Dirac delta function, $FT\{\cdot\}$ is the opration of Fourier transform, and

$$\omega\varphi(\xi, \eta) = \tfrac{1}{2}[\Delta\varphi_1(\xi, \eta) + \Delta\varphi_2(\xi, \eta)]. \quad (10.2.17)$$

This is a spectrum with a zeroth-order diffraction at the plane centre and two first-order diffractions at a distance of f_0 from the origin of the plane. Letting only the first-order diffraction pass through the filter plane, the complex amplitude in the image plane (x_i, y_i) can be expressed as

$$U_i(x_i, y_i) = K \cos \left[\frac{\Delta\varphi_1(x_i, y_i) - \Delta\varphi_2(x_i, y_i)}{2} \right]$$
$$\times \exp\{-j[2\pi f_0 x_i + \varpi\varphi(x_i, y_i)]\}. \tag{10.2.18}$$

The intensity of the image can thus be written as

$$I_i(x_i, y_i) = K^2 \cos^2 \tfrac{1}{2}[\Delta\varphi_1(x_i, y_i) - \Delta\varphi_2(x_i, y_i)]. \tag{10.2.19}$$

Dark fringes appear at points where the following equation is satisfied:

$$\frac{\partial u(x_i, y_i)}{\partial x_i} = \frac{1}{\sin\theta} \frac{(2n+1)\lambda}{4\Delta x_i} \qquad (n = 0, \pm 1, \pm 2, \dots). \tag{10.2.20}$$

In a similar way, $\partial u/\partial y$ can also be obtained if we place the glass wedge G at an angle α with respect to the z axis in the y–O–z plane to shear the diffraction wavefront in the η direction. When the illumination on the grating is in the y–O–z plane rather than in the x–O–z plane, the displacement derivative patterns of $\partial v/\partial x$, and $\partial v/\partial y$, can also be achieved in real time.

10.2.3.3   An incoherent technique for gradient patterns

In the previous section, the specimen grating adhering to the contact region is illuminated by two beams with orthogonal polarization, produced by splitting a laser beam into two parts and inserting a quartz crystal rotator into one beam to turn the polarization plane 90°. There is no any 'virtual grating' to superimpose the specimen grating but the coupling between the in-plane and out-of-plane displacement derivatives is eliminated due to the shearing of the wavefronts. In this section, a more practical method is presented. It utilizes the temporal correlation of the laser instead of its polarization property, to produce interference between the diffractions from the beams without time coherence [14].

The optical arrangement of this experimental method is illustrated in figure 10.2.5. A He–Ne laser ($\lambda = 6328$ Å) with a coherence length of $l = 30$ cm is used as the light source. After the laser beam is split into two beams by a beam-spilter, BS, a portion of the light is reflected and then it illuminates the specimen grating SG at an angle θ_1, directly after it is expanded by E_1 and collimated by the lens L_1. Reflected by the mirror M_1 and M_2, another portion of the beam from BS is also collimated by E_2 and L_2 to illuminate the grating. It is important that the pathlength of these two beams, or the distance $M_1 - M_2$, is about 100 mm, which is much larger than the coherence length of the laser.

To produce a uniform shift in the diffraction wavefront so that the optical differentiation is kept constant over the whole field, two lens, L_{01} and L_{02}, are placed on the normal axis to receive the diffracted images. The lens L_{01} converges the images into smaller sizes and L_{02} collimates the light into parallel rays. A glass wedge G is used as in the previous section to shear the diffraction rays

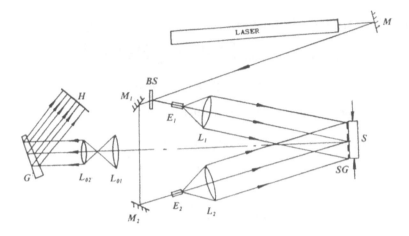

Figure 10.2.5. Optical arrangement of an incoherent illumination and diffraction shearing.

laterally. In the recording plate H, the complex amplitude of the wavefronts from the front and rear surfaces of the wedge G can be expressed as

$$U_{1f}(\xi, \eta) = a_1 \exp\{j\varphi_1(\xi, \eta)\} \qquad (10.2.21a)$$
$$U_{2f}(\xi, \eta) = a_2 \exp\{j\varphi_2(\xi, \eta)\} \qquad (10.2.21b)$$

and

$$U_{1r}(\xi, \eta) = a_1' \exp\{j\varphi_1(\xi + \Delta, \eta) + 2\pi f_0\xi\} \qquad (10.2.22a)$$
$$U_{2r}(\xi, \eta) = a_2' \exp\{j\varphi_2(\xi + \Delta, \eta) + 2\pi f_0\xi\} \qquad (10.2.22b)$$

where a_1, a_2, a_1' and a_2' are the amplitudes of the wavefronts.

Since the optical path difference of the dual beam is longer than the coherent length, the mutual coherence $\Gamma(\tau)$ of these four wavefronts U_{1r}, U_{2r}, U_{1f} and U_{2f} becomes null or

$$\Gamma_{U_{1r}U_{2r}}(\tau_0) = \Gamma_{U_{1r}U_{2f}}(\tau_0) = \Gamma_{U_{1f}U_{2f}}(\tau_0) = \Gamma_{U_{1f}U_{2r}}(\tau_0) = 0 \qquad (10.2.23)$$

where τ_0 is the time delay of the two illuminating beams. Therefore, the intensity distribution in the image plane can be expressed by two mutual correlation functions given by

$$\begin{aligned} I(\xi, \eta) &= \langle U_{1f}U_{1r}\rangle + \langle U_{2f}U_{2r}\rangle \\ &= a_1^2 + a_1'^2 + 2a_1a_1'\cos[\Delta\varphi_1(\xi, \eta) + 2\pi f_0\xi]|\gamma_{11}(\tau_d)| \\ &\quad + a_2^2 + a_2'^2 + 2a_2a_2'\cos[\Delta\varphi_2(\xi, \eta) + 2\pi f_0\xi]|\gamma_{22}(\tau_d)| \end{aligned} \qquad (10.2.24)$$

where $\Delta\varphi_1(\xi, \eta)$ and $\Delta\varphi_2(\xi, \eta)$ are given in equations (10.2.7a) and (10.2.7b), and the correlation coefficients $\gamma_{11}(\tau_d), \gamma_{22}(\tau_d)$ are defined by

$$\gamma_{11}(\tau_d) = \frac{\Gamma_{U_{1f}U_{1r}}(\tau_d)}{\Gamma_{U_{1f}U_{1r}}(0)} = \frac{\langle U_{1f}(t)U_{1r}^*(t + \tau_d)\rangle}{\langle U_{1f}(t)U_{1r}^*(t)\rangle} \qquad (10.2.25a)$$

Figure 10.2.6. Fringes of displacement derivatives $\partial v/\partial x$ obtained by optical shearing.

$$\gamma_{22}(\tau_d) = \frac{\Gamma_{U_{2f}U_{2r}}(\tau_d)}{\Gamma_{U_{2f}U_{2r}}(0)} = \frac{\langle U_{2f}(t)U_{2r}^*(t+\tau_d)\rangle}{\langle U_{2f}(t)U_{2r}^*(t)\rangle} \qquad (10.2.25b)$$

whereas τ_d is the relative time delay between the shearing wavefronts reflected by the front and rear surfaces of the glass wedge. Since this time delay τ_d is much smaller than the coherent time τ_c of laser, both $|\gamma_{11}(\tau_d)|$ and $|\gamma_{22}(\tau_d)|$ can be regarded as unity. Moreover, to obtain a higher visibility for the modulated fringe carrier, the rear surface of the glass wedge is coated with a full reflector and the front one with partial transmission film, respectively. This makes the intensities of the emerging light beams almost identical so that the amplitudes are identical, i.e. $a_1 = a_2 = a'_1 = a'_2 = a$. Therefore, the intensity distribution $I(\xi,\eta)$ becomes

$$I(\xi,\eta) = I_0 + 4a^2 \cos\left(\frac{\Delta\varphi_1 + \Delta\varphi_2}{2} + 2\pi f_0\xi\right) \cos\left(\frac{\Delta\varphi_1 - \Delta\varphi_2}{2}\right). \quad (10.2.26)$$

This is the same as equation (10.2.15); thus, the in-plane displacement gradient patterns can be obtained by optical filtering of Fourier transform as before. Figure 10.2.6 shows an example of the fringes $\partial v/\partial x$ near the contact region of a plate, loaded by another body with a large deformation at both ends of the distributed contact stress, as illustrated along the contact interface of figure 10.2.7.

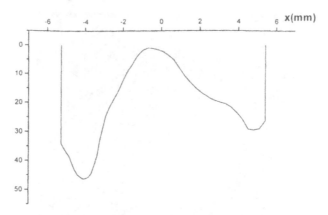

Figure 10.2.7. Variation of the slope $\partial v / \partial x$ along the contact interface.

10.2.4   The hybrid solution of the contact stresses

In section 10.2.2, we found that the arbitrarily distributed normal and tangential tractions in the contact zone are the sum of the infinitesimal concentrated normal and tangential forces. By moiré interferometry, as shown in the previous section, the normal and tangential displacements, or the strains and slopes of the contact profile, can be obtained by various methods. These measured data provide the boundary conditions by which the displacements or displacement derivatives are specified over the contact interface of two plane bodies. As the unknown terms, the normal and tangential tractions are involved in the coupling integral equations of (10.2.3) and (10.2.4).

To eliminate the constants in the displacement expressions in (10.2.3), the relative displacements are often utilized. These relative values are obtained by subtracting the displacement at a distant point or at the original point of the coordinates from those in equations (10.2.3). For example, when the contact surface of the two plates is only loaded by normal traction $p(x)$ distributed in a zone $0 \leq x \leq c$, the relative deformation $\Delta v_{x0}(x, 0) = v(x, 0) - v(0, 0) = v'_{x0}(x, o)\Delta x$ can be expressed as

$$v'_{x0}(x, 0) = \frac{2}{\pi Et} \int_0^c \int_0^x \frac{p(\xi)}{x - \xi} \, dx \, d\omega \qquad (10.2.27)$$

where t is the thickness of the specimen over which $p(x)$ is uniformly pressed. In this equation, the logarithmic operation has been replaced by an integration in terms of x. The displacement derivatives are processed in the same way. The solution of the normal and tangential stresses is associated with the solution of two singular integral equations (10.2.4a) and (10.2.4b) where a singularity exists

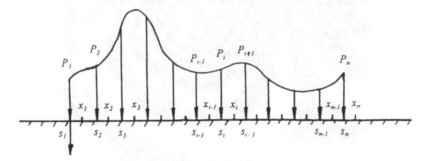

Figure 10.2.8. Division of the contact area by the points where the fringe orders and contact stresses are determined.

at $x = \xi$. The equations (10.2.4a) and (10.2.4b) can be rewritten as

$$\frac{\partial u(x,0)}{\partial x} = -Ap(x) - B \int_{-b}^{a} \frac{q(\xi)}{x - \xi} \, d\xi \tag{10.2.28a}$$

$$\frac{\partial v(x,0)}{\partial x} = -B \int_{-b}^{a} \frac{p(\xi)}{x - \xi} \, d\xi + Aq(x) \tag{10.2.28b}$$

where $A = (1 - 2v)(1 + v)/E$ and $B = 2(1 - v^2)/(\pi E)$.

To obtain the distributed contact stresses, numerical integration based on boundary conditions from experimental measurements of the displacement or displacement derivatives using moiré interferometry are utilized. To avoid a singularity in the integration, the contact zone $0 \le x \le c$ is divided into n subintervals, in which the discrete points s_i $(i = 1, 2, \ldots, n)$, where the tractions p_i and q_i are applied do not coincide with the point x_j $(j = 1, 2, \ldots, n)$ where the displacement derivatives $\Delta u_j / \Delta x$, $\Delta v_j / \Delta x$ are taken from the moiré fringes, as shown in figure 10.2.8. Therefore, equations (10.2.28a) and (10.2.28b) can be expressed in the form of finite summations:

$$\left.\frac{\partial u}{\partial x}\right|_{x_1} = -Ap(s_1) - B \sum_{i=1}^{n} \frac{q(s_i)}{x_1 - s_i} \Delta s$$

$$\left.\frac{\partial u}{\partial x}\right|_{x_n} = -Ap(s_n) - B \sum_{i=1}^{n} \frac{q(s_i)}{x_n - s_i} \Delta s \tag{10.2.29}$$

$$\left.\frac{\partial v}{\partial x}\right|_{x_1} = -B \sum_{i=1}^{n} \frac{p(s_i)}{x_1 - s_i} \Delta s + Aq(s_1)$$

$$\left.\frac{\partial v}{\partial x}\right|_{x_n} = -B \sum_{i=1}^{n} \frac{p(s_i)}{x_n - s_i} \Delta s + Aq(s_n)$$

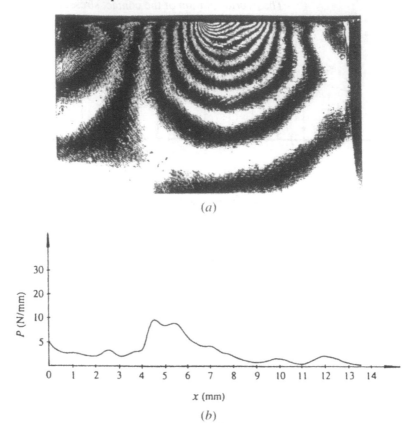

(a)

(b)

Figure 10.2.9. Fringe patterns (*a*) and the stress distributions (*b*) for different contacts for the distributed loads.

where $\Delta s = s_{i+1} - s_i$. In this way, the contact stress components $p(s_i)$ and $q(s_i)$ are the solution of the linear algebraic equations

$$\begin{bmatrix} -[\bar{A}] & [\bar{B}] \\ [\bar{B}] & [\bar{A}] \end{bmatrix} \begin{Bmatrix} \{p(s_i)\} \\ \{q(s_i)\} \end{Bmatrix} = \begin{Bmatrix} \{\Delta u/\Delta x|_{x_i}\} \\ \{\Delta v/\Delta x|_{x_i}\} \end{Bmatrix} \qquad (10.2.30)$$

where the contact strains $\{\Delta u/\Delta x|_{x_i}\}$ and the surface slopes $\{\Delta v/\Delta x|_{x_i}\}$ have been obtained from experimental measurements and the matrices $[\bar{A}]$ and $[\bar{B}]$ are

$$[\bar{A}] = \frac{(1-2v)(1+v)}{E} \begin{bmatrix} 1 & 0 & \cdot & 0 \\ 0 & 1 & \cdot & \cdot \\ \cdot & \cdot & \cdot & 0 \\ 0 & 0 & \cdot & 1 \end{bmatrix} \qquad (10.2.31)$$

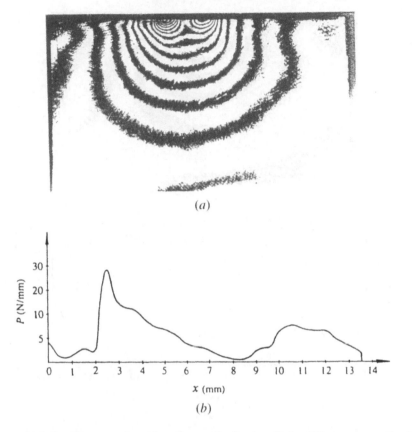

Figure 10.2.10. Fringe patterns (*a*) and stress distribution (*b*) for different contact for the distributed load (see text for details).

$$[\bar{B}] = \begin{bmatrix} \dfrac{1}{x_1 - s_1} & \dfrac{1}{x_1 - s_2} & \cdot & \cdot & \dfrac{1}{x_1 - s_n} \\ \dfrac{1}{x_2 - s_1} & \dfrac{1}{x_2 - s_2} & \cdot & \cdot & \dfrac{1}{x_2 - s_n} \\ \cdot & \cdot & \cdot & \cdot & \cdot \\ \cdot & \cdot & \cdot & \cdot & \cdot \\ \dfrac{1}{x_n - s_1} & \dfrac{1}{x_n - s_2} & \cdot & \cdot & \dfrac{1}{x_n - s_n} \end{bmatrix} . \qquad (10.2.32)$$

Some patterns for a large plate pressed by a smaller plate are presented in figures 10.2.9(*a*), 10.2.10(*a*) and 10.1.11(*a*), respectively, to show the displacement fringes near the contact interface. Correspondingly, the solved distribution of the normal stress $p(x)$ are illustrated in figure 10.2.9(*b*), 10.2.10(*b*) and 10.2.11(*b*), respectively.

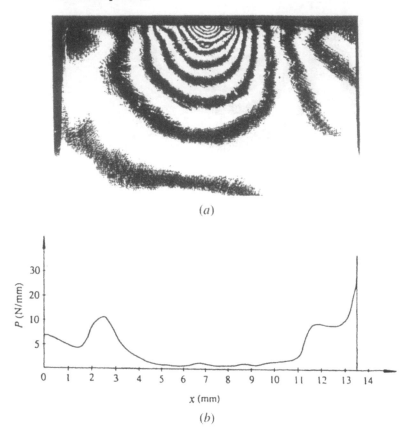

(a)

(b)

Figure 10.2.11. Fringe pattern (a) and stress distribution (b) for different contact for the distributed load (see text for details).

These results show the different contact situations by both the fringe patterns and the stress curves. In figure 10.2.9(a), most contact loads are applied gently onto the plane edge. Despite being small, they can be solved accurately as shown by figure 10.2.9(b). In figure 10.2.10(a), the two bodies are mainly in touch in two small zones within the contact region. The concentrated force in the left-hand part is relatively higher than that in the right-hand part, as illustrated in figure 10.2.10(b). At the right-hand end of the contact area in figure 10.2.11(a), a singularity exists at the end point. Meanwhile, as indicated in figure 10.2.11(b), there is also some stress concentration inside the contact region near the left-hand end.

10.2.5   Remarks

In contact mechanics, evaluating the deformation profile and determining the interface stresses are two major research subjects.   The hybrid method of combining moiré interferometry and numerical computation presented here offers a tool to obtain both of them for a half-plane subjected to arbitrarily distributed contact loads.

With the same sensitivity as holographic interferometry or speckle interferometry, moiré interferometry can be used to obtain the in-plane displacement field near the plane contact region directly, without either the necessity of separating the deformation components as in holographic inteferometry or the speckle noise in speckle inteferometry. In fact, the high quality of the fringes near the contact zone enables accurate measurement of the deformation of the contact surface, which is applicable to the solution of the contact stresses even when the contact region is subjected to a slight load. Meanwhile, moiré interferometry has an extends the limit for large deformation measurements, which makes it possible to apply the method to cases where the stress concentration is high at some local contact points or a stress singularity exists either inside the contact area or at the ends of the contact zone.

References

[1] Hertz H 1882 Ueber die Beruehrung fester elastische Koerper *J. Reine Ang. Meth.* **92** 156–71
[2] Boussinesq J 1885 *Application des Potentials a Letude de l'Equilibre et du Mouvement des Solides Elastiques* (Paris: Gauthier-Villars)
[3] Johnson K L 1985 *Contact Mechanics* (Cambridge: Cambridge University Press)
[4] Dai F L and Zhong G C 1978 Application of holographic photoelasticity *J. Qinghua University* **3** 25–9
[5] Theocaris P S 1973 Stress singularity at concentrated loads *Exp. Mech.* **13** 511–18
[6] Karami G A 1989 *Boundary Element Method for Two-Dimensional Contact Problem* (Berlin: Springer)
[7] Theocaris P S and Razem C I 1977 Deformation boundary determination by the method of caustics *J. Strain Anal.* **12** 223–32
[8] Kobayashi A S 1993 Hybrid experimental-numerical stress analysis *Handbook of Experimental Mechanics* ed A S Kobayashi (Englewood Cliffs, NJ: Prentice-Hall) ch 17
[9] Fang J, Qi J and Pan K L 1993 A hybrid method of combining speckle interferometry with boundary element method to solve plane contact problem *Proc. SEM 50th Anniversary on Experimental Mechanics* (Dearborn) (Bethel, CT: SEM) pp 460–1
[10] Fang J 1990 Application of moiré interferometry to a contact problem *J. Strain Anal.* **25** 229
[11] Flamant 1892 *Compt. Rendus* **114** 1465
[12] Post D 1987 Moiré interferometry ch 7 *Handbook of Experimental Mechanics* ed A S Kobayashi (Englewood Cliffs, NJ: Prentice-Hall)

[13] Dai F L and Fang J 1988 Polarized shearing moiré interferometry *Opt. Commun.* **65** 411

[14] Fang J and Dai F L 1990 A new real-time method of moiré interferometry for displacement derivative fringe patterns *Opt. Lasers Eng.* **13** 51

Chapter 10.3

Moiré interferometry for measurement of strain near mechanical fasteners in composites

Gary Cloud
Michigan State University, USA

10.3.1 Case study: Whole-field measurement of strain concentrations in composites

The project described here involved analytical and experimental study of the mechanics of fasteners in glass-fibre-reinforced epoxy laminates.

Fibre-reinforced composite materials are considerably weakened when bolt or rivet holes pierce the laminate to form a mechanical joint. The strength and stiffness of the joint are adversely affected by stress concentrations, stress channelling effects, free-edge effects and large bearing stresses. Still, requirements for assembly ease, economy, repairability and ultimate dependability, especially in hostile environments, often dictate the use of mechanical fastenings instead of adhesives.

Areas of study included multiple-fastener arrays, load-spreading washers of three types, bolt pre-load effects, strain-relief inserts and some field measurements [1].

Characteristics of the experimental aspects of the problem include:

- sensitivities of a few microstrain are required;
- analysis must extend over the whole field in the region of the fastener, requiring a field size of at least 75 mm in the case of multiple fastener arrays;
- the complete strain state at each point in the field must be established, requiring two orthogonal normal strains and the shear strain;
- an eventual requirement was that measurements be made at a testing machine or in the field rather than on an optical bench; and

- the large test matrix implied that a large amount of data would be generated and computer-based data reduction was highly desirable.

Clearly, a whole-field optical technique operating at interferometric sensitivities was required. Moiré interferometry was chosen in lieu of electronic speckle interferometry which, at that time, was not highly enough developed (this situation has changed in the mean time). Holographic interferometry was rejected because of the difficulty in obtaining displacement components precise enough in magnitude and location to compute all three strain components accurately.

To this end, a three-axis, six-beam moiré interferometer having a relatively large field was designed and constructed. In the following sections, some general considerations relevant to the design of moiré interferometry apparatus are presented, then the construction and adjustment of a particular device are described.

10.3.2   Approaches to design

Discussion of the theory of moiré interferometry, beginning with the three-dimensional diffraction equation, is not appropriate for this case study. It may be reviewed in the extensive literature on the subject (e.g. [2–4, ch 13]). Attention is focused here on the design, construction and use of a moiré apparatus that is easy to build and which serves very well for studies of the type being discussed.

There are three general design steps in the physical realization of a device for performing interferometry, including the moiré variety. First, the needed special capabilities and tolerable limitations must be listed. Second, an optical arrangement which will satisfy the requirements must be designed. Finally, optical components must be obtained and arranged to perform the desired tasks.

Various optical arrangements for moiré interferometry have been employed [2, 3, 5–10]. Discussion of all the possibilities would be somewhat tedious and it might be counterproductive in that attention is distracted from the development process. Once the process is understood, the practitioner can review literature on the designs that have been employed and modify them or create new ones to fit the problems and resources at hand. For that reason, attention here is focused on the design and construction of a general-purpose six-beam instrument which has useful capabilities for experiments which required full knowledge of the surface strain fields in different types of specimens [1, 4, 9, 11, 12].

The main objectives pursued in the construction of this moiré interferometer were:

- Rigid connections between the optical elements and the specimen being tested were to be avoided in order to gain flexibility and allow the instrument to be used in several different experimental set-ups.
- Incorporate capability to perform measurements in three different directions over the whole area of interest so that the complete surface strain field could

be measured. This 'strain rosette' approach was preferred over the more common practice of determining shear strain by using the cross-derivatives of the displacement for reasons having to do with error propagation.

- Obtain a good efficiency of light utilization so that a low-power laser would suffice.
- The experimental set-up should be suitable for performing measurements in environments less ideal than an optics laboratory.
- Incorporate some sort of unitized construction so that critical adjustments were simple to perform and, once completed, would be fixed until circumstances dictated a controlled change.
- Facilitate the creation of positive and negative pitch mismatch at desired levels to fit the problem under study.
- Be compatible with photographic and video fringe recording and digital data reduction.
- Fundamental grating frequency is easily changed to suit the measurement problem.

10.3.3 Design, construction and alignment of interferometer

A system which uses three pairs of beams for measurement of strain in three directions is described here. This device is properly called either a three-axis interferometer or a six-beam interferometer, even though only three physical beams are needed to form the desired three pairs of interfering beams. The instrument meets all the criteria listed previously.

Many practitioners seem not to realize that moiré interferometry is totally a process of two-beam interference coupled with diffraction. First, the gratings are formed by two-beam interference. Then, two-beam interference creates the fringe pattern at the observing location. To illustrate consider the case where only two incident beams fall on a single specimen grating. The beams that are diffracted from the specimen surface travel back along the instrument axis with slight angular deviations that depend on local variations of grating spatial frequency caused by the local surface strain. These angular deviations are preserved (modified by the lenses, of course) right up to the film plane in the fringe recording camera. At any point of the film, two-beam interference forms a local fringe system that is dependent on the angular spread between the two beams. That this interference pattern is the same as would be produced by geometric moiré is not immediately obvious [4].

Figure 10.3.1 is a photograph showing an overall view of the experimental apparatus including a small loading frame with a load transducer. Only part of the imaging system is included so that the optical elements will not be obscured.

Figure 10.3.2 is a conceptual sketch of the optical system that illustrates how the required three beams are created and directed to the specimen. Recall that this device utilizes three pairs of two beams each to obtain the strain measurements

Figure 10.3.1. A three-axis six-beam moiré interferometer.

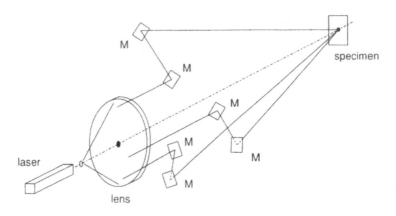

Figure 10.3.2. Conceptual optical scheme of a three-beam six-axis moiré interferometer not including fringe imaging optics.

along three directions. The sketch is distorted to clarify the idea, and the fringe imaging system is not included in this rendition.

Figure 10.3.3 shows a two-dimensional schematic that establishes the basic plan and function of the optical system. A device built according to this plan could be used to measure one strain component, which might be adequate for demonstration purposes or for certain types of measurement applications.

Referring to figure 10.3.3, the basic functions of the device are as follows: a 20 mW helium–neon laser 1 produces a beam of coherent light having a wavelength of 632.8 nm (24.913 × 10⁻⁶ in). The two front surface mirrors 2

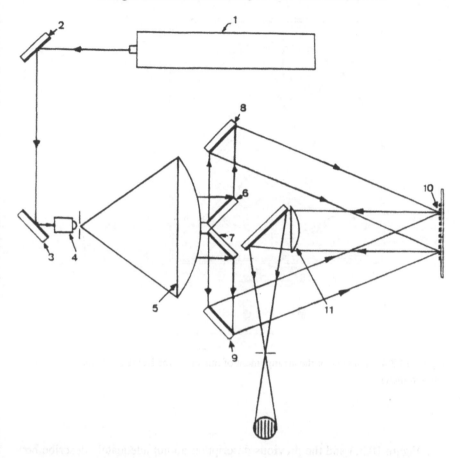

Figure 10.3.3. Two-dimensional representation showing the basic plan of each of the two-beam interferometers contained in the three-axis moiré interferometer.

and 3 direct the beam to the spatial filter where it is filtered and expanded with a 40× microscope objective and a pinhole 4. A collimating lens 5 (a 13 in diameter plano-convex lens, focal length $= 1.0$ m) changes the expanding beam into a parallel beam. This collimated beam reaches two first-surface flat mirrors 6 and 7 oriented at 45° with respect to the optical axis of the lens. Each of these two mirrors directs portions of the parallel rays to another two mirrors 8 and 9 which can be oriented easily to direct the light towards the specimen grating 10 and to obtain the desired angle of interference between the two beams. These two beams incident onto the specimen grating are diffracted. The diffracted rays pass through a converging lens 11 (focal length $= 75$ cm) and are directed by a front surface mirror, oriented at 45° with respect to the normal of the grating, to the observing system or photographic camera. The beam narrows and passes through an aperture before reaching the camera.

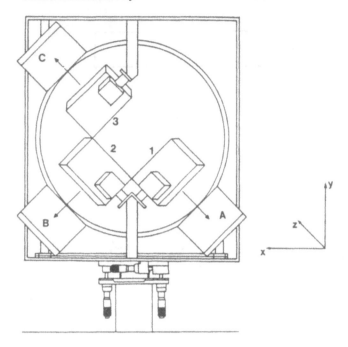

Figure 10.3.4. Plan view of the arrangement of mirrors in the frame of the three-axis moiré interferometer.

Figure 10.3.3 and the previous description do not adequately describe how the mirrors must be oriented in three dimensions to create the required three pairs of beams for measurement of the three strain components. For that purpose, consider figure 10.3.4, which shows a plan view (looking along the main optic axis) of the arrangement of the mirrors in their mounting frame. This is the heart of the system. This sketch also shows how the mirror frame is mounted on a base that has micrometre adjustments for tilt, rotation and position. The individual mirrors are mounted on micrometre adjustable bases also but these are not shown.

The setting-up and adjusting of an optical interferometer can be frustrating to the novice, because seemingly minute changes cause such large changes in the output. This six-beam moiré apparatus can be especially tricky, because each beam is used in two different two-beam interferometers. The mirror adjustments are all interdependent. Space limitations preclude discussion of the alignment procedure, so the interested reader is referred to the literature [4, 9]. Suffice it to say here that a thorough understanding of the instrument is helpful; minimal practice reduces the alignment time to a few minutes; and, barring laboratory disasters, the procedure needs to be performed only occasionally.

10.3.4   Specimen gratings

The creation of specimen gratings that are appropriate for the problem at hand is one of the most important aspects of moiré interferometry and one that has absorbed considerable effort.

It is important to realize that the specimen gratings (actually a two-way grating or a grid) for a three-axis moiré interferometer of this type must be oriented at $\pm 45°$ to the vertical axis of the instrument. This is necessary so that all the useful beams that are diffracted by the specimen grating will travel back along the horizontal optical axis to the central receiving mirror. The fringe patterns still represent the displacements parallel and perpendicular to the specimen axes and at 45° to the specimen axes (that is u_x, u_y, and u_{45}).

There are three aspects of the problem. In the first place, some sort of master grating at the desired frequency must be created. Then, gratings suitable for transfer to the specimen must be manufactured. Finally, the gratings be somehow fastened to the specimen. Walker [13] outlines several different approaches that have been pursued through recent years. Post *et al* [3] also describe various techniques.

An important grid transfer method has been developed by Basehore and Post [3, 7, 14]. This technique was adopted with only minor variations. In a way, the approach is reminiscent of the transfer moulding idea and the old lithographic stripping film method for creating specimen gratings for geometric moiré. In brief, a phase-type relief grating is made by exposing a photoplate in the interferometer. The emulsion side is metallized and then glued to the specimen. The glass substrate with the emulsion is pulled away, leaving a metal-coated relief grating attached to the specimen. While simple in concept, the process is tricky to carry out at first and much credit is due the developers.

For the work described here, specimen gratings having spatial frequencies of 640 lines/mm were made by exposing holographic photoplate (Agfa 8E75) to only two beams of collimated light in the interferometer. As previously mentioned, only a two-way grating is needed. After development and drying, a thin film of aluminum was vacuum-deposited on the emulsion to create a reflective phase grating. The specimen is then positioned in a simple fixture that aligns the grating axes at $\pm 45°$ to the specimen axes. A small amount of cement, in this case PC-10C photoelastic coatings cement from Measurements Group, Raleigh, NC, USA, is poured onto the specimen. Next, the aluminum-coated grating is placed onto the specimen, aluminum side inward. Excess cement is squeezed out by applying pressure of approximately 10 psi by means of a pad and distributed weights. After about 12 h, the glass photoplate is separated from the specimen by a combined twisting and prying action. The thin film of aluminum is thus transferred to the epoxy cement layer, which now carries a replica of the lenticulation grooves that were formed in the photoemulsion. The result is a highly reflective phase-type diffraction grating on the specimen surface. Details of the grating manufacture and transfer processes are given by Herrera Franco [9] and Cloud [4].

10.3.5 Recording fringes

After construction and adjustment of the interferometer and the production of specimens having appropriate gratings, the collection and reduction of data are routine. With the specimen in the loading frame, the recording camera, whether photographic or video, is mounted with its iris plane at the focus of the decollimating lens. The camera is focused in the plane of the image of the specimen surface as seen through the decollimating lens. After final checks for alignment, baseline patterns at small load are recorded. One of the interferometer beams is blocked and a pattern is recorded with the two remaining beams. This process is performed three times, blocking each of the three beams in turn. The result is a set of three fringe patterns for gratings at 0°, 45° and 90° from the specimen axes.

The specimen is then loaded and the cycle of recording three fringe patterns is repeated.

Two such fringe patterns are offered for illustration in figure 10.3.5.

10.3.6 Data reduction

The reason for recording an initial or baseline fringe data set is that moiré investigations, like other experiments, are best conducted in differential mode. Here, one is always looking at the difference of fringe order caused by an increment of load, and the data reduction scheme is arranged to handle fringe patterns in pairs. This approach allows automatic elimination of pitch mismatch effects, whether accidental or deliberate, as well as the reduction or total removal of other errors, such as would be caused by film shrinkage, lens distortions and the like.

For the studies discussed here, the fringe patterns are printed or rear-projected onto a screen or digitizing tablet. The intersections of fringes with the chosen axes are located and measured relative to the fiducial marks that have been placed on the specimen surface. The distances between fiducial marks are also measured in order to establish the final scaling of the fringe records relative to specimen space.

The measurements from a pair of fringe patterns are entered into a computer program that uses cubic-spline curve fitting and finite-difference differentiation to determine displacements and strains along the chosen axes [1, 15]. The results are in the form of plots of displacement and strain along a set of axes, an example of which may be seen in figure 10.3.6.

10.3.7 Closing remarks

The moiré interferometer and procedures described here have been used extensively to investigate the strain field in the vicinity of mechanical fastenings in

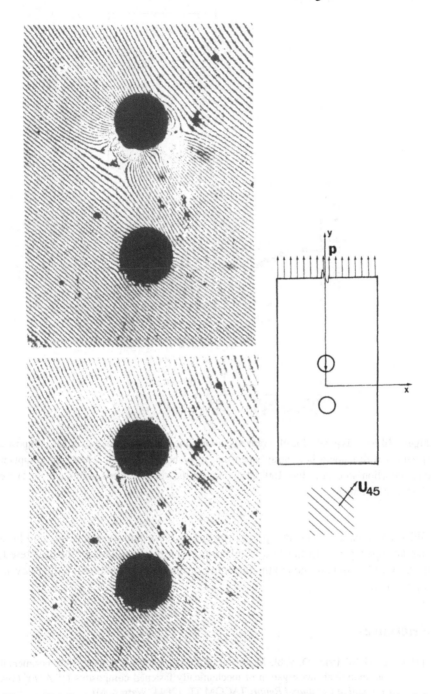

Figure 10.3.5. Moiré fringe patterns for displacements at 45° to direction of load for composite specimen with tandem hole array when upper hole is loaded.

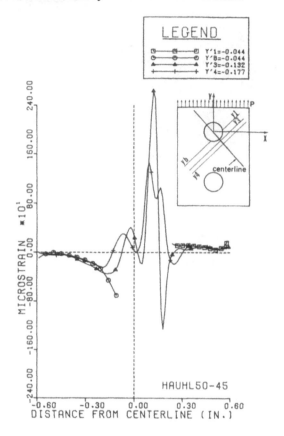

Figure 10.3.6. Typical plot of strain ε_{45} along lines at 45° to load axis for composite specimen with tandem hole array when the upper hole is loaded. Note; the computer data reduction software plots 100 data points in the span. Plotted points are for curve identification only.

GFRP composites. Once set up, it is very easy to use and its sensitivity and field size are appropriate to the task. Only representative results have been offered; more complete data related to the fastening problems are contained in the relevant report [1, 12]

References

[1] Cloud G, Sikarskie D, Vable M, Herrera Franco P J and Bayer M 1987 Experimental and analytical investigation of mechanically fastened composites *US Army Tank Automotive Command Report* TACOM TR 12844, Warren, MI
[2] Walker C A and McKelvie J 1978 A practical multiplied moiré system *Exp. Mech.* **8** 316–320

[3] Post D, Han B and Ifju P G 1994 *High Sensitivity Moiré: Experimental Analysis for Mechanics and Materials* (New York: Springer)

[4] Cloud G 1995 *Optical Methods of Engineering Analysis* (New York: Cambridge University Press)

[5] McDonach A, McKelvie J and Walker C A 1980 Stress analysis of fibrous composites using moiré interferometry *Opt. Lasers Eng.* **1** 85–105

[6] Post D 1980 Optical interference for deformation measurements—classical, holographic and moiré interferometry *Mechanics of Nondestructive Testing* ed W W Stinchcomb (New York: Plenum)

[7] Post D 1987 Moiré interferometry *Handbook on Experimental Mechanics* ed A Kobayashi (Englewood Cliffs, NJ: Prentice-Hall) ch 7

[8] Post D and Baracat W A 1981 High sensitivity moiré interferometry—a simplified approach *Exp. Mech.* **21** 100–4

[9] Herrera Franco P J 1985 A study of mechanically fastened composite using high sensitivity interferometric moiré technique *PhD Dissertation* Michigan State University, E Lansing, MI

[10] Cloud G and Herrera Franco P J 1986 Moiré analysis of multiple hole arrays of composite material fasteners *Proc. Spring 1986 Conference on Experimental Mechanics* (Bethel, CT: Society for Experimental Mechanics)

[11] Herrera Franco P J and Cloud G L 1988 Strain-relief inserts for composite fasteners *J. Composite Mater.* **26** 751–68

[12] Cloud G, Herrera Franco P J and Bayer M H 1989 Some strategies to reduce stress concentrations at bolted joints in FGRP *Proc. 1989 SEM Spring Conf. on Experimental Mechanics* (Bethel, CT: Society for Experimental Mechanics)

[13] Walker C A 1994 A historical review of moiré interferometry *Exp. Mech.* **34** 281–99

[14] Basehore M L and Post D 1984 High frequency, high reflectance transferable moiré gratings *Exp. Tech.* **8** 29–31

[15] Cloud G L 1978 Residual surface strain distributions near fastener holes which are coldworked to various degrees *Air Force Material Lab Report* AFML-TR-78-1-53 (Ohio: Wright Aeronautical Labs)

CHAPTER 11

OUT-OF-PLANE DEFORMATION, FLATNESS AND SHAPE MEASUREMENT

Chapter 11.1

Shadow moiré interferometry applied to composite column buckling

Peter G Ifju, Xiaokai Niu and Melih Papila
University of Florida, USA

The post-buckling behaviour of an I-shaped laminated graphite/epoxy column was investigated using shadow moiré interferometry [1]. The purpose of the experiment was to verify a theoretical model [2] and to obtain experimental insight into the post-buckling behaviour of such structures. The out-of-plane deflections of both the web and the flange were measured as a function of load.

Since the maximum out-of-plane displacement was expected to be quite high (approximately 6 mm), relative low-frequency linear gratings, called *grilles* were used. The reference grilles were made by printing an image created with drawing software onto transparent acetate material, using a 600 dpi laser printer. The acetate material was bonded to a rigid transparent acrylic plate with acetone. Since it was desired to measure the out-of-plane deflections simultaneously on both the web and the flange of the column, two sets of reference grilles were produced. Deflection was monitored throughout the entire test, up to failure. Since the range of deflections was high and accuracy at each load was required, two grille frequencies were used. A frequency of 3.94 lins/mm (100 lines/inch) was used at the early stages of loading and 1.97 lines/mm (50 lines/inch) at later stages. The experimental set-up is illustrated in figure 11.1.1. The grilles were mounted on adjustable bases so that their positions relative to the specimen could be adjusted easily. Mirrors were used to direct the light to the reference grilles. This was necessary because the loading machine and fixtures would have blocked a large portion of the specimen if direct illumination had been used. Figure 11.1.1 shows the mirror/grille assemblies separated from the specimen for illustration purposes. During testing, both grilles were moved to nearly contact the specimen. The grille used for the web was sized to fit between the flanges of the specimen. Two clear glass incandescent light bulbs, mounted on tripods, were used for the light sources. The angle of illumination was 45° and the viewing angle was

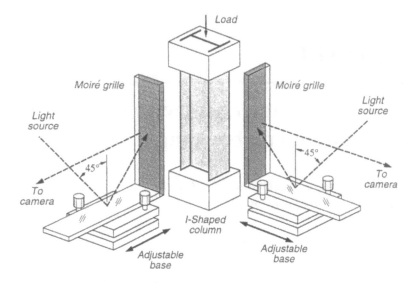

Figure 11.1.1. Set-up for shadow moiré experiment on buckled composite column.

normal to the specimen surface. Both the light source and camera position were located at the same perpendicular distance from the specimen surface. With this configuration, the contour interval of the moiré fringes is equal to the pitch of the reference grille. The specimen was sprayed with flat white paint to provide a matte surface. Images were recorded by two 35 mm cameras using technical pan film.

The specimens were made from graphite epoxy prepreg in an autoclave oven. The plies in the web and flange were arranged in a cross-ply configuration where, for each specimen, the ply sequence differed. The ends of the specimens were potted in a polyester/sand mixture to simulate the fixed end condition of the model. A total of five specimens were tested in a screw-driven universal testing machine, operated in displacement control.

In addition to shadow moiré other experimental techniques were incorporated. LVDTs were used to monitor the deflection of the web at two arbitrary locations along the centre line on the back side of the web. This information was used to determine the absolute deflection of the specimen in order to obtain a baseline for the shadow moiré information. Otherwise, a live count of the fringes passing through a reference point during load application would be required for absolute displacement data. Strain gauges were also mounted in the centre of both sides of the web to monitor the load at which buckling occurred. Thin lead wires were used in order to minimize their appearance on the shadow moiré fringe patterns.

Figure 11.1.2 illustrates examples of the shadow moiré fringe patterns on the web and flange of the specimen. Nearly the entire web and flange were observed.

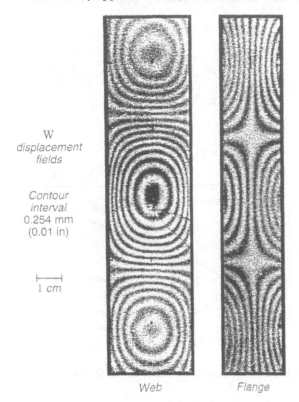

W
displacement
fields

Contour
interval
0.254 mm
(0.01 in)

⊢——⊣
1 cm

Web   Flange

Figure 11.1.2. Shadow moiré fringe patterns on the web and flange of the composite columns.

From these patterns, the out-of-plane deflections were determined as a function of load. A plot of the out-of-plane deflection *versus* position along the vertical centre line of the web for various load levels is shown in figure 11.1.3. These results were then compared to those obtained from numerical modelling to guide and validate the model.

The average buckling load for the five specimens was around 8000 N (1800 lb) while the average failure load for the specimens was around 22 000 N (5000 lb). All of the specimens failed at locations which coincided with zero curvature in the web. These locations are the inflection points where the interlaminar shear stresses are maximum. From the moiré tests, we gained valuable insight into the failure mode as well as validation of our model.

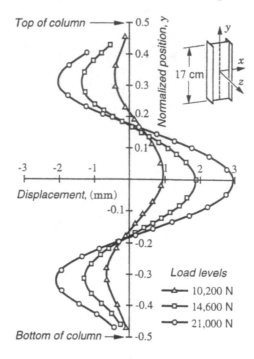

Figure 11.1.3. Deflection along the centre line of the web at various load levels.

References

[1] Post D, Han B T and Ifju P G 1994 *High Sensitivity Moiré: Experimental Analysis for Mechanics and Materials (Mechanical Engineering Series)* (New York: Springer)
[2] Papila M 1995 Buckling and post-buckling analysis of composite I—sections *PhD Thesis* Middle East Technical University

Chapter 11.2

Shadow moiré interferometry with enhanced sensitivity by optical/digital fringe multiplication

Bongtae Han
University of Maryland, USA

Although fringe shifting and image processing are popular for extracting fractional fringe orders, few methods are well adapted for geometric moiré patterns. The main reasons are (1) the intensity distribution across geometric moiré fringes is a complicated function of fringe order, rather than the simple harmonic function of classical interference fringes; and (2) in practice, the function is not fixed but varies from region to region in the fringe pattern. In the idealized case of equal bar and space widths of moiré gratings, the (averaged) intensity distribution is triangular [1]. In real systems, the finite aperture of the camera lens acts as a filter that distorts the triangular distribution by rounding its corners: the rounding of corners by filtering is much stronger for closely spaced fringes than widely spaced fringes and, therefore, the intensity distribution is not a fixed function. In addition, the function varies in practice because the ratio of bar and space widths of the moiré grating usually varies from region to region across the field.

The optical/digital fringe multiplication (O/DFM) method was originally developed to enhance the sensitivity of moiré interferometry further [2]. However, it is a very robust scheme [3] and it is not affected by the conditions enumerated earlier. It is compatible with diverse intensity distributions, which can vary from region to region in the field and it copes with variations in bar-to-space ratios and optical noise. Consequently, the O/DFM method of image processing is well suited to shadow moiré. This chapter reviews the O/DFM method and demonstrates its applicability to shadow moiré [4].

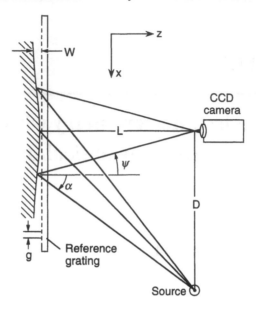

Figure 11.2.1. Shadow moiré arrangement for constant sensitivity across the field.

11.2.1 Optical configuration and fringe shifting

The most popular and practical shadow moiré arrangement is illustrated in figure 11.2.1, utilizing oblique illumination and normal viewing [1]. Normal viewing is preferred to avoid foreshortening and distortion of the image, even though measurement sensitivity can be increased with oblique viewing. With this arrangement, where the source and camera lie at the same distance from the plane of the specimen, the relationship between W and fringe order N_z is

$$W = \frac{g}{\tan \alpha + \tan \beta} N_z = \frac{gL}{D} N_z \qquad (11.2.1)$$

where W and N_z apply to each x, y point in the field. It is important to note that although the incidence and viewing angles, α and ψ, respectively, vary across the field, the sum of their tangents is a constant, which provides a constant sensitivity across the field.

Fringe shifting is accomplished by moving the reference grating in the z direction to increase W uniformly throughout the field. By equation (11.2.1), N_z is changed by one fringe order when W is increased by g. Accordingly, a series of β fringe patterns can be recorded for successive shifts of g/β of the reference grating.

11.2.2   Optical/digital fringe multiplication (O/DFM)

The O/DFM method utilizes phase stepping of β equal steps to achieve fringe multiplication by a factor of β. The patterns are recorded by a CCD camera and the intensity at each pixel is converted to a digital value on a 256 grey scale. Figure 11.2.2 provides a physical explanation of the algorithm. A detailed mathematical description can be found in [1, 2]. Figure 11.2.2(a) is a graphical representation of the intensity distributions of a shadow moiré pattern and its complementary pattern that is fringe-shifted by π (the broken curve). The two patterns are complementary patterns, a name given to patterns in which the phase at each point differs by π and consequently the maximum and minimum intensities are interchanged. The O/DFM algorithm subtracts the intensities at each point, as illustrated in figure 11.2.2(b). Then it inverts the negative portions by taking absolute values, as illustrated in (c). The algorithm proceeds by truncating the data near $|I_r| = 0$ and binarizing by assigning intensities of zero and one to points below and above the truncation value, respectively: the result is graphed in (d).

The result is a sharpened contour map that has twice as many contour lines as the number of fringes in the initial pattern. The sharpened contours occur at the crossing points of the complementary graphs, where the intensities of the complimentary patterns are equal. Note that any noise or other factor that affects the two patterns equally has no influence on the locations of the crossing points. The crossing points and resulting fringe contours occur at odd multiples of $N/4$, as seen in figure 11.2.2(a).

The final step of the O/DFM method is illustrated in figure 11.2.2(e), which represents a case of $\beta = 6$. If additional shadow moiré patterns are recorded with phase steps of $\pi/6$, $2\pi/6$, $4\pi/6$ and $5\pi/6$, they can be processed in complementary pairs to provide intermediate fringe contours. The result is portrayed in (e), where the fringe contours are combined in a single computer output map. Six fringe contours are produced for every moiré fringe in the original pattern, providing fringe multiplication by a factor of six.

11.2.3   Applications

11.2.3.1   Warpage of electronic device

A plastic ball grid array (PBGA) package is composed of an active silicon chip on an organic carrier board called a substrate, which is usually a glass/epoxy composite. The warpage of the substrate is attributed to a large mismatch between the coefficents of thermal expansion of the chip and substrate. In the subsequent assembly process, electrical and mechanical connections are made by solder balls between the substrate and a printed circuit board (PCB). If the bottom side of the substrate warps significantly at the solder reflow temperature, it yields an uneven height of solder interconnections, which could cause premature failure

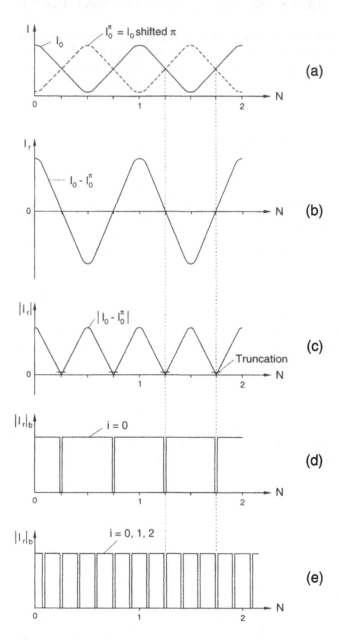

Figure 11.2.2. O/D fringe multiplication: (*a*) intensities of fringe-shifted complementary patterns; (*b*) subtracted; (*c*) absolute values; (*d*) truncated and binarized; and (*e*) combined patterns for multiplication by six.

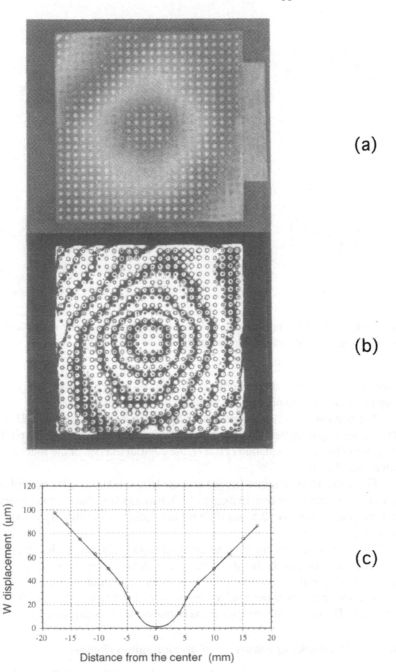

Figure 11.2.3. Warpage of an electronic package by the O/DFM algorithm: (*a*) shadow moiré fringes; (*b*) the O/DFM pattern for multiplication by four, providing 12.5 μm per contour; and (*c*) plot of the global deflection along the upper-left to lower-right diagonal.

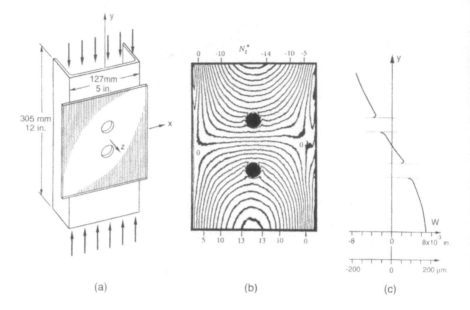

Figure 11.2.4. (a) Experimental arrangement; (b) fringe contours for multiplication by six; (c) displacement W along vertical centreline.

of the assembly. Detailed knowledge of this out-of-plane deformation is essential to optimize design and process parameters for reliable assemblies.

The shadow moiré configuration of figure 11.2.1 was used with $D = 2L$. The reference grating frequency was 10 lines/mm (254 lines/in) and its pitch was 100 μm (0.004 in). By equation (11.2.1), the contour interval of the shadow moiré pattern was 50 μm.

The initial shadow moiré fringes and the fringe contours after multiplication by a factor of four are shown in figures 11.2.3(a) and (b). The contour interval of the multiplied pattern is 12.5 μm (0.0005 in). Enhanced sensitivity is evident in the multiplied pattern.

The small circles result from an array of raised soldering pads. The method is capable of resolving the height of these fine details. The global displacement along the diagonal (upper-left to lower-right) is plotted in figure 11.2.3(c). Although the specimen is nominally symmetrical, the warpage deviates from pure symmetry. The deviations are typical, as corroborated by many examples.

11.2.3.2　Pre-buckling behaviour of an aluminium channel

The specimen was an aluminium channel with two 19 mm (0.75 in) diameter holes, loaded in compression as illustrated in figure 11.2.4(a). A 300 lines/in. ($\sim$ 12 lines/mm) reference grating was positioned ahead of the specimen. The shadow moiré configuration of figure 11.2.1 was used with $D = L$. In this

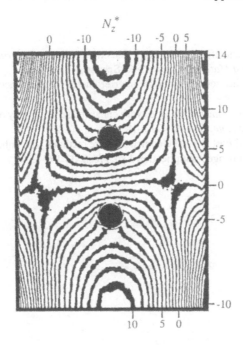

Figure 11.2.5. Fringe contours at a higher load level with a fringe multiplication factor of 10.

configuration, the contour interval of the moiré fringe pattern is equal to the pitch of the reference grating.

The pattern in figure 11.2.4(*b*) shows the fringe contours for multiplication by a factor of six. The left- and right-hand sides of the specimen were stiffened by the flanges of the channel and the out-of-plane distortions at the sides were very small. The web, or face of the channel experienced a second bending mode wherein the upper portion deflected inwards and the lower portion deflected outwards. In the immediate vicinity of the holes, the multiplied pattern shows a ridge on the left- and right-hand side of each hole; the ridge was in the positive z direction for both holes. At the top and bottom of each hole, the displacements were in the negative direction. The displacements along the vertical centre line are plotted in figure 11.2.4(*c*). The contour interval for the multiplied pattern is g/β, i.e. 14.1 μm/fringe.

Figure 11.2.5 shows the deformation at a different load level, but here the multiplication factor is $\beta = 10$. In this case, the reference grating was inclined to the face of the specimen, as revealed by the fringe gradients along the left- and right-hand sides. Again the mode 2 bending is evident. Here, the contour interval is 8.5 μm/fringe, equivalent to that of shadow moiré with a grating of 118 lines/mm. The initial patterns corresponding to those shown in figures 11.24 and 11.2.5 can be found in [1, 2].

References

[1] Post D, Han B and Ifju P 1994 *High Sensitivity Moiré: Experimental Analysis for Mechanics and Materials* (New York: Springer)
[2] Han B 1992 Higher sensitivity moiré interferometry for micromechanics studies *Opt. Eng.* **31** 1517–26
[3] Han B 1993 Interferometric methods with enhanced sensitivity by optical/digital fringe multiplication *Appl. Opt.* **32** 4713–18
[4] Han B, Ifju P and Post D 1993 Geometric moiré methods with enhanced sensitivity by optical/digital fringe multiplication *Exp. Mech.* **33** 195–200

Chapter 11.3

Shadow moiré interferometry with a phase-stepping technique applied to thermal warpage measurement of modern electronic packages

Yinyan Wang
Electronic Packaging Services Ltd Co, Atlanta, USA

11.3.1 Introduction

Thermally induced warpage of electronic substrates is a big concern in the manufacturing process. Electronic package miniaturization has led to wide usage of BGAs (ball grid arrays) and other surface mounted technologies that require a greater flatness for the chip carrier than the older through-hole connection techniques. The ability to evaluate the substrate behaviour as a function of temperature is particularly important to the manufacturing engineers.

Common out-of-plane deformations of BGAs are on the order of dozens of micrometres. This amount of deformation is too large for high-sensitivity measurement methods, for example, Twyman–Green interferometry, are too small for ordinary shadow moiré or projection moiré methods. The sensitivity of optical interferometric methods is one-half wavelength of light. For a BGA 27 mm × 27 mm, if the maximum out-of-plane deformation is 25 μm, there will be approximately 80 fringes in the field. Resolving this number of fringes can be difficult and error prone. In addition, surface irregularities will create difficulties in extracting useful global deformation information.

If one uses the ordinary shadow moiré method to measure the same out-of-plane deformation, only one fringe appears in the whole field, even with a reference grating of 40 lines/mm. Much information on surface characteristics can be lost. A method that can fill the gap between high- and low-sensitivity

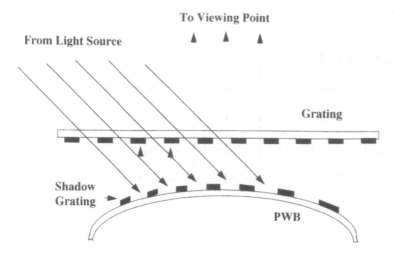

Figure 11.3.1. Schematic diagram of shadow moiré.

methods is necessary for the measurement of thermally induced deformation of BGAs.

As has been pointed out [1], the sensitivity of ordinary shadow moiré can be improved significantly by using all of the light-intensity information in the field or view and not just fringe centre locations. It should be noted that the increment in the sensitivity mentioned is here the displacement for a change of one fringe order. Among the two categories of digital fringe pattern analysis that make use of grey-scale intensity information, the

phase-stepping technique is more attractive for applying to the fringe pattern images of shadow moiré than the Fourier transform method [2, 3]. The fringe gradient often changes its sign across the field for shadow moiré fringe patterns. Using a set of shifted fringe patterns, the phase-stepping technique can judge the increasing or decreasing order of a given fringe automatically. The major advantages include: (1) the sensitivity can be increased by up to two orders of magnitude; (2) the ordering of fringes for a given image can be performed automatically, thus eliminating a significant source of analysis error; and (3) an out-of-plane displacement matrix is obtained directly from a set of original fringe pattern images.

11.3.2 Application of phase stepping to shadow moiré fringe patterns

Shadow moiré fringes are generated by the geometric interference of the real grating and the shadow grating. As illustrated in figure 11.3.1, when an illuminating source is projected through the real grating at a given angle, the

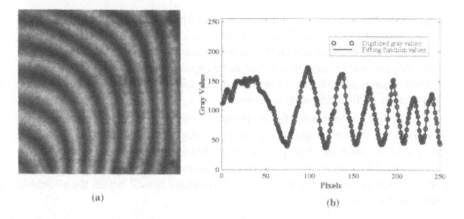

Figure 11.3.2. The light intensity distribution of a typical shadow moiré fringe pattern: (*a*) a fringe pattern of an electronic substrate and (*b*) the digitized and the sinusoidal function fitted grey scale values of the fringes across a horizontal line.

shadow of the grating will be cast onto the surface of the sample. Shadow moiré fringes are isothetics, i.e. the loci of the points of equal distance between the sample surface and the grating. The fundamental assumption that the phase-stepping technique makes is that the distribution of the light intensity of a fringe pattern can be described using a sinusoidal function. The distribution of the light intensity of a shadow moiré fringe pattern is in principal piecewise linear. However, when the observing resolution is low, the triangular distribution of the light intensity is rounded off at the maximum and minimum [4]. When rounding happens, the sinusoidal approximation is allowed. Then, the following function can be used to describe the light intensity of a fringe pattern image.

$$I(x, y) = I_0(xy) + A(x, y)\cos[\varphi(x, y)] \qquad (11.3.1)$$

in which $I_0(x, y)$ is the dc component (background light), $A(x, y)$ is the modulation of the fringes and $\varphi(x, y)$ is the phase. The unknown values $I_0(x, y)$, $A(x, y)$ and $\varphi(x, y)$ can be determined from three or more shifted fringe patterns using a least-square technique [5–7]. Figure 11.3.2(*a*) is a typical shadow moiré fringe pattern of an electronic package substrate and figure 11.3.2(*b*) shows the digitized grey values across a horizontal line of the pattern and the least-square fitting values using the sinusoidal function. Since there are three unknowns, at least three fringe patterns are needed while only one, which is $\varphi(x, y)$, needs to be solved explicitly.

    Many studies have been performed on error analysis for the optimum number of images to use in solving for φ [5, 8]. The more measurements are taken, the less error there is in the result. However, in practice, the fewer measurements taken, the higher the data acquisition rate is and the lower the storage memory required. In the measurement of the thermally induced warpage of BGAs, the sample is heated according to a temperature profile that stimulates the real-world manufacturing process. The fringe pattern images need to be recorded as quickly as possible so as to avoid any change in temperature between image acquisitions. Three strips of fringe shifting are used in this application. Considering the three-step technique is sensitive to the error in the shifting mechanism, a calibration of the system is performed before each test.

In the system that is used for the measurement of thermally induced warpage of BGAs, the reference grating is set on a stage that is driven by a stepper motor and the number of steps per fringe order is determined. The average precision is 2%. According to the error analyses done by Creath [5], the error due to the stepper motor for the three-step technique is well within the allowable experimental error limits.

11.3.3   Phase calculation

Three separate images of the fringe pattern are recorded by a CCD camera as the fringe pattern is shifted by 0, 1/3 and 2/3 fringe cycles. This is achieved by translating the reference grating by $1/3W$ steps for each image recorded, where W is the displacement per fringe order, so that a phase change of $2\pi/3$ is added to the phase of the fringe patterns. Thus,

$$I_1 = I_0 + A\cos[\varphi(x, y)]$$
$$I_2 = I_0 + A\cos[\varphi(x, y) - \tfrac{2}{3}\pi] \qquad\qquad (11.3.2)$$
$$I_3 = I_0 + A\cos[\varphi(x, y) - \tfrac{4}{3}\pi]$$

Solving for the phase term, $\varphi(x, y)$

$$\varphi(x, y) = \arctan\frac{\sqrt{3}(I_2 - I_3)}{(2I_1 - I_2 - I_3)}. \qquad\qquad (11.3.3)$$

The CCD camera measures I_1, I_2 and I_3 at each pixel, which allows the phase and fractional fringe order to be calculated throughout the field. Due to the nature of arctangent calculations, $\varphi(x, y)$ can only be determined in the range 0–2π. The phase ambiguities are removed by an unwrapping process. Any gap between the adjacent elements larger or smaller than π is searched and filled by adding or subtracting 2π. The displacement matrix is obtained simply from the phase distribution as

$$w(x, y) = \frac{\varphi(x, y)}{2\pi}W. \qquad\qquad (11.3.4)$$

Creath [5] has a brief description of three-, four- and five-step algorithms. N-step phase calculation is discussed by Greivenkamp [6] and Morgan [7].

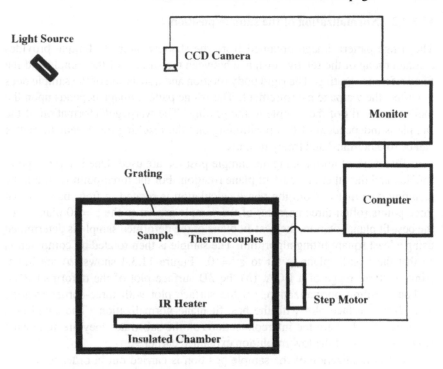

Figure 11.3.3. Set-up for the measurement of thermal warpage of BGAs.

11.3.4   On-line measurement of thermal warpage of modern electronic packages

11.3.4.1   The experimental system

The system for temperature-dependent shadow moiré analysis of electronic packaging materials was described by Stiteker and Ume [9] and it was improved by applying the
phase-stepping technique so that the system is able to analyse the fringe patterns automatically. The heating system is capable of simulating a variety of soldering processes, such as wave soldering and infrared reflow soldering. Figure 11.3.3 is a diagram of the system. The optical grating is set on a stage that is driven by a computer-controlled stepper-motor. The test sample is set inside of an insulated chamber and its temperature is measured by multiple thermocouples interfaced through signal conditioning circuitry to a desktop computer. The signal of a master thermocouple is compared with a user-entered temperature profile and the heater is controlled accordingly. The fringe pattern images are recorded in real time by a CCD camera and the recorded images are later analysed.

11.3.4.2   Normalization of the sample position

The fringe pattern image obtained from the shadow moiré technique provides a contour map of the relative distance between the surface of the sample and the fixed reference grating. The rigid body rotation and translation of the sample does not affect the warpage measurement. The fringe pattern image depends upon the inclination or tilt of the sample to the grating. The warpage/deformation of the sample is independent of this positioning and the resulting displacement matrix can be rotated/translated freely in space.

Two ways of normalizing the sample position are used. One is a three-point rotation and the other is a best-fit plane rotation. For the three-point rotation, the resulting data matrix from the phase calculation is rotated in free space so that three points (often three corners) of the sample are set to the $z = 0$ plane. For the best-fit plane rotation, the best-fit plane of the deformed sample is determined using a least square fitting algorithm. The sample is then rotated by computation so that the best-fit plane is set to $z = 0$. Figure 11.3.4 shows (*a*) the initial fringe pattern image of a BGA, (*b*) the 3D surface plot of the deformed BGA without position normalization, (*c*) the surface plot with three-corner rotation and (*d*) the surface plot with the best-fit plane normalization. The small dots in figure 11.3.4(*a*) are the interconnections on the substrate: they are somewhat fuzzy as a result of the low resolution explained earlier.

The normalization of the sample position is carried out in order to present the deformation of the sample more clearly. It is also helpful when the relative warpage needs to be determined. The relative warpage is the total deformation minus the initial warpage. The relative warpage provides the change of the sample caused by the temperature change.

11.3.5   Application examples

11.3.5.1   A PBGA through a heating–cooling cycle

Figure 11.3.5 shows the results of an on-line measurement of thermally induced warpage of a plastic ball grid array (PBGA) when it experienced a heating–cooling cycle. The size of the PBGA was 27 mm × 27 mm. The sample was heated and cooled inside the insulated chamber illustrated in figure 11.3.3. A designated temperature profile was followed as shown in figure 11.3.5. The fringe pattern images were recorded at different temperature points. The temperature of the sample was monitored by two thermocouples, one of which was designated as a master. In order to ensure that the sample was heated uniformly, the temperature was held constant for about 100 s at each point of data acquisition.

The frequency of the reference grating used in the measurement was 11.8 lines/mm (300 lines/in). The fringe sensitivity of the set-up was 84.7 μm (3.33 mils) per fringe. The measurement was made on the die side (top side of the package). The data matrix of the sample was normalized by best-fit plane

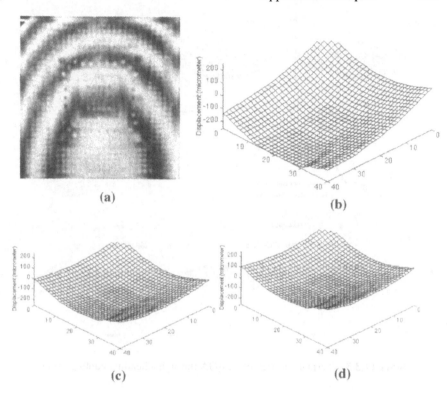

(a)

(b)

(c)

(d)

Figure 11.3.4. Normalization of the sample position: (*a*) initial fringe pattern, (*b*) surface plot without rotation, (*c*) surface plot with three-corner rotation and (*d*) surface plot with best-fit-plane rotation.

rotation. As seen in figure 11.3.5, the sample started initially with a slightly concave shape. When it was first heated, the type of deformation remained the same but the magnitude increased. Between 100 and 200 °C, the magnitude of the warpage decreased. At 200 °C, the sample was relatively flat and at the peak temperature point it experienced significant deformation (convex). Its warpage decreased when it was cooled.

The packaging temperature of the PBGA was 178 °C and it was expected to be stress free at this temperature. Results showed the minimum warpage when the sample was near this temperature. There were a total of three samples tested and each behaved similarly. At peak temperature, the maximum deflection of the sample was as large as 161 μm. This magnitude of deformation could indicate that de-lamination had occurred just prior to reaching the peak temperature.

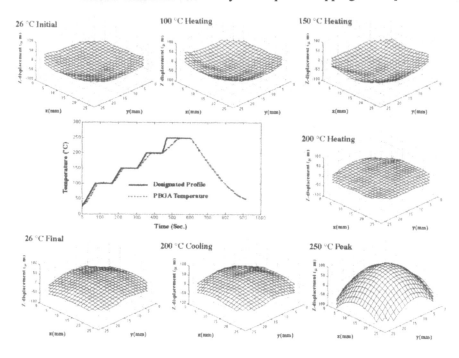

Figure 11.3.5. Thermal warpage of a PBGA through a heating–cooling cycle.

11.3.5.2   Analysis of a BGA and its seating printed wire board

A BGA (size 44 mm × 44 mm) and its seating printed wiring board (PWB) were measured separately though simulated reflow cycles. For the PWB, the measurement was made on the area where the BGA would be present in the package. Each sample was heated inside the insulated chamber and the fringe pattern images were recorded at various temperature points. The sensitivity was again configured for 84.7 μm (3.33 mils) per fringe. At peak temperature, 225 °C, the PWB was bowed downward while the BGA substrate was bowed upward. The maximum gap between the two substrates was as large as 428 μm.

Figure 11.3.6 plots the deformed BGA and PWB at the peak temperature. Each sample was normalized by best-fit plane rotation and then the BGA was moved up along the z axis so that there is no overlap between the BGA and the PWB. Figure 11.3.6(b) is a 2D plot of the deformed samples along one diagonal (from top-left to bottom-right corners). The maximum gap is the difference between the maximum z value of the BGA and the minimum z value of the PWB. Such a large change of the gap between the interconnect points of the BGA and its PWB caused by their respective thermally induced warpages would very likely create problems during the reflow processing.

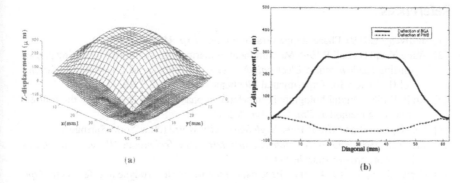

Figure 11.3.6. Thermal warpage of BGA and its seating PWB: (*a*) 3D surface plot of the deformed BGA and PWB at peak temperature 225°C; and (*b*) deformation along the diagonal of the BGA and the seating site of the PWB.

11.3.6   Conclusion

The sensitivity of the analysis of fringe patterns of shadow moiré has been significantly increased by applying the phase-stepping technique. Making use of the grey-scale information contained in the fringe pattern images, the phase-stepping technique resolves the fractional fringe orders. Since the technique assigns the grey scale level of the light intensity in 256 steps, it can, theoretically, resolve the fractional fringe to 1/256. In practice, a resolution of 0.1–0.01 fringe order has been accepted for various different applications [1, 10, 11]. The most important advantage of the technique is that the fringe ordering is fully automated. This advantage makes the phase-stepping analysis most attractive for electronic packaging systems because monotonic deformation is rarely seen there.

As seen in the test of the PBGA, very few fringes (one or two total) appeared across the field for several temperature points. For example, the maximum deflection of the sample was only 14.1 μm at 200 °C. If the phase-stepping technique had not been applied, it would have been difficult to obtain useful information regarding the surface topography. Had a high-sensitivity method been applied, such as ESPI (electronic speckle pattern interferometry) or Twyman–Green interferometry (the fringe sensitivity of these methods is about 0.3 μm), the fringe pattern image analysis might have been impossible for the peak temperature point (maximum deflection of the sample was 161 μm). Another advantage of the phase-stepping technique is that it remains relatively insensitive to the environment as a result of the lower resolution associated with the shadow moiré technique. Shadow moiré with phase stepping is a good tool to investigate the thermal mechanical behaviour of modern electronic packages.

References

[1] Sullivan J L 1991 Phase stepped fraction moiré *Exp. Mech.* **31** 373–81

[2] Wang Y and Hassell P 1998 Measurement of thermal deformation of bga using phase-shifting shadow moiré *Electronic/Numerical Mechanics in Electronic Packaging* vol II (Society for Experimental Mechanics) (Bethel, CT: SEM) pp 32–9

[3] Kreis T 1986 Digital holographic interference-phase measurement using the fourier-transform method *J. Opt. Soc. Am.* A **3** 847–55

[4] Post D, Han B and Ifju P 1994 *High Sensitivity Moiré* (New York: Springer)

[5] Creath K 1988 *Phase-Measurement Interferometry Techniques (Progress in Optics 26)* (Amsterdam: Elsevier) ch 5

[6] Greivenkamp J E 1984 Generalized data reduction for heterodyne interferometry *Opt. Eng.* **23** 350–2

[7] Morgan C J 1982 Least-squares estimation in phase-measurement interferometry *Opt. Lett.* **7** 368–70

[8] Cheng Y Y and Wyant J C 1985 Phase shifter calibration in phase-shifting interferometry *Appl. Opt.* **24** 3049-52

[9] Stiteler M and Ume C 1997 System for real-time measurements of thermally induced warpage in a simulated infrared soldering environment *ASME J. Electron. Packaging* **119** 1–7

[10] Chang M and Ho C S 1993 Phase-measuring profilometry using sinusoidal grating *Exp. Mech.* **33** 117–22

[11] Chang M, Hu C P, Lam P and Wyant J C 1985 High precision deformation measurement by digital phase shifting holographic interferometry *Appl. Opt.* **24** 3780–3

Chapter 11.4

On-line system for measuring the flatness of large steel plates using projection moiré

Jussi Paakkari
VTT Electronics, Oulu, Finland

11.4.1 Motivation

Flatness and dimensional criteria are included in the current quality standards for many sheet and plate products, e.g. steel plates, aluminium plates, door leaves, insulating products for building applications and floor coverings. It is evident that there are many reasons why one would want to know and control the shapes and dimensions of such products. The consumers of steel plate and strip products are increasingly requiring tighter dimensional tolerances and better profiling and shaping. The following motivations for flatness measurement can be listed in the case of steel plates:

- Analysis of the flatness of the product assists in process development and diagnostics.
- Feedback from flatness measurements to automatic process control using numerical and user-independent information is needed.
- There is a need for extremely flat plates for use in automatic welding and plate-handling machines.
- Plate products need to fulfill quality standards [1, 2].
- Plate products can be classified according to their flatness and prices can be fixed according to this classification.
- Mishandling of the plate in lifting, transferring and cutting can locally or totally destroy the flatness of properly processed plates. The steel company can avoid product replacement due to the customer complaints by identifying the plate and recovering its flatness data.

11.4.2   Measurement specifications

The measurement site at Rautaruukki Oy Raahe Steel plate mill is at the end of the mechanical cutting line, before the cold leveller. The standard flatness specification at this measurement site is 6 mm m^{-1} (height/length of the cord) and a special flatness value of 3 mm m^{-1} is often required.

The technical objectives for designing the measurement system were:

- depth resolution 0.3 mm (coordinate direction z)
- depth of field at least 350 mm,
- minimum length of the wave interval is 300 mm (see length of the cord),
- maximum measurement time 1 s,
- maximum plate temperature 100 °C,
- operating temperature +5–+40 °C and
- spatial point resolution on the plate 10 mm × 10 mm (coordinate directions x and y)

The plates to be measured are up to 3.5 m wide and may be from 4 m to several tens of metres long. They may be either stationary or moving on the production line at a speed of 1 m s^{-1}. The plates have both diffuse and specular surface properties, which can vary between plate products. The surfaces may also have markings, paintings, dust, water, oil or other kinds of anomalies.

The system is expected to achieve high operational reliability in a hostile plate mill environment and should measure automatically 99% of the plate products passing the measurement site. It must withstand stray light, dust, vibrations and electromagnetic fields.

11.4.3   Design of the flatness measurement system

11.4.3.1   System overview

The system consists of three main parts: an optomechanical measurement unit, a digital sensor processor (DSP) frame grabber and a host PC with a user interface (see figure 11.4.1). Control and data signals are transferred between these three main units.

The system is connected to the mill's upper level computer system and process control PC computers via a LAN (local area network). The measurements are synchronized with the material flow on the production line and the steel plates to be measured are identified by plate numbers. The measurements can be seen in real time on the user interface. The flatness map of the plate is saved on the hard disc of the host PC for later diagnostic use and its maximum flatness deviation is sent to the mill's database via the LAN. Plate movements on the production line are followed by an ultrasonic switch and rotating encoder. Other necessary interfaces of the system are the power supply and the air pressure connection to cool the illuminator of the optomechanical measurement unit and to make a slight

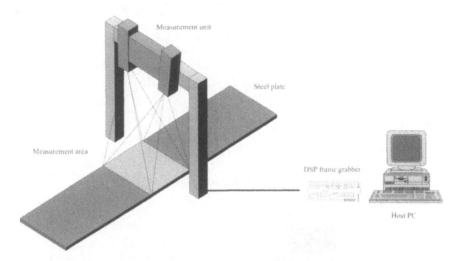

Figure 11.4.1. The main parts of the flatness measurement system.

excess pressure around the optical components to eliminate dust. The external interfaces of the system can be seen in figure 11.4.2.

The optomechanical measurement unit produces the moiré fringes. The measurement unit is mounted over the production line on a steel frame, the basis of which is isolated from the production line. The installation of the measurement unit at the plate mill is shown in figure 11.4.3. Since the variations in background illumination depend on the time of day and season of the year, a shade was designed around the measurement area (see figure 11.4.3).

11.4.3.2   Optomechanical design of the measurement unit

The most critical details of the design work from the optical point of view were the high depth resolution requirement, the large measurement area, sufficient lighting power and mechanical stability.

The projection moiré system (see figure 11.4.4) can be divided into three subsystems: the projection system, the illumination system and the imaging system. The projection system was designed first, followed by the imaging system and, finally, the necessary illumination system.

The projection moiré method normally uses two gratings with a similar period for the generation of moiré fringes. The master grating is projected on the target and it is imaged through a lens onto the submaster grating and the moiré fringes are seen by the camera as a superimposition of the two gratings.

The sensitivity and measurement range of the projection moiré depend on the pitch of the gratings, the illumination and viewing angles as well as on the dimensions of the system. Following equation describes the relations between the

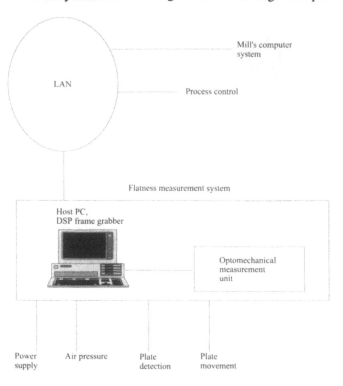

Figure 11.4.2. The external interfaces of the flatness measurement system.

sensitivity, the grating pitch on the object, the illuminating angle and the viewing angle:

$$\Delta z = P/[\tan(\Psi) + \tan(\phi)] \qquad (11.4.1)$$

where P = grating pitch on the object, ϕ = the angle of incidence of the projected grating, Ψ = the viewing angle and Δz = the sensitivity per fringe.

The depth resolution needed is 0.3 mm and it was decided that the grating period on the steel should be 1 LP mm^{-1}. This is equal to a single line width of 0.5 mm. If the grating pitch on the unfinished steel plate falls below 1 mm, there is an increased likelihood that the grating lines will be blurred and distorted by the grain structure of the steel. It was also assumed that 1/20 of the fringe period can be solved by the image-processing algorithm. The imaging angle of the system was set at 0 degrees during the design phase. Applying equation (11.4.1), it can be calculated that with a system angle of 20 degrees the fringe interval will be 2.7 mm. Thus the image-processing algorithm is capable of a resolution of 0.14 mm. a 20 LP mm^{-1} bar and the space grating is used in the projection system. It is magnified by the projection system on the steel plate.

Figure 11.4.3. Installation of the measurement unit in the plate mill.

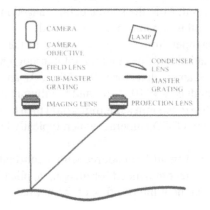

Figure 11.4.4. Typical construction of the projection moiré device.

11.4.3.2.1 Projection system

The projection system consists of the grating and the projection optics. The periodic structure of the grating is imaged and magnified onto the surface to be measured. Two wide-angle enlargers with short focal lengths were tested in the course of the design work for the projection optics. Both wide-angle lenses gave results in experiments that showed a significant degradation in performance across the area needed. A normal enlarger with longer focal was appropriate to this application and resulted a standoff distance as great as 4.5 m or more.

11.4.3.2.2   Imaging system

The imaging system images the object surface together with the projected grating onto a second detection grating. The imaging system is usually a twin of the projection system.

11.4.3.2.3   Illumination system

A large measurement area needs to be illuminated sufficiently well and to be flat enough. The illumination system is based on the traditional projecting system, with a concave mirror reflecting back to the light source and condenser optics in both the illumination module and the camera module, as seen in figure 11.4.4. One critical component is a small high-power light source. A lifetime of one month or more was specified. The light source should be available as a stock product, because it can then be replaced easily and cheaply during maintenance. Also, the need of a special power supply should be avoided. The spectral distribution of the source must fit the spectral responsivity of the camera and the light must be directed and focused through the optical system. A maximum light throughput will result from the source being imaged into the aperture of the projection lens and, for this, a point source is needed.

The power throughput from the point source to the camera detector was calculated. The calculation indicates that the total throughput of the system from the point source to the camera detector is 9×10^{-8}. As an example, a 10 000 lm point source would produce 9×10^{-4} lm, and the illumination on the retangular 2/3 in CCD detector of size 6.6 mm $\times$ 8.8 mm would then be 9.5 lx (lm m^{-2}). This is enough for commercial CCD cameras, which typically have a light sensitivity of about 0.5 lx.

A halogen lamp is a small light source with a constant light output over its lifetime. A dc voltage is recommended for imaging applications. The throughput calculation suggests that the luminous flux of the normal halogen projector lamp is enough.

11.4.3.2.4   Design of the optomechanics

A stable optomechanical construction is necessary in order to achieve high performance and long-term stability under plate mill conditions. It is necessary to adjust the optical components during system composition, after which they should not move beyond their permitted tolerances. Questions of maintenance and environmental conditions also had to be considered in the mechanical design.

The most critical component positions in the system with respect to the projection and imaging lenses are the gratings. If thermal expansion or mechanical instability causes the optical distance from the projection grating to the projection lens to exceed the depth of focus at the grating, the system will not produce moiré contours and a similar situation is also encountered on the imaging side of the system. The stability of the depth of focus was estimated to be kept to 50 μm.

The high-power halogen light source generates heat and some arrangement is necessary to protect the system from heat deformation. Air cooling is used for this purpose, and the whole illumination box and all the components are air cooled in order to stabilize the temperature.

11.4.3.3 Image analysis software

The main objectives of the fringe interpretation algorithm and its implementation were a resolution of 1/20, real-time operation and robustness. The algorithm must be capable of interpreting the fringe image despite variation of surface properties within plate products.

The fringe image of the system includes bias fringes (see figure 11.4.5) achieved by tilting the optomechanical unit and allowing a rotational mismatch between gratings. The purpose of this was to assist both fringe interpretation and visual inspection of the plate products via a monitor image. If needed, the operator can look at the monitor image and observe the straight lines in the case of a flat plate. The use of bias fringes also simplifies the image processing, because mainly vertical and open fringes are imaged with a known carrier frequency.

By tilting the system for bias fringes it is also possible to avoid or reduce the glare from the metal surface. The dynamic range of the camera is set to cover all the main types of plate products that pass the measurement area. A number of images are taken of each plate as it passes, the imaging process being synchronized with the plate movement to prevent overlapping images. These sequential images are later combined to form a set of flatness data for the entire plate.

Phase-shifting techniques [3–6], a Fourier method [7] or a phase-locked loop (PLL) technique [8] were not appropriate for this type of application. An intensity-based phase calculation was chosen as the basic method for fringe interpretation. Intensity methods can be expected to have an accuracy of 1/20 of a fringe interval [9], which is enough for this application. The computation of plate flatness is based on an intensity method which takes advantage of the pixel surroundings, i.e. local image areas. The effects of image distortions on the fringe computation and final flatness values are kept local.

The variation in surface properties between steel plates means that noise is added to the moiré signal. After bounding the plate area, the image-processing algorithm includes functions for noise filtering, correction of the signal profile and adjustment of the signal level.

A calibration value is used to convert phase information to relative flatness information, a concept that implies that, after imaging the plate area, the flatness information is computed according to its surface with no attention paid to the absolute distance between the plate and the optical unit or to the 3D orientation of the plate.

11.4.4 Performance

The performance of the system was studied under both laboratory and factory conditions. The tests were performed according to prearranged plans.

11.4.4.1 Quality of the moiré signal

Two steel plates of size 2 m × 2.5 m were placed in parallel in the measurement area, together covering an area of 4 m × 2.5 m in the laboratory hall. An example of the moiré image and the profile of the moiré signal are seen in figure 11.4.5. The horizontal profile was taken from the middle of the image. The unknown to be extracted is the phase information, not the fringe contrast or intensity.

The visual interpretation of the image suggests that a depth deviation of less than 1.0 mm will be seen as a curved fringe. The moiré image can be used by the operators as a visual aid.

11.4.4.2 Calibration

The system is calibrated using a specially manufactured reference object having two plane levels. This is placed in four locations at equal intervals along the width of the measurement area of size 3.5 m × 1.0 m (width × length) and 0.5 m up from the lower edge of the area. The fringe deformations on the reference object are interpreted by the algorithm used for the flatness measurements. The average of these four calibration values indicates that a +1.0 mm depth deviation corresponds to a fringe deformation of 9.06 pixels.

11.4.4.3 Repeatability and depth resolution of flatness measurement

Repeatability was assessed by measuring the calibration reference object in different locations within the measurement area at the steel mill. The object has a diffuse surface without disturbances. The repeatability of the flatness measurement is 0.046 mm (2σ) and 95.5% of the random measurements will lie between -2σ and $+2\sigma$. The depth resolution, i.e. the smallest flatness deviation which can be reliably detected, can be deduced from the normal distribution of the random measurements. If the flatness deviation in the object is 4σ, which amounts here to 0.09 mm, it can be reliably detected.

The practical performance of the system was evaluated under production conditions using steel plates passing the measurement site. Eight steel plates of varying thickness and surface properties were selected and measured manually as a reference, their flatness being measured using a straight edge along the

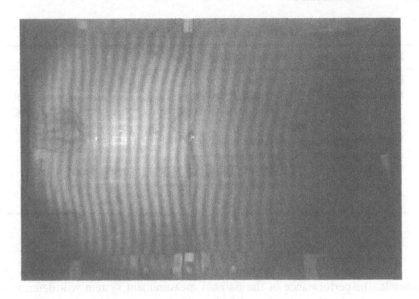

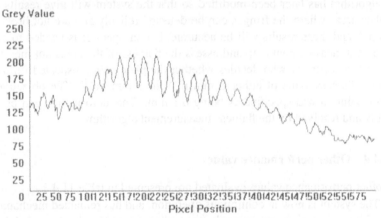

Figure 11.4.5. Example of a moiré image and a profile as obtained in the laboratory.

longitudinal axis. The discrepancies between these manual measurements and the flatness results based on moiré contours were in the range 0.3–2.0 mm or 0.5–1.5 mm in the case of three plates with dark, non-reflective surfaces. Conversely, it was the one plate with a highly reflective surface in which the difference was 2.0 mm. A plate with white markings on it was also measured, and this, too, gave a difference of 2.0 mm. The use of manual measurements as a reference is somewhat problematic, however, as the exact positions of the supporting points and measurement point of the gauge are not always known and, in any case, longitudinal measurements fail to detect the transverse curvature of the plate.

Table 11.4.1. Performance of the flatness measurement system.

Performance factor	Specification	Result
Depth of field	350 mm	520 mm
Measurement area	3.5 m wide and from 4 m to several tens of metres long	3.5 m × 1 m, sequential non-overlapped images are captured
Spatial point resolution on the plate surface	10 mm × 10 mm (x and y directions)	5.6 mm × 5.3 mm
Usability	99% of plate products must be measured	100%, based on the discussion with the production personnel

If the moiré contours are distorted by markings or bad surface quality, for example, the performance of the flatness measurement system will deteriorate. The algorithm has later been modified so that the system will give results only for plate areas where the fringes can be detected reliably and are not broken or scattered, and these results will be accurate, but an operator is needed to check the flatness measurement map and assess the flatness of the areas not measured. It is thus the operator who decides whether the plate being inspected fulfils the specified flatness value of either 6 mm m^{-1} or 3 mm m^{-1}. The objective for depth resolution was specified as being 0.3 mm. This deformation can be seen visually and resolved by the flatness measurement algorithm.

11.4.4.4   Other performance values

The other performance values evaluated are presented in table 11.4.1.

The system is now in continuous operation, and has remained mechanically stable for years. No optical or mechanical adjustments have been needed during operation and no dust has penetrated inside the illumination or camera module.

11.4.5   Conclusion

The system has achieved the confidence of the operators and utilization is high. The process is now controlled by only one operator (earlier two were used) and since flatness control is now based on numerical values, the difference between shifts in routing plates to the levelling process is significantly reduced. This results in better standards of quality assurance control. The company Spectra-Physics Visiontech Oy has modified the first pilot system and commercialized it. It is now called FM3500. Projection moiré topography is a feasible and competitive technique for measuring the flatness of steel plates. After this successful development project, we expect to see new applications for moiré-based real-time 3D measurements in other fields of industry.

Acknowledgments

The development work was undertaken jointly by the follwing organizations: VTT Electronics, Spectra-Physics Vision Tech Oy and Rautaruuukki Oy Raahe Steel.

References

[1] ASTM designation: A 568/A 568M-96 1996 Standard spesification for steel, sheet, carbon, and high-strength, low-alloy, hot-rolled and cold-rolled, general requirements. ASTM American Society for Testing and Materials, 22pp

[2] SFS-EN 10 029. 1991. Hot rolled steel plates 3 mm thick or above *Tolerances on Dimensions, Shape and Mass* (Helsinki: Finnish Standards Association (SFS)) 12pp

[3] Kujawinska M 1993 The architecture of a multipurpose fringe pattern analysis system *Opt. Lasers Eng.* **19** 261–8

[4] Schmit J 1996 Spatial phase measurement techniques in modified grating interferometry (strain, error reduction) *Phd Thesis* University of Arizona

[5] Sough D 1993 Beyond fringe analysis *Interferometry (San Diego, CA) (Proc. SPIE 2003)* (Bellingham, WA: SPIE) pp 208–23

[6] Harding K G, Coletta M P and VanDommelen C H 1988 Color encoded moiré contouring Optics, *Illumination and Image Sensing for Machine Vision III (Cambridge, MA) (Proc. SPIE 1005)* (Bellingham, WA: SPIE) pp 169–78

[7] Perry K E and McKelvie J 1993. A comparison of phase shifting and fourier methods in the analysis of discontinuous fringe patterns *Opt. Lasers Eng.* **19** 269–84

[8] Servin M, Rodriguez-Vera R and Malacara D 1995 Noisy fringe pattern demodulation by an iterative phase locked loop *Opt. Lasers Eng.* **23** 355–65

[9] Kozlowski J and Serra G 1997 New modified phase locked loop method for fringe pattern demodulation *Opt. Eng.* **36** 2025–30

Chapter 11.5

Optical triangular profilometry: unified calibration

Anand Asundi and Zhou Wensen
Nanyang Technological University, Singapore

Optical profilometry measurement [1–8], an efficient tool in scientific, industrial, medical and other applications, has been extensively studied. In most of the applications, a structured pattern, such as a spot, single line or grating is projected onto the surface to be measured and the distorted pattern caused by the depth variation of the surface is recorded by a CCD camera at an angle to the illumination direction. This is optical triangular profilometry. On the basis of different structured light illuminations, optical triangular profilometry is classified under three types: spot inspection, a single-line system or a projection grating system. Although the configurations of three types are quite different, their distance mapping algorithms [10–12] between depth variation and deformed signal are inherently the same. The ray tracing can start from either the projection or image ray.

11.5.1 Projection ray tracing

A schematic diagram of the triangular profilometry based on projection ray tracing is shown in figure 11.5.1(a). Points I_1 and I_2 are the projection and imaging centre, respectively. First, a projection ray $I_1 O_R$ illuminates a reference point O_R and the image O_R' is recorded by a camera. When an object point A is measured, the image O_R' is moved to A' due to the height variation $O_R A$. Point O is a reference point of the recorded signal. In 360° profilometry, the reference point O_R is the rotational centre of the turntable, while in $2\frac{1}{2}$D profilometry, the point O_R is a point on the reference plane.

There are two types of recorded signals: position and phase.

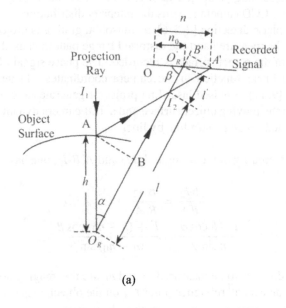

(a)

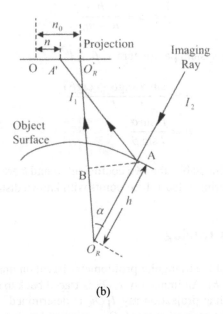

(b)

Figure 11.5.1. Schematic diagram of the unified triangular profilometry: (*a*) projection ray tracing and (*b*) image ray tracing.

- In position sensing [1, 3, 4], a laser spot or single line is projected onto the object and a CCD camera records the intensity distribution.
- In spatial phase detection [2, 5, 6], a sinusoidal grating is projected onto the surface to be measured and the distorted fringe pattern caused by the depth variation of the object surface is recorded. The phase signal of a sinusoidal wave is a linear function in the camera coordinates. In temporal phase detection [9, 11], a pulsed laser line projector generates a temporal grating pattern on the moving object surface and a TDI camera dynamically records the deformed fringe pattern line by line.

From the similarity of the triangle ABI_2 and $A'B'I_2$, one has

$$\frac{BI_2}{BA} = \frac{B'I_2}{B'A'}$$

$$\frac{l - h\cos\alpha}{h\sin\alpha} = \frac{l' + (n - n_0)\cos\beta}{(n - n_0)\sin\beta} \qquad (11.5.1)$$

where h is the distance to be measured. n and n_0 are the imaging positions of the point A on the object and reference point O_R on the object, respectively.

The distance mapping formula can, thus, be written as

$$\frac{1}{h} = a + \frac{b}{n - n_0} \qquad (11.5.2)$$

where a and b relate to system parameters α, β, l, l'.

$$a = \frac{\sin\alpha(\text{ctg}\alpha + \text{ctg}\beta)}{l}$$

$$b = \frac{l'\sin\alpha}{l\sin\beta}$$

When the system remains stable, the two coefficients a and b are constant and can be calculated by measuring at least three points with known distances.

11.5.2 Image ray tracing

A schematic diagram of the triangular profilometry based on image ray tracing is shown in figure 11.5.1(b). An image ray $I_2 O_R$ is traced back to a reference point O_R and its corresponding projection ray $I_1 O_R$ is determined. When an object point A is measured, the projection ray $I_1 O_R$ is changed to $I_1 A$ due to the height variation $O_R A$. From schematic symmetry, the mapping algorithm for projection ray tracing can be readily used in image ray tracing by substituting the image variable to projection variable.

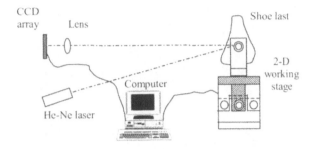

Figure 11.5.2. Set-up for spot inspection.

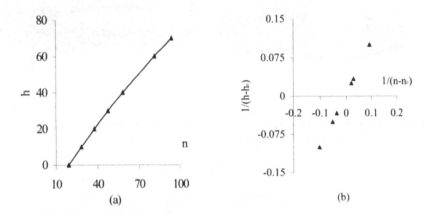

Figure 11.5.3. Calibration of spot inspection: (*a*) sampling data at different distances; and (*b*) inverse linear mapping relationship.

11.5.3   Choice

The choice of either projection or image ray tracing is determined by two criteria: tracing validity and computational complexity. Projection ray tracing is valid for spot inspection and single-lines and projection grating systems since the distance variation can be recorded. But this method is time consuming due to the complexity in matching the distorted signal in the image and reference rays. Image ray tracing is only valid in the projection grating system. The projection ray is recorded as a phase value by a CCD pixel and the determination of signal distortion in the projection is simplified by subtracting the phase values of the CCD pixel. Therefore, projection ray tracing is the only configuration where image ray tracing can speed the measurement.

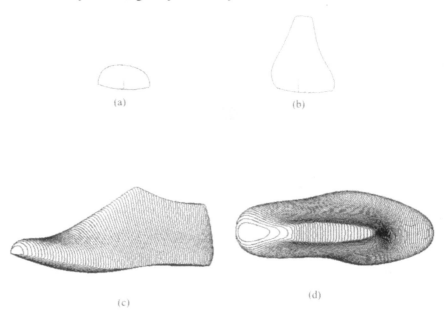

Figure 11.5.4. Profile measurement of a shoe: (*a*), (*b*) section displays; and (*c*), (*d*) 3D displays of the shoe from two viewpoints.

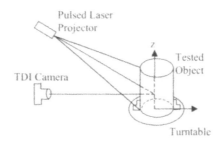

Figure 11.5.5. Set-up for pulsed single-line system.

11.5.4 Spot inspection

A schematic diagram of spot inspection [10] is shown in figure 11.5.2. A 3D object is mounted on a 2D working stage to enable 3D acquisition. A He–Ne laser illuminates a point on the surface of the object and a CCD camera records the displacement of the light spot.

To calibrate the measuring system, seven points at different distances are measured. The data are shown in figure 11.5.3(*a*). Then the point with 30 mm is chosen as the reference point. An inverse linear relationship is obtained as shown in figure 11.5.3(*b*) and the regression coefficients *a*, *b* and inter-related coefficient

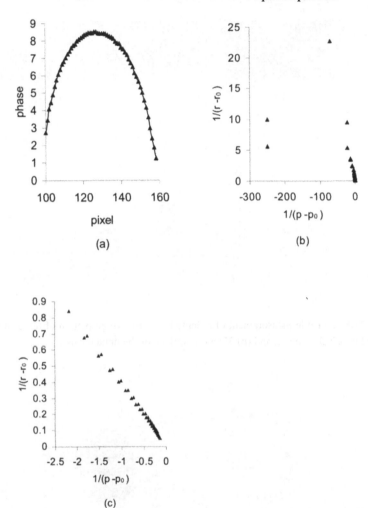

Figure 11.5.6. Calibration of a single-line system (*a*) phase distribution, (*b*) inverse linear distribution and (*c*) inverse linear distribution after the exclusion of some points near r_0.

r are:

$$a = 0.0034 \qquad b = 1.0146 \qquad r = 0.9999$$

For the purpose of demonstration, one point with a given height 50 mm is measured and this is compared with the calculated height 49.93 mm. Applications of the approach are valid with the shoe last shown in figure 11.5.4.

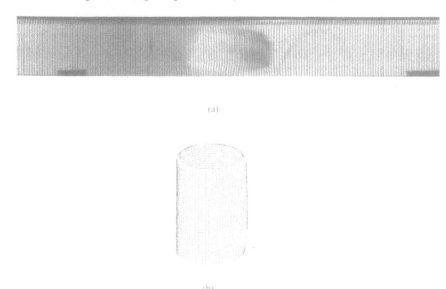

(a)

(b)

Figure 11.5.7. Profile measurement of a dented can: (*a*) fringe pattern of the dented can recorded by a TDI camera; and (*b*) 3D mesh surface of the dented can.

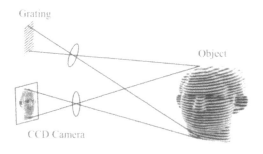

Grating

Object

CCD Camera

Figure 11.5.8. Set-up for projection grating system.

11.5.5 Single-line system

The pulsed single-line system shown in figure 11.5.5 uses a narrow pulsed line of light to generate a temporal grating pattern on the surface of moving (rotating) object and a TDI camera dynamically records the deformed fringe pattern line by line [9]. The fringe pattern can be considered analogous to many images recorded for a standard single-line system sequentially and compiled to form a fringe pattern.

The calibration gauge is a plane with a distance 20.47 mm away from the rotation axis. From the experimental data shown in figure 11.5.6(*a*), one can

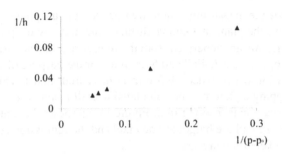

Figure 11.5.9. Inverse linear distribution in projection grating system.

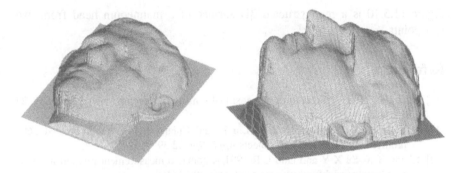

Figure 11.5.10. Reconstructed 3D surface of a mannequin head from two viewpoints.

easily find that errors in the phase measurement cause many points to deviate from the expected linear distribution as shown in figure 11.5.6(*b*), particularly points near p_0, which make the denominators close to zero. Figure 11.5.6(*c*) shows a good linear relationship after the exclusion of 20 points near p_0.

$$a = 0.0101 \qquad b = -0.3802 \qquad r = -0.9983.$$

A rectangular object with known distance 51.05 mm is used for verification. The calculated overall mean of distances is $\mu = 51.07$ mm and the standard deviation $\sigma = 0.11$ mm. An example of the TDI imaging system to measure a dented can is shown in figure 11.5.7. The fringe pattern of the dented can is shown in figure 11.5.7(*a*). Demodulation, processing and reconstruction provide the 3D profile of the dented can shown in figure 11.5.7(*b*).

11.5.6 Projection grating system

The set-up for the grating method [5, 12] is shown in figure 11.5.8. When a sinusoidal grating pattern is projected onto a 3D diffusion object, the distorted grating image is recorded by a CCD camera.

To calibrate the measuring system we first measure the phase of the reference plane, then move the plane at a known distance and measure the phase again. The mapping coefficients are generated from the reference plane and the planes at the distances of 9.6, 19.5, 39.7, 49.7 and 59.8 mm. For the purpose of demonstration, a plane with a known distance 29.3 mm is measured, and is calculated by its phase and mapping equation. The calculated overall mean is $\mu = 29.4$ mm and the standard deviation is $\sigma = 0.1$ mm. Figure 11.5.9 shows the linear relationship between the phase and the height at one pixel and the regression coefficients a, b and inter-related coefficient r are:

$$a = -0.0027 \qquad b = 0.3939 \qquad r = 0.9999.$$

Figure 11.5.10 is a reconstructed 3D surface of a mannequin head from two viewpoints

References

[1] Strand T C 1985 Optical three-dimensional sensing for machine vision *Opt. Eng.* **24** 33–40
[2] Halioua M, Krishnamurthy R S, Liu H and Chiang F P 1983'Automated 360° profilometry of 3-D diffuse objects *Appl. Opt.* **22** 3977–82
[3] Cheng X X, Su X Y and Guo L R 1991 Automated measurement method for 360° profilometry of diffuse objects *Appl. Opt.* **30** 1274–8
[4] Ohara H, Konno H, Sasaki M, Suzuki M and Murata K 1996 Automated 360° profilometry of a three dimensional diffuse object and its reconstruction by use of the shading model *Appl. Opt.* **35** 4476–80
[5] Srinivasan V, Liu H C and Halioua M 1984 Automated phase-measuring profilometry of 3-D diffuse objects *Appl. Opt.* **23** 3105–8
[6] Takeda M and Mutoh K 1983 Fourier transform profilometry for the automatic measurement of 3-D object shapes *Appl. Opt.* **22** 3977–82
[7] Saint-Marc P, Jezouin J L and Medioni G 1991 A versatile pc-based range finding system *IEEE Trans, Robotics and Automation* **7** 250–6
[8] Valkenburg R J and Mclor A M 1998 Accurate 3D measurement using a structured light system *Image and Vision Comput.* **16** 99–110
[9] Asundi A, Chan C S and Sajan M R 1994 360° prodilometry: new techniques for display and acquisition *Opt. Eng.* **33** 2760–69
[10] Su X Y and Zhou W 1990 Algorithm for the generation of look-up range table in 3-D sensing *SPIE* **1332** 355–7
[11] Sajan M, Tay C, Shang H and Asundi A 1998 Improved spatial phase detection for profilometry using a TDI imager *Opt. Commun.* **150** 66–70
[12] Zhou W and Su X Y 1994 A direct mapping algorithm for phase-measuring profilometry *J. Mod. Opt.* **41** 89–94

Chapter 11.6

Grating diffraction method

Anand Asundi and Bing Zhao
Nanyang Technological University, Singapore

Grating diffraction techniques are a special class of moiré/grating techniques. They directly measure strain, i.e. the strain is obtained without differentiation of the displacement field. The strain measurement is independent of the rigid body motion of the grating. The grating diffraction method uses only the specimen grating with no master grating. The basic principle of measurement is illustrated in figure 11.6.1. A diffraction grating bonded to the surface of the specimen follows the deformation of the underlying specimen. Consider a thin monochromatic collimated beam that is normal to the line-grating plane and illuminating a point on the specimen grating. The diffraction is governed by the equation

$$P \sin \theta_n = n\lambda \qquad n = 0, \pm 1, \pm 2, \ldots. \tag{11.6.1}$$

Here P is the grating pitch, n is the diffraction order, λ is the light wavelength and θ is the diffraction angle (see figure 11.6.1). After the grating is deformed, the pitch changes to P^*, with a corresponding change in θ according to equation (11.6.1). Thus, the strain along the X direction (figure 11.6.1) at the illuminated point on the grating can be determined as [1]

$$\varepsilon_{xx} = -\frac{P* - P}{P} = \frac{X_n^* - X_n}{X_n} \cos \theta_n \tag{11.6.2}$$

where X_n and are the centroids of the undeformed and deformed nth-order diffraction spot. This centroid can be determined precisely with subpixel accuracy [1, 2]. If a cross grating is used instead of a line grating, the strain along the Y direction, ε_{yy}, can be similarly obtained. The shear strain ε_{xy} can be evaluated with reference to figure 11.6.2, which depicts a grating before and after deformation and their corresponding first-order diffraction patterns. From the coordinates of the diffraction spots, α and β defined in this figure can be

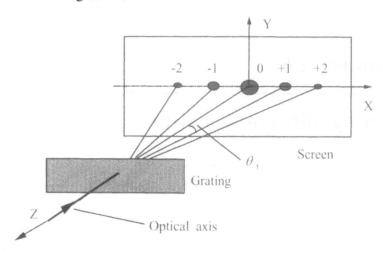

Figure 11.6.1. Sketch of a one-dimensional grating diffraction. The grating illuminated by a collimated beam diffracts the beam into discrete orders based on the pitch of the grating.

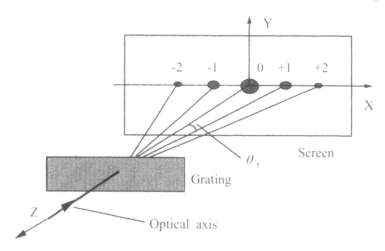

Figure 11.6.2. First-order diffraction pattern of undeformed and deformed specimen gratings.

computed as

$$\varepsilon_{xy} = \frac{\alpha + \beta}{2}.$$

(11.6.3)

This technique possesses several advantages [1]:

(1) no reference grid is needed;
(2) the sign of the deformation is known [1, 2];
(3) the system is simple and it integrates easily; and

(4) no high-resolution recording system is needed.

In 1956, Bell [3] initiated the use of the 'diffraction grating strain gauge'. He analysed the grating diffraction pattern during the test by measuring changes in the grating frequency at various loads and calculated the strain due to this spatial frequency variation of the grating. Since then, this technique has advanced steadily and is widely applied. These are reviewed in the works of Moulder and Cardenas-Garcia [4], Sevenhuijsen *et al* [5] and Bremand *et al* [6]. In 1965, Douglas [7] modified the use of gratings to an application on a flat surface, which allowed the study of the crack problem. Pryor and North [8] later developed a diffractographic strain gauge using the diffraction of light from a single aperture to measure strain. Boone [9] used a laser beam to illuminate the specimen grating for a full field measurement. Sevenhuijsen [10] used an optoelectronic device to calculate the strain distribution automatically. With recent developments in CCD cameras and image-processing techniques, the grating diffraction method is being redeveloped [4, 11–13]. Bremand *et al* [1] and Sevenhuijsen *et al* [5] used a numerical grating spectrum method to measure the local strain directly. Qin *et al* [14] and Asundi [2,15] developed a pointwise strain sensor with high sensitivity by using a high-frequency grating diffraction. Compared to other photomechanics techniques, the grating diffraction technique avoids the difficulty in fringe pattern interpretation.

Some new advances in grating diffraction technique are introduced here.

11.6.1 A strain microscope with grating diffraction method [16]

The development of a compact microscope system for direct strain measurement involves the grating diffraction method coupled with microscopy and an image-processing technique. A Leitz optical transmitting microscope with a white light source is reconstructed by developing a loading and recording system. Gratings with densities from 40–200 lines/mm are used. The specimen is made of transparent or semi-transparent materials such as plastic film, PMMA, photoelastic materials etc. The grating and the specimen are between the condenser and the objective of the microscope (see figure 11.6.3). The loading system which is capable of providing a uni-axial tensile load is fixed on the mechanical stage of the microscope.

The important step in the measurement is to obtain the grating diffraction pattern. Using an optical microscope where the image of the object is examined, its Fourier spectrum (or diffraction pattern) is not usually obtained. According to the microscopic image-forming principle (Abbe's theory [17]), the transition from object to image involves two stages. First, the grating-like object when illuminated gives rise to a Fraunhofer diffraction pattern in the back focal plane of the objective. Then, every point in the focal plane may be considered to be the centre of a coherent secondary disturbance. The light waves that proceed from

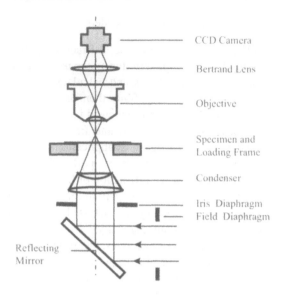

Figure 11.6.3. Sketch of the optical system and principle.

these secondary sources will then interfere with each other and give rise to an image of the object in the image plane of the objective.

Thus, the diffraction pattern is formed in the back plane and image of the grating is formed at the image plane. Hence, in order to get a video image of the grating diffraction pattern through a microscope, the back focal plane of the objective should be extended to the CCD sensor plane. This is realized using a Bertrand lens. This lens is generally placed between eyepiece and objective and is often used to observe the interference or polarization patterns. With this lens, one can see clearly the image formed in the back focal plane of the objective. The image-forming procedure from objective to Bertrand lens is illustrated in figure 11.6.4. On gradually adjusting the focusing screw of this lens which moves the lens up and down, the transformation from the grating image to the grating spectrum pattern is observed through the eyepiece or the video monitor.

The developed microscope system possess several features:

(1)   a high spatial resolution of 0.18 mm,
(2)   real-time loading and measurement,
(3)   variable sensitivity,
(4)   the field of view (examined area) can be easily changed, as the specimen can rotate and move in the X and Y directions;
(5)   the set-up is compact and the optical components are easily adjustable; and
(6)   the image quality is very good—the diffraction pattern approximates to a binary image.

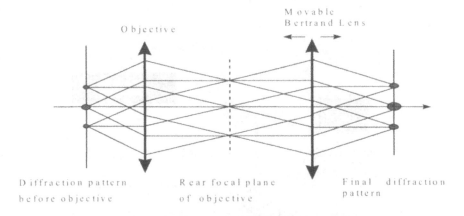

Figure 11.6.4. Illustration of image forming from objective lens to Bertrand lens.

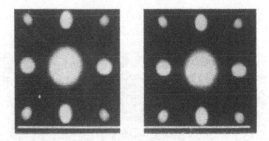

Figure 11.6.5. Diffraction patterns of an 80 line/mm cross grating before and after loading. The white line is a reference one.

A test example is demonstrated in figure 11.6.5, where the diffraction patterns of an 80 line/mm cross grating before and after loading are shown. The specimen has an overall displacement of 0.6 mm and the calculated average strain is 12 000 $\mu\varepsilon$. The white reference line in this figure clearly shows the shift in the diffraction spots.

11.6.2 Grating diffraction strain sensor [18]

In this section, developments in the instrumentation of the grating diffraction strain sensor are introduced. A new compact system for directly measuring in-plane strain using a high-frequency grating and two position sensitive detectors (PSDs) is demonstrated. The system is shown in figure 11.6.6(*a*). A focused laser beam illuminates the grating with a frequency of 1200 lines/mm attached on the surface of a specimen. The spatial resolution for strain measurement, i.e. the illuminating area on the specimen, is about 0.4 mm. Two PSDS connected to a personal computer automatically determine the centroids of diffracted beam

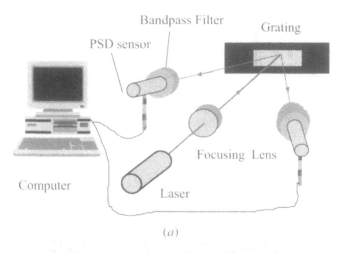

(*a*)

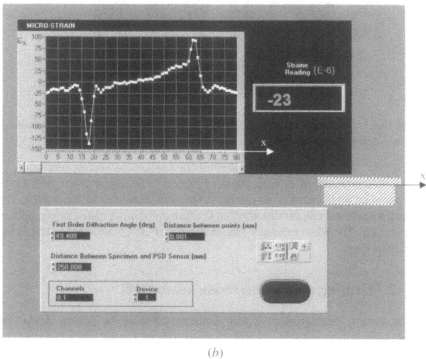

(*b*)

Figure 11.6.6. (*a*) Grating diffraction system with PSD sensors. (*b*) Interface of the diffraction strain sensor.

spots from the grating. The shift in the diffracted beam spots due to the specimen deformation is then detected and related to strain. Figure 11.6.6(*b*) shows a typical interface of the strain sensor depicting the thermal strains in electronic packaging. The beam size, which corresponds to the gauge-length of the strain sensor, is 0.5 mm and recordings are made at 1 mm intervals.

This optical strain sensor has a sensitivity comparable to a strain gauge but with the advantages of non-contact measurement along a continuous strip of the sensor with a variable gauge-length. It also provides much higher resolution than existing devices. The system specifications are:

(1) sensitivity of 1 μm strain,
(2) variable spatial resolution from 0.4 mm to 2 mm,
(3) large range of strain up to 15%,
(4) automatic data acquisition and processing,
(5) user-friendly visual instrument interface,
(6) any location of the grating area can be measured,
(7) high-speed data processing and continuous measurement,
(8) dynamic strain measurement and
(9) compact size

References

[1] Bremand F, Dupre J C and Lagarde A 1992 Non-contact and non-disturbing local strain measurement methods, I. Principle *Eur. J. Mech. A/Solids* **11** 349–66
[2] Asundi A 1996 Moiré interferometric strain sensor *Final Report for Project on Moiré Interferometry Strain* ensor, submitted to committee on research and conference Grants, University of Hong Kong
[3] Bell J F 1956 Determination of Dynamic plastic strain through the use of diffraction gratings *J. Appl. Phys.* **27** 1109–13
[4] Moulder J C and Cardenas-Garcia J F 1993 Two-dimensional strain analysis using a video optical diffractometer *Exp. Tech.* September/October 11–16
[5] Sevenhuijsen P J, Sirkis J S and Bremand F 1993 Current trends in obtaining deformation data from grids *Exp. Tech.* May/June 22–6
[6] Bremand F, Dupre J C and Lagarde A 1992 Non-contact and non-disturbing local strain measurement methods, I. Principle *Eur. J. Mech. A/Solids* **11** 349–66
[7] Douglas R A 1965 Strain-field investigations with plane diffraction gratings *Exp. Mech.* **5** 233–8
[8] Pryor T R and North W P 1971 The diffractographic strain gauge *Exp. Mech.* **11** 565–8
[9] Boone P M 1911 A method for directly determining surface strain fields using diffraction gratings *Exp. Mech.* **11** 481–789
[10] Sevenhuijsen P J 1978 The developpment of a laser grating method for the measurement of strain distribution in plane, opaque surfaces *VDI Berichte* **313** 143–7

[11] Cardenas-Garcia J F and Wu M S 1989 Further development of the video optical diffractometer for strain measurement *Proc. 1989 SEM Spring Conf. (Cambridge, MA)* (Bethel, CT: SEM) pp 73–9

[12] Bremand F and Lagarde A 1986 A new method of optical strain measurement with application *Proc. 1986 SEM Spring Conf. (New Orleans)* (Bethel, CT: SEM) pp 686–94

[13] Bremand F and Lagarde A 1989 Two methods of large and small strain measurement on small area *Proc. 1989 SEM Spring Conf. (Cambridge, MA)* (Bethel, CT: SEM) pp 80–6

[14] Qin Y W, Wang Y Y and Hung Y Y 1988 Grating strain method *Proc. SPIE Industrial Laser Interferometry* **955** 63–8

[15] Asundi A 1989 Developments in the moire interferometric strain sensor *Proc. SPIE Holographic Interferometry and Speckle Metrology* **1163** 63–8

[16] Zhao B and Asundi A 2001 Microscopic grid methods—resolution and sensitivity *Opt. Laser Eng.* **36** 437–50

[17] Born M and Wolf E 1983 *Principles of Optics, Electromagnetic Theory of Propagation, Interference and Diffraction of Light* 6th edn (Oxford: Pergamon)

[18] Asundi A and Zhao B 1998 Moiré interferometric strain sensor *Proc. Int. Conf. on Optics and Optoelectronics (Dehradun)* (Bellingham, WA: SPIE)

CHAPTER 12

REAL TIME AND DYNAMIC MOIRÉ

Chapter 12.1

Real-time video moiré topography

Joris J J Dirckx and Willem F Decraemer
University of Antwerp, Belgium

12.1.1 Introduction

In traditional projection moiré topography [1, 2], a transmission ruling is imaged onto the surface of the object. The projected grid lines are modulated by the object shape. This deformed ruling is then imaged by a second lens onto another identical ruling. The interference between the deformed and non-deformed grid lines produces moiré fringes, which can be interpreted as contours of equal object surface height. Deformation measurements have to be made by obtaining moiré topograms of the object in its original state and in its deformed state, followed by reconstruction of the object shape from these topograms and finally by calculating the difference between these two shape measurements.

In this chapter we will describe a projection moiré system which is combined with video techniques. The system does not use a second optical ruling for demodulation but an electronic recording of a reference grid. This allows one to obtain in real time both contours of equal height and contours of equal deformation. The apparatus was developed for *in vitro* shape and deformation studies of the eardrum. By changing the set-up geometry, the apparatus can be adapted for the study of the deformations of any object.

12.1.2 Theory of fringe formation

12.1.2.1 Modulation and demodulation

Because the fringe formation is somewhat different from traditional moiré topography, in the sense that fringes are not produced by an optical multiplication process but rather by an electronic subtraction process, we will give a concise overview of the theory behind the technique.

Let us consider the projection set-up represented schematically in figure 12.1.1. The light from projector P is passed through the grid G and projected by lens L_1 onto the surface of the object. The projected grid lines are modulated by the object shape and are imaged by a second lens L_2 onto the camera C. The focal point of the projection and imaging system coincide at the origin of an orthogonal coordinate system. The z axis of the coordinate system is chosen along the viewing direction and the x axis in the plane formed by the projection and imaging axes.

For a sinusoidal transmission ruling with grid lines parallel to the y axis, the modulation function T is given by

$$T = I_0 \left[\frac{1}{2} + \frac{1}{2} \sin \left(\frac{2\pi}{p} x' \right) \right] \qquad (12.1.1)$$

where I_0 is the intensity of the incoming light (the projector), p is the period of the grid and x' is the x axis coordinate of the point on the grid where the light emerges. The y axis coordinate does not appear in the equation since the grid lines are positioned parallel to the y axis so that for a given value of x' the modulation is identical for all values of y.

From a geometric derivation, we find that the camera will observe a light distribution I on the object given by

$$I = I_0 \left\{ \frac{1}{2} + \frac{1}{2} \sin \left[\frac{2\pi}{p} \left(\frac{(D-x)(L+L'-z)}{L-z} + x \right) \right] \right\}. \qquad (12.1.2)$$

In an ordinary projection moiré set-up, this light distribution is demodulated by multiplying it with the transmission function of a second ruling. In practice, this multiplication is obtained by passing the light through a second ruling positioned in the focal plane of L_2. In our set-up, there is no second grid and the light distribution given by (12.1.2) is projected directly by L_2 onto the camera sensor, without demodulation.

After recording the intensity distribution given by (12.1.2), a second intensity distribution is recorded for an object with a different shape (which may be the original object after it has undergone a deformation). On the point with coordinates (x, y), this object has a height which we will denote as z', so the recorded intensity distribution will be once again given by equation (12.1.2) but with z' substituted for z.

To obtain the moiré interference, these two intensity distributions are now subtracted from one another, yielding the difference I_d. If we take the set-up dimensions D and L large in comparison to the values of x and z for all points on the object, I_d is given by

$$I_d = I_0 \cos \left[\frac{2\pi}{p} x + \varphi \right] \sin \left[\frac{2\pi}{p} \frac{D(z'-z)}{2L} \right] \qquad (12.1.3)$$

where

$$\varphi = \frac{2\pi}{p} \left(\frac{D(L+L')}{L} + \frac{D(z-z')}{2L} \right) \qquad (12.1.4)$$

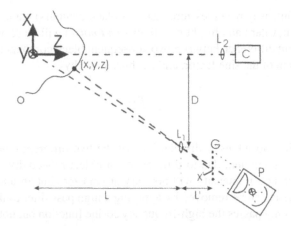

Figure 12.1.1. Schematic diagram of the optical set-up, showing projector (P), grid (G), projection lens (L₁), object (O), imaging lens (L₂) and camera (C).

Equation (12.1.3) contains the product of a cosine factor, with the same frequency as the projected ruling, and a sine factor, which is only dependent on the difference in surface height $(z' - z)$ between the two objects. In the image , the cosine factor is seen as high spatial frequency grid lines or *grid* noise. The sine factor can be regarded as an amplitude modulation of the cosine factor and it describes contours of equal height displacement. The term φ in the cosine argument can be regarded as a phase offset.

12.1.2.2 Shape and deformation measurement

If we use, as second object, a flat plate positioned in the origin and perpendicular to the viewing direction, so that z' equals zero for all values of (x, y), equation (12.1.3) reduces to

$$I_s = I_0 \cos \left[\frac{2\pi}{p} x + \varphi \right] \sin \left[\frac{2\pi}{p} \frac{Dz}{2L} \right] \tag{12.1.5}$$

In equation (12.1.5), the sine factor describes contours of equal surface height, so from I_s we can determine the shape of the object.

If we do not replace the object by a flat plate but leave it in place and make it undergo some kind of deformation, then equation (12.1.3) describes fringes of equal displacement along the z axis. I_d can then be used to measure the deformation of the object along the z axis.

12.1.2.3 Fringe characteristics and fringe enhancement

The fringes described by (12.1.3) and (12.1.5) are not dark and bright fringes as we are used to see in interferometry. In zones where the sine argument is zero, the observed intensity becomes zero (seen as black contours) but elsewhere high spatial frequency dark and bright grid lines (*grid noise*) will be seen. The height or displacement difference between two subsequent black contours is determined by the argument of the sine factor, and is, thus, given by

$$\lambda = \frac{pL}{D}.$$ (12.1.6)

In practice, the two intensity distributions of the two different objects may not have the same amplitude, due to differences in object reflectivity. If this is the case, the difference in intensity will not vary around zero but around some offset value. This offset can be removed by applying a high pass filter to the subtracted image, which only passes the high-frequency cosine function but not a DC offset.

Negative intensities can, of course, not be visualized, so half of the cosine signal in equation (12.1.3) is lost. We make the signal all positive by taking its absolute value, which is done, in practice, by performing a full wave rectification of the video signal. The effect on fringe visibility is demonstrated in the results section.

12.1.3 Apparatus

12.1.3.1 Optical set-up

The optical set-up is essentially the same as for any other projection moiré set-up, and was schematically represented in figure 12.1.1. The light of a 100 W halogen lamp projector is passed through a 300 lines per 25.4 mm transmission grid (G) and is projected by a 100 mm focal length photographic lens (L_1) onto the surface of the object (O). A second identical lens (L_2) images the deformed grid lines onto a CCD camera (Sony AVE-D7CE) (C). In contrast to a normal projection moiré set-up, there is no second grid in the focal plane of the imaging lens: the projected grid lines are not optically demodulated but are directly recorded by the camera.

The projection and imaging optics are mounted on optical rails which can pivot in the x–y plane around the focal point. In this way, the angle between the projection and imaging direction and, hence, the fringe plane distance or sensitivity of the set-up can be easily changed while the object is in place and remains in focus of both the imaging and the projection systems. The camera and imaging lens are mounted on a translation rail and the imaging lens is coupled to the camera with a bellows, so that the field of view of the CCD camera can be adjusted to the object size. In our experiments the grid-to-object distance and the camera to object distance is of the order of 40 cm and the angle between the projection and imaging direction is about 35 degrees.

12.1.3.2 Electronic image processing

The image-processing electronics are represented schematically in figure 12.1.2. The camera signal is digitized by an analogue-to-digital converter (A/D) with a resolution of 8-bit, at a rate of 25 images per second. One complete frame (two interlaced frames) of digitized video data is stored in a 512 by 512 pixel memory matrix. The memory of this frame store consists of 250 kB of inexpensive standard computer dynamic random access memory. When the frame store is set to *free run* mode, the previous video frame is cleared and replaced by the next video frame coming from the camera. When the frame store is set to *hold* mode, the last incoming video frame is stored and remains in the memory until the frame store is again released.

The data in the frame store are continuously read at a speed of 25 image frames per second. These data go into the subtraction unit which can be set to either *direct* or *subtraction* mode. In *direct* mode, the frame store data are simply passed on to the output of the unit. In *subtraction* mode, the digitized real-time image data coming from the camera are subtracted from the data coming from the frame store. This subtraction process is also performed in real time, at a rate of 25 images per second, using a commercially available subtraction chip (model reference 74181). The data at the output of the subtraction unit are converted back to an analog video signal by a digital-to-analogue converter (D/A).

If the subtraction unit is set to *direct* mode, a monitor connected to point A displays the stored image when the frame store is set to hold. The same monitor can be used to look at the direct image by switching the frame store to free run. Monitoring the direct camera image is necessary when positioning and focusing an object in the set-up. If an image is captured in the frame store and the *subtraction* mode is chosen, the monitor at point A will show the image described by equation (12.1.3). We will show such an image in the results section.

The signal coming from the subtraction unit contains the difference between the reference image and direct camera image. The signal is fed through a high pass filter to remove offset and full wave rectified by a bridge rectifier to take the absolute value. Because the camera image is read in horizontal direction, one line after the other, this processing of the video time signal is equivalent to filtering the image data along the x axis. By performing the high pass filtering and rectification on the analog video signal, no complicated and fast digital signal processor is needed to obtain the processed image in real time. The moiré fringes are finally displayed on a monitor connected to point B at a rate of 25 new interferograms per second.

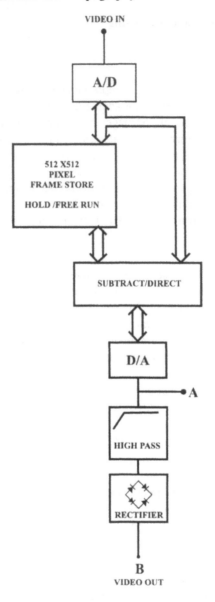

Figure 12.1.2. Block diagram of the image-processing electronics.

12.1.4 Measurement results

12.1.4.1 Adjustment and calibration

For studies on objects of different size, such as tympanic membranes of different species, we use different magnification factors. The field of view can be adjusted to the object size by using different lens-to-object distances. By using different rulings or a different projection to observation angle, the fringe plane distance can easily be adapted.

To measure the shape or deformation of an object accurately from the moiré fringe pattern, the fringe plane distance λ has to be known precisely. Because it is impractical to pinpoint the distances L and D with high accuracy, the fringe plane distance is obtained by direct calibration, rather than by calculating it from the geometrical parameters. To perform the calibration measurement, an object is mounted on a calibrated translation table with the translation axis of the stage parallel to the viewing direction. The grid-line image on the object is stored in the frame store and the processed subtraction image is monitored. When the object is moved backward along the z axis using the translation stage, the processed image turns alternating bright and dark each time as the displacement grows with one fringe plane distance. First, the object is positioned so that it turns completely dark. Then the object is translated over one or more fringe plane distances until it is again dark. The local minimum in the intensity is easily judged by eye on the monitor image. The localization of this minimum is repeatable within a few micrometres. The fringe plane distance is calculated from the translation distance, read from the micrometre screw on the translation stage and divided by the number of fringe planes encountered in the translation. To obtain high accuracy, we perform the calibration by measuring the translation distance for 10 fringe orders. As we explained in the previous section, the fringe plane distance varies slightly with object height. Calculating λ from the distance of ten fringe planes gives an average fringe plane distance within this depth-measuring range.

For the measurements presented in this section the fringe plane distance was set at 0.082 mm. This setting was also used for our studies on gerbil tympanic membranes [4, 6]. With projection and lens apertures of $f/16$ the set-up gives good fringe contrast over a depth range of about 25 fringe orders or 2 mm. The field of view for the processed image was 3.5 mm horizontally and 2.9 mm vertically. This is somewhat smaller than the actual camera image, because the frame store does not capture the video image over its full width.

12.1.4.2 Measuring resolution

The measuring resolution was tested on a steel bearing ball with a spherical surface precision better than 1 μm. The diameter of the ball is (6.750 ± 0.005) mm. Because diffuse reflection is needed in projection moiré topography, the surface of the ball was coated with white Chinese ink.

Figure 12.1.3(a) shows the direct image of the grid projected onto a flat surface positioned in the x–y plane: the projected grid lines are straight and equidistant. Figure 12.1.3(b) shows the direct image of the grid projected onto the surface of the sphere. We see the projected grid lines bending due to the spherical surface shape. At the right-hand side, the image is darker due to shadow effects

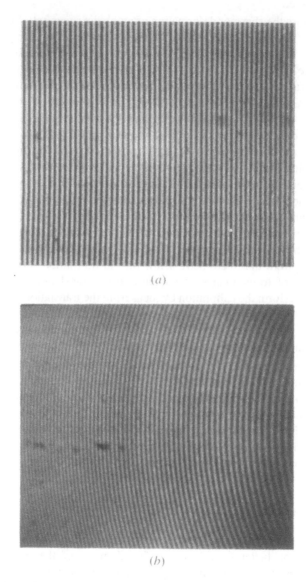

(*a*)

(*b*)

Figure 12.1.3. (*a*) Direct image of the grid lines projected on a flat plate positioned in the *x–y* plane. (*b*) Direct image of the grid lines projected onto the surface of a sphere.

caused by the strong curvature of the ball surface and the large angle between illumination and observation direction.

The moiré fringes are generated by subtracting the two images of figure 12.1.3. The image shown in figure 12.1.3(*b*) is captured in the frame store and then the sphere is replaced by the flat plate positioned parallel to the $x-y$ plane. Figure 12.1.4(*a*) shows a photograph of the image displayed by a monitor connected to point A of figure 12.1.2, when the set-up is switched to subtraction mode. We recognize circular bands in this image which are alternating grey and striped. These bands are the fringes described by equation (12.1.5). The electronics of the display monitor automatically generate an offset so that the video signal becomes completely positive. In this way, the negative intensity minima of the grid stripes ($I_s = -I_0$ in equation (12.1.5)) are put at zero level (black) and, as a consequence, the dark fringes (sine factor in equation (12.1.5) equal to zero) become grey. Although it is perfectly possible to follow the fringes, their visibility is not very good, especially in the zones where the fringes are close to one another. Figure 12.1.4(*b*) shows the result after low-pass filtering and half-wave rectification of the signal, as displayed on a monitor connected to point B shown on the electronics scheme presented in figure 12.1.2. In this image we have clear bright and dark fringes. Because the fringes are obtained in real time, any motion of the sphere can be followed at a rate of 25 interferograms per second.

Instead of recording a flat plate as a reference image, we could record any other shape (such as the steel ball itself in some deformed state) and obtain fringes which directly give contours of equal height difference between the two shapes. We will demonstrate this application in the next section.

The fringes seen in figure 12.1.4 are contours of equal height difference between the spherical surface and the flat surface positioned parallel to the $x-y$ plane, and thus are contours of equal height of the spherical surface. We determined the (x, y) coordinates of a number of points in the darkness centre of the fringes, and we attributed a relative z value to each point as the fringe plane distance multiplied by the order of the fringe. Then we fitted a best sphere through this set of three-dimensional data points. The diameter of the best fitting sphere was 6.96 mm (with a fitting tolerance of 1/1000), which deviates less than 3% from the actual diameter of the steel ball. The maximal difference between the data points and a circle fitted through the fringe darkness centres along a horizontal section, is less than 5 μm. The total depth range of the measurement was 0.41 mm, so the maximal deviation of the data points is 1.2% of the depth-measuring range. Measurements on a sphere of 3 mm diameter yield similar values for accuracy and precision [3].

12.1.5 Application example: measurements on the eardrum

We developed this moiré interferometer for *in vitro* studies of the effect of static pressure loading on the eardrum. The high sensitivity allows accurate

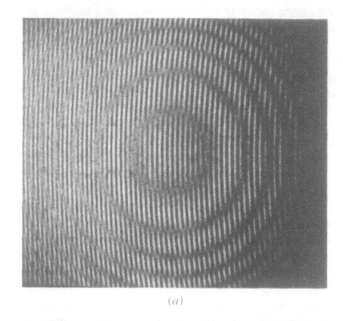

(*a*)

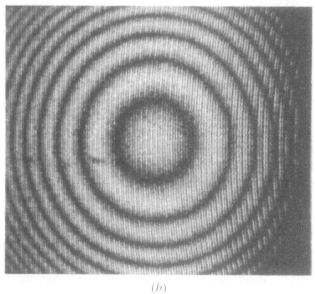

(*b*)

Figure 12.1.4. (*a*) Image displayed by a monitor connected to point A in figure 12.1.2. The image is obtained by subtraction of the images shown in figures 12.1.3(*a*) and 12.1.3(*b*). (*b*) The image shown in 12.1.4(*a*) after low-pass filtering and rectification, as obtained on a monitor connected to point B in figure 12.1.2.

measurement of the shape and the deformation of tympanic membranes under small static pressures, as they appear in the ear in everyday life.

Figure 12.1.5(*a*) shows the shape moiré interferogram obtained with our apparatus on an *in vitro* preparation of the tympanic membrane of a Mongolian gerbil [4]. The middle ear has been removed so that the membrane can be seen from the middle ear side. The ear canal was filled with white Chinese ink to improve light reflectance. This approach is possible because the gerbil eardrum is nearly completely transparent. The fringes seen in figure 12.1.5(*a*) represent contours of equal surface height, and demonstrate the conical shape of the tympanic membrane. They are the result of taking the difference between the grid image obtained on a flat plate and the grid image obtained on the eardrum. The fringe plane distance, and thus the height difference between successive fringe orders, was 0.082 mm.

Through the fluid ink in the ear canal, static pressures can be applied to the eardrum. Figure 12.1.5(*b*) shows the moiré interferogram of such a pressure-induced deformation. For this measurement the difference was made between the projected grid-line image for the tympanic membrane in rest state and the image obtained for the membrane under an overpressure of 0.2 kPa. Now the fringes represent contours of equal displacement. The fringe plane distance was the same as in the measurement presented in figure 12.1.5(*a*), so the displacement increases with 0.082 mm steps per contour order. The fringe pattern shows the characteristic two-lobed shape which was also obtained by time-averaged holography measurements on eardrums vibrating at low frequencies [5].

The spot on the right-hand side of the image is caused by specular reflections which saturate the imaging system. The shape measurement, shown in figure 12.1.5(*a*), and the deformation measurement, represented in figure 12.1.5(*b*), were obtained on the same membrane, in exactly the same position. Both figures are photographs of the display monitor, showing the result which is obtained in real time without any additional image enhancement.

We also use the apparatus to study the *pars flaccida* of the gerbil tympanic membrane [6]. The *pars flaccida* is a small soft zone located at one side of the eardrum and it is of particular interest to the pressure-regulating mechanisms of the ear. In the gerbil, the membrane has a circular border of about 1.6 mm diameter and it takes the shape of a sphere cap when it is deformed by static pressure. This results in circular fringes on the moiré topograms, as shown in figure 12.1.6. Using a mathematical sphere cap model, the fringes are used to calculate the volume displacement of the *pars flaccida* as a function of middle ear pressure. Because the interferograms are generated at a high rate, many measurements at different pressure values can be made in a short time, so that post mortem artifacts can be eliminated.

The video moiré topogram images are recorded on a VCR, so that the interferograms can be analysed after the measurements are finished. To obtain the fringe location data, images from the VCR recording are captured in a PC using a frame grabber board. The location of the fringes is then determined

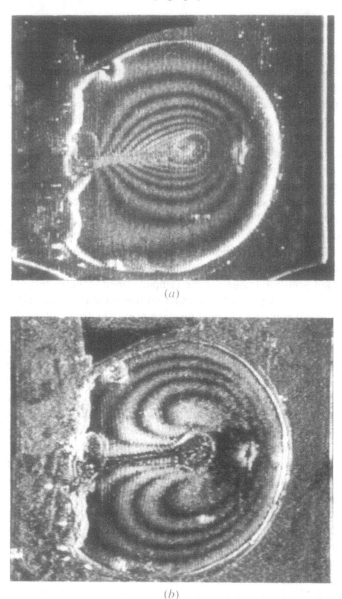

(a)

(b)

Figure 12.1.5. (a) Moiré interferogram obtained on a gerbil tympanic membrane, by taking the difference of the grid-line images on a flat plate and on the eardrum. Fringes represent contours of equal surface height, with a fringe plane distance of 82 μm. (b) Moiré interferogram obtained on the same membrane as in 12.1.5(a), but now using the difference between the grid-line images obtained on the deformed and undeformed membrane. Fringes now represent contours of equal displacement, again with a fringe plane distance of 82 μm.

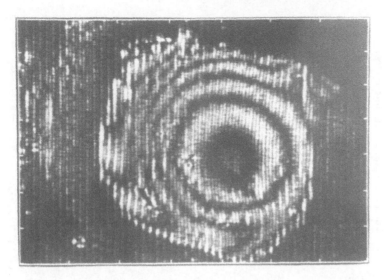

Figure 12.1.6. Real-time moiré topogram obtained with our apparatus on a gerbil tympanic membrane *pars flaccida* which is deformed by 400 Pa middle ear pressure.

using an interactive computer program. The crosses in figure 12.1.6 show the data points which are used to perform the volume calculation. From such measurements, several parameters can be calculated. For middle ear research, knowledge of elasticity and volume displacement as a function of middle ear pressure are important parameters. The high resolution of the fringes allows us to measure volume displacement of less than 0.5 μl to an accuracy of better than 10%. Within a physiological relevant range of pressures, deformation topograms are recorded at many intermediate pressure steps in a cycle of increasing and decreasing pressure. For each pressure step, volume displacement is calculated. Figure 12.1.7 shows the resulting curve of volume displacement *versus* applied pressure. The figure demonstrates that volume measurement resolution is high enough to determine hysteresis and nonlinearity.

12.1.6   Discussion and conclusions

A major advantage of the apparatus lies in the fact that both shape and deformation measurements can be performed in immediate succession, without altering the optical set-up and without touching the specimen. The demonstration application showed how, by a simple reset of the frame store, the set-up allows both shape and displacement fringes to be obtained from the object sitting in the same position. We use the shape measurement to determine the exact position of the membrane in the coordinate system

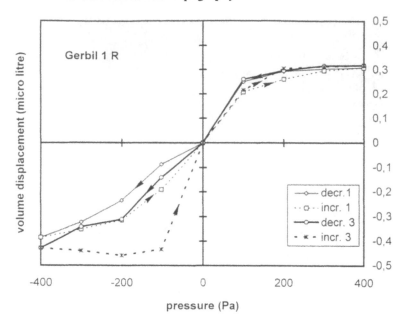

Figure 12.1.7. Volume displacement as a function of applied pressure, calculated from deformation moiré topograms obtained on a 1.6 mm diameter *pars flaccida* of the gerbil eardrum.

By storing a new reference image, deformation can be followed starting from any reference state. In studies of viscoelastic behaviour, for instance, the reference image can be stored immediately after applying the pressure step. Displacement fringes will then appear if the pressure step is followed by viscoelastic creep.

Because the apparatus uses standard video signals, it is also possible to perform the image subtraction and processing on images which have been recorded on video tape. In our experiments we often use this post-processing capability. During a measurement, the direct image of the camera is constantly recorded on a VHS video recorder. If it turns out afterwards that some more information is needed, the video recorder signal is fed to the input of the frame store and any image on the tape can be chosen as the new reference image to study deformations with respect to that object shape.

Using a fringe plane distance of 82 μm , a precision of better than 5 μm is obtained in a measuring field of 3.5 mm by 2.9 mm by 0.41 mm. Measurements on a calibrated spherical surface show an absolute measuring accuracy of better than 3%. At least 25 fringe orders can be clearly visualized, so the maximal depth measuring range is about 2 mm.

The interferometer was especially designed for shape and deformation studies of the tympanic membrane. It allows the acquisition of valuable information on the mechanical behaviour of the eardrum and the middle ear under static pressure, both in healthy and diseased ears. By changing the set-up geometry, the image field can be adapted for the study of objects of varying size. For research in otology, the real-time optoelectronic moiré device offers interesting possibilities. Because interferograms are obtained in real time, it is possible to obtain a large number of measurements in a short time, so one can study the response of the membrane to pressure with a large number of small pressure steps. A short measurement time is necessary, because the dissected middle ear and the eardrum tend to dehydrate and to stiffen quickly. This set-up also allows us to study the time course of a deformation such as viscoelastic-induced creep following a sudden pressure step.

References

[1] Der Hovanesian J and Hung Y Y 1971 Moiré contour-sum contour-difference and vibration analysis of arbitrary objects *Appl. Opt.* **10** 2734–8

[2] Halioua M, Krishnamurthy R S, Liu H and Chiang F P 1983 Projection moiré with moving gratings for automated 3-D topography *Appl. Opt.* **22**

[3] Dirckx J J J and Decraemer W F 1997 Optoelectronic moiré projector for real-time shape and deformation studies of the tympanic membrane *J. Biomed. Opt.* **2** 176–85

[4] von Unge M, Decraemer W F, Bagger-Sjöbäck D and Dirckx J J 1993 Displacement of the gerbil tympanic membrane under static pressure variations measured with a real-time differential moiré interferometer *Hear. Res.* **70** 229–42

[5] Tonndorf J and Khanna S M 1972 Tympanic membrane vibrations in human cadaver ears studied by time-averaged holography *J. Acoust. Soc. Am.* **52** 1221–33

[6] Dirckx J J J, Decraemer W F, von Unge M and Larsson Ch 1997 Measurement and modelling of boundary shape and surface deformation of the Mongolian gerbil pars flaccida *Hear Res.* **111** 153–64

Chapter 12.2

Tensile impact delamination of a cross-ply interface studied by moiré photography

L G Melin[1], J C Thesken[2], S Nilsson[1] and L R Benckert[3]
[1] *The Aeronautical Research Institute of Sweden, Sweden*
[2] *Second Engineering AB, Sweden*
[3] *Luleå University of Technology, Sweden*

12.2.1   Introduction

It is well known that impact and dynamic damage evolution may severely impair the structural performance of carbon fibre/epoxy laminates. Faced with the inability to predict the initiation and evolution of such damage accurately, designers of aircraft composite structures must be satisfied with conservative dimensions which do not fully utilize the material's specific stiffness. The problem is challenged by a host of analytical and numerical approaches which attempt to predict damage [1–4] but progress is hampered. Despite a few notable experimental and phenomenological studies [5, 6], fundamental rate-dependent fracture behaviour is not well understood and material data remain in shortage. A further review of the literature in this area is given by Thesken [7].

Perhaps the principal experiment difficulty associated with the study of dynamic damage evolution in polymer composites is the accurate measurement of boundary and crack-tip field data during the application of a rapid loading to a thin laminate. The present work demonstrates that moiré measurement procedures combined with analysis may be used to overcome these difficulties. As in fracture studies made by Huntley *et al* [8, 9], high-resolution moiré imaging and high-speed photography are an essential feature of the experiments. The resulting fringe patterns are converted to full-field displacement maps to verify specimen performance and to examine details of the complex deformation fields. These measurements have a clear advantage over strain gauges which integrate strain over a length and have a stiffness which might interfere with the deformation of the specimen. These advantages are well exemplified in the applications of moiré interferometry to quasi-static delamination in double cantilever beam specimens by Arakawa *et al* [10].

The example treated here is for delamination along a cross-ply interface due to tensile impact. Using sandwich specimen construction procedures similar to [7] and [11], a double-edge notch specimen with a single 0° ply sandwiched between bulk layers of 90° plies was made. The cross-ply interface appears frequently in laminate structures: cross-ply laminates have been the focus of several impact damage modelling studies [12–15]. The tensile impact loading is made using an instrumented split-Hopkinson bar: analysis of strain signals acquired from the input and output bars provide the load, displacement and displacement rate. Since a uniform tensile loading applied normally to the place of the laminate is difficult to achieve, special care was taken to manufacture accurately aligned specimens.

A quasi-static linear elastic three-dimensional analysis of the specimen was made using a self-adaptive p-version of the finite-element method [16, 17]. The code is equipped with advanced extraction methods to compute edge and vertex stress intensity factors for 3D cracked bodies. These methods are fully generalized to evaluate the energy release rate for bimaterial interface cracks. The purpose of the analysis is to gain some insight into the behaviour of the specimen and to provide a means to compare quasi-static and rapid deformation rate results. Comparisons of the experimental and the computed displacement fields provide an estimate of the specimen alignment, stiffness and linearity up to the point of failure. Once calibrated, the finite-element model gives insight into the through-thickness energy release rate distribution and its mode I, II and III components present along the crack front.

This chapter is based on [18]. More detailed description of the experiments may be found in [19] and [20].

12.2.2 Experimental methods

12.2.2.1 Optical set-up

The moiré method used here can be placed in the range between geometrical moiré and moiré interferometry. It is based on the high-resolution moiré photography method developed by Burch and Forno [21] in 1975. A development of this technique has been made by Huntley and Field [22] for measurements on transparent specimens. The specimen grating is imaged on a static reference grating making the moiré fringes visible in real time while single grating lines do not have to be resolved by the camera. This enables high-speed image convertor cameras, which have a relatively low spatial resolution, to be used.

A further development was made by Huntley *et al* [8] and Whitworth and Huntley [9] allowing opaque specimens to be tested. Essentially the same set-up has been employed in these experiments, see figure 12.2.1. Our set-up was built using Spindler & Hoyer Macrobench ensuring good mechanical stability.

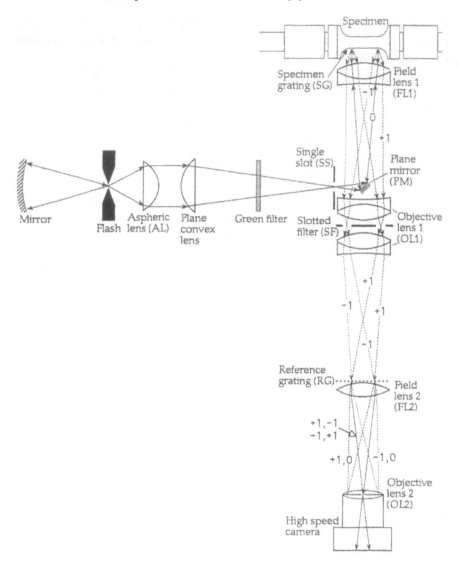

Figure 12.2.1. Optical set-up with a specimen inserted. Light rays are numbered according to their diffraction order at SG and RG.

Light from the flash is directed onto the specimen by a small plane mirror (PM) in front of the objective lens. A beam-splitter was used instead of a mirror in earlier work [8, 9]. The advantage with the mirror is that theoretically no light is lost, compared to the beam-splitter which gives a 75% loss. This is an important improvement in high-speed photography where a low light level often is a problem.

A reflection phase grating with 75 lines/mm, SG in figure 12.2.1, is attached to the specimen. This grating is imaged by the objective lens OL1 onto the reference grating RG, a transmission phase grating with approximately the same pitch as SG, giving rise to moiré fringes. Carrier fringes parallel to the grating lines, the so-called carrier fringes of extension, are introduced by setting the magnification slightly larger than 1:1. The carrier fringes are needed for the quantitative fringe analysis. In the centre of OL1 a slotted filter (SF) is inserted. This filter only accepts the +1 and −1 diffraction orders and blocks the zeroth order, giving a doubled effective grating frequency of 150 lines/mm. Together with filtering at the camera aperture at OL2, this frequency multiplication doubles the sensitivity for in-plane displacements [9].

The light is made quasi-monochromatic by a green filter. It is not necessary to use a filter to achieve moiré fringes: they also appear in white light. The problem with white light is to filter the desired diffraction orders at OL1 and OL2. The diffraction orders are then spread out over wider angles. To ensure correct filtering, it is important that only the +1 and −1 orders are accepted at OL1 and similarly at OL2. The single slot (SS) determines the width of the diffraction orders for each wavelength. SS is imaged at magnification 1:1 onto the SF by the field lens FL1 in front of the specimen. Wide slots are desired in order to increase the amount of light available but the width must be limited to make sure that the surrounding diffraction orders are completely blocked. The slots should also allow for a specimen tilt. Another function of FL1 is to make the system telecentric, i.e. to make the magnification of the specimen insensitive to out-of-plane movements. Three camera lenses (Schneider Symmar 1:5.5/300) were used as FL1 and as the two lenses in OL1. Initially achromats were tested but these were replaced with the camera lenses due to problems with distortion.

The grating on the specimen is replicated from a master grating made with interferometric techniques [23] on which an aluminium film has been evaporated. An amount of epoxy resin (PC10C, Measurements Group Inc.) was applied between the specimen and the master grating cured under pressure. When the specimen is separated from the master grating after curing, the aluminium is transferred. The specimen grating must be thin in order not to reinforce the specimen or cause shear lag effects. By applying a load of 30–40 kPa as the epoxy cures during the moulding, grating thicknesses of about 10 μm have been achieved.

12.2.2.2 Tensile split-Hopkinson bar

In a Hopkinson bar apparatus, mechanical waves are transmitted through thin bars to load specimens at high strain rates. The theory of the tensile split-Hopkinson bar apparatus has been presented by Nicolas [24] and by Li *et al* [25]. Figure 12.2.2 shows the design of the Hopkinson bar used in these experiments with the specimen screwed into the steel bars and tightened using locking nuts. The projectile is accelerated by compressed air and forms a tensile wave when

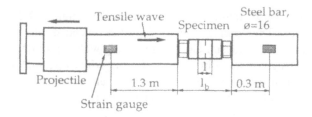

Figure 12.2.2. Tensile split-Hopkinson bar with double-edge notch specimen installed.

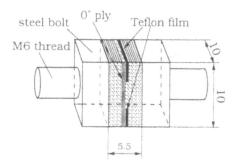

Figure 12.2.3. Double-edge notch specimen.

hitting the end of the input bar. The strain gauges, which were mounted in pairs on opposite sides of the bars, measured the strain of the incident, the reflected and the transmitted waves. Based on one-dimensional wave propagation theory, the force and the elongation of the specimen can be obtained from the strain gauge signals.

12.2.2.3   Specimen design and manufacturing

Figure 12.2.3 shows a schematic diagram of the double-edge notch specimen geometry. The specimen is loaded through square steel bolt end tabs bonded to the composite. The steel bolts are threaded to transfer load from the tensile split-Hopkinson bar. The laminate used here was prepared from Ciba-Geigy HTA/6376C prepeg containing 65% volume fraction of carbon fibres embedded in a toughened epoxy matrix. The lay-up sequence for the laminate was $(90^\circ_{20}/0^\circ/90^\circ_{20})$. The double-edge notches are created by inserting two strips of 25 μm Teflon film parallel with the 90° and spaced 4.0 mm apart. After sectioning, the specimen was bonded to the steel bolts using the high-strength adhesive FM 300K *Cyanamid). To alleviate sliding during bonding the bolt heads were machined with a 0.2 mm pin in the centre which holds the centre alignment of the specimen.

The available engineering in-plane elastic ply properties for this material are $E_{11} = 145.6$ GPa, $E_{22} = 10.5$ GPa, $G_{12} = 5.25$ GPa, $v_{12} = 0.30$. From double cantilever beam tests of $(0°_{12}/90°/0°_{12})$ sandwich specimens, Thesken *et al* [11] found that the global mode I initiation fracture toughness of $90°/0°$ interfaces were about the same (200–300 J m^{-2}) or lower than initiation values for conventional unidirectional DCB specimens with $0°/0°$ interfaces. However, after the initiation of growth in the $(0°_{12}/90°/0°_{12})$ sandwich specimens, an extremely large bridging zone developed leading to toughness values exceeding 1000 J m^{-2}. This was not expected in these experiments because the reverse order of the stacking sequence cannot evolve into intralaminar growth of the central cross-ply. The quasi-static failure load for this type of specimen has been measured to be about 859 N, corresponding to a fracture toughness of about 73 J m^{-2}.

12.2.2.4 Procedure

Recordings of images were made by an Hadland Imacon 790, giving a sequence of eight frames on a Polaroid film. A framing rate of $200\,000$ frames/s was used. The flash and the camera are triggered from the arrival of the incident tensile wave strain gauge pair in the input bar. To get reference points for the analysis of the resultant frames, scratches were made in the grating at the interfaces between the specimen and the steel bolts.

12.2.2.5 Fringe analysis

Two image sequences have been captured for the experiment: one reference sequence of the pre-set carrier fringe pattern before the loading pulse arrives; and one sequence during the dynamic event. A CCD camera has been used to digitize the frames from the photographs for the quantitative analysis. The fringe analysis [26] has been made with the Fourier transform fringe method. The resulting phase maps are unwrapped using an algorithm developed by Huntley [27]. The phase is then scaled according to the sensitivity, in this case a phase difference of 2π is equal to a displacement of 6.67 μm $= 1/150$ mm. The reference frames are analysed in the same way as the dynamic frames and the displacement fields are obtained by subtracting the results from corresponding frames eliminating the influence of the carrier fringes. This procedure also minimizes the influence of camera distortion.

12.2.3 Results

A sequence of moiré fringe patterns from an experiment is shown in figure 12.2.4(a). The tensile wave arrives from the left-hand side. The rise time of the wave is relatively long, about 30 μs, compared to the propagation time across the specimen, about 2 μs, so the difference in strain due to the propagating wave is expected to be small. In the second frame 5 μs later the fringes starts to bend

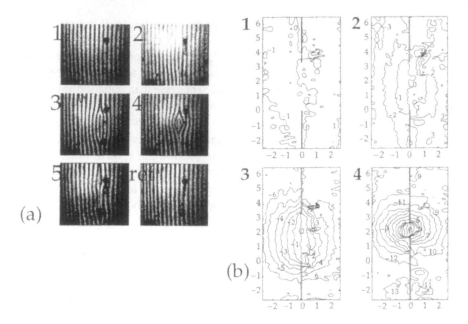

Figure 12.2.4. Moiré results: (*a*) the first five moiré fringe patterns from the dynamic experiment and a reference pattern captured before loading; (*b*) displacement fields from frames 1–4. The displacement component normal to the crack plane is plotted as contours with 1 μm interval. Length scales are in mm.

around the two vertical cracks which means that delamination along the Teflon films has started. In frame 4 the cracks have started to propagate and in frame 5 the whole specimen is fractured. A few defects can be seen disturbing the pattern, in those areas where the grating has been damaged during the grating replication process.

The resultant in-plane displacement component over the composite specimen area of frames 1–4 is presented in figure 12.2.4(*b*) as contours with an interval of 1 μm. Only the relative displacement is obtained while the zero displacement contour is placed arbitrarily. In the fourth frame the upper crack has propagated 0.7 mm and the lower 1.7 mm: the corresponding average crack-tip velocities between these two frames are, respectively, 140 and 340 m s^{-1}. The actual velocities might be higher since it is not known exactly when the propagation starts.

The time history of force F and elongation of the specimen u, calculated from the strain gauge signals, is shown in figure 12.2.5(*a*). Also plotted in figure 12.2.5(*a*) are the optical measurements of u taken from the displacement maps in figure 12.2.4(*b*). These values appear to be in good agreement with the measurements determined using the strain gauge data. In figure 12.2.5(*b*) force is plotted against both displacement measurements. These force displacement

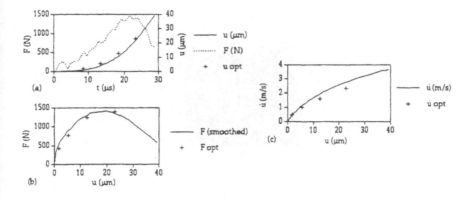

Figure 12.2.5. Time histories measured with strain gauges (lines) and the corresponding moiré measurements (crosses). (*a*) Force (dotted line) and displacement (full line) histories. The moiré values of specimen elongation are taken from figure 12.2.4(*b*). (*b*) Force *versus* displacement. The force signal has been smoothed compared to (*a*).

curves are nonlinear with a high initial tangent stiffness that decreases with increasing load.

Since the load pulse duration is much longer than the pulse travel time across the specimen, the specimen behaviour can be considered quasi-static. In such a case, an elastic analysis with displacement boundary conditions would probably characterize the specimen's displacement fields quite well. Trends in calculated stresses and crack energy release rate should, at least, be qualitatively correct. Therefore, a three-dimensional elastostatic analysis has been made using a self-adaptive version of the finite-element method formulated in the code STRIPE [16, 17]. Details of the calculation may be found in [18].

Figure 12.2.6(*a*) shows an example of the finite-element mesh. Figure 12.2.6(*b*) shows a displacement map from the finite-element results in comparison to experimental results for image 3. The finite-element solution has been scaled by setting the relative displacement from the left to the right of the area in figure 12.2.6(*b*) equal to the displacement experimentally measured over the same distance. The experimentally determined stiffness for this point in time (18 μs) agrees well with the stiffness of the finite-element model 94.5 kN mm^{-1}. Figure 12.2.7 compares crack opening displacement profiles for the second and third frames to the scaled finite-element solutions. The profile of frame 2 is not opened as wide as the finite-element solution and no opening at all can be seen in frame 1 but the results for frame 3 compare well to the finite-element model. Presently it is unknown whether this is a viscoelastic effect or an effect of tractions across the Teflon film. Indeed, during recent quasi-static experiments followed at high magnification, it was observed that Teflon films do not open until a certain load level is reached [28]. Thus the nonlinearity in the experimental load displacement data may be due to the behaviour of the Teflon film as an elastic inclusion.

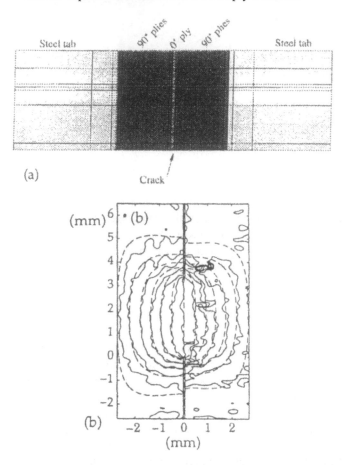

Figure 12.2.6. Finite-element calculations compared to experiment: (*a*) finite-element mesh on a half-plane; (*b*) finite-element displacement contours (broken lines) and corresponding contours from frame 3 (full lines) with 1 μm interval.

The finite-element analysis was also used to evaluate the energy release rate and find out the influence of the shearing modes. Figure 12.2.8 gives the through-thickness distribution for the normalized total energy release rate, G/G_0, where G_0 is the total energy release rate at the centre of the specimen ($x_3/w = 0$). Defining G_1 as the portion of the total energy release rate due to the work of normal tractions at the crack-tip, it is possible to show the small influence of shearing tractions in this interface by comparing the ratios G_1/G_0 and G/G_0 in figure 12.2.8. Clearly for this sandwich specimen construction, the work of fracture is dominated by normal tractions everywhere except near the vertex ($x_3/w = 1$) where boundary layer and bimaterial effects bring other deformation

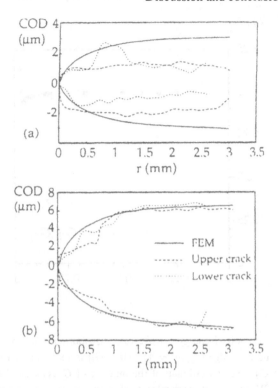

Figure 12.2.7. Profiles of the crack face opening displacement (COD) *versus* distance from the crack-tip (*r*) for (*a*) frame 2 and (*b*) frame 3.

modes into play. At the failure load of 1410 N, G_0 is equal to 196 J m^{-2}: this value is about 2.7 times greater than the quasi-static value of 73 J m^{-2}.

12.2.4 Discussion and conclusions

Experimental procedures have been developed to investigate rate-dependent effects on fracture mechanisms in carbon fibre/epoxy laminates. The optically measured displacement has been compared with data from conventional strain gauge measurement on the Hopkinson bar with acceptable agreement. A quasi-static linear elastic 3D finite-element analysis was made to evaluate the specimen and the experimental results. Computed displacement fields and specimen stiffness compared well to experimental measurements for a point in time just prior to failure. This is not the case for measurements made earlier in the loading history. A comparison of the optically and numerically determined crack opening displacements gives some indication that the Teflon film flaws may play a role in this observation. Early in the loading history, the crack does not open to the extent linear elasticity predicts. Tractions across these films are expected to account for

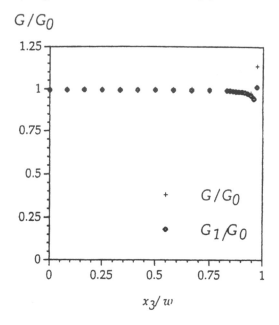

Figure 12.2.8. Normalized energy release rate distribution from the specimen centre $(x_3/w = 0)$ to the free edge $(x_3/w = 1)$. Here G_0 is the total energy release rate at the centre while G_1 gives an estimate of the amount of G associated with the work or normal tractions. When it is not identical to G shearing modes are present.

this behaviour and raises questions about the further use of films in impact fracture testing of polymer composites.

   In conclusion, the experimental procedures and analysis presented here were found to have good potential for extensive use in the determination of rate-dependent material properties of carbon fibre/epoxy composites. The combination of measurement techniques and analysis provided several checks on the results which raises confidence in the approach. However, it is recognized that the Hopkinson bar system should be tuned to provide constant strain rate pulses and work is underway to make this improvement. The particular example of a cross-ply interface delamination clearly shows that 0° fracture data cannot be applied indiscriminately to other interfaces.

Acknowledgments

The tensile Hopkinson bar equipment and the optical equipment was developed at Luleå University of Technology (LTU) with initial funds from the Swedish Defence Material Administration (FMV); the work by G Melin is presently supported by The Swedish National Board for Technical Development (NUTEK).

The authors thank Dr J M Huntley (Loughborough University) for giving us permission to use his fringe analysis program and for valuable discussions about the optics and Dr K-G Sundin (LTU) for help with the construction of the Hopkinson bar apparatus.

Specimen development and finite-element analysis were made at the FFA with finds from the FMV and the Swedish National Aeronautical Research Program (NFFP, BG 2.36). Mr F Brandt and Mr B Thörnquist assisted with specimen manufacturing and Mr L Nystedt helped with quasi-static testing. A special thanks is due to Dr B Andersson for assistance with the self-adaptive p-version of the finite-element method. J C Thesken would like to acknowledge helpful discussions with Dr A E Giannakopoulos and the valuable comments of an unknown reviewer.

References

[1] Choi H Y and Chang F K 1992 A model for predicting damage in graphite/epoxy laminated composites resulting from low-velocity point impact *J. Composite Mater.* **26** 2134–69

[2] Besant T, Davies G and Hitchings D 2001 Finite element modelling of low velocity impact damage in composite sandwich panels *Composite A* **32** 1189–96

[3] Olsson R 1992 Impact response of orthotropic composite plates predicted from a one-parameter differential equation *AIAA J.* **30** 1587–96

[4] Liu S 1994 Quasi-impact damage initiation and growth of thick-section and toughened composite materials *Int. J. Solids Structures* **31** 3079–8

[5] Daniel I M, Hamilton W G and Labedz R H 1982 Strain rate characterization of unidirectional graphite/epoxy composite *Composite Materials: Testing and Design (Sixth Conf.) ASTM STP* ed I M Daniel (American Society for Testing and Materials) (Conshohocken, PA: ASTM) pp 393–413

[6] Friedrich K, Walter R, Carlsson L A, Smiley A J and Gillespie J W Jr 1989 Mechanisms for rate effects on interlaminar fracture toughness of carbon/epoxy and carbon/PEEK composites *J. Mater. Sci.* **24** 3387–8

[7] Thesken J C 1994 *Theoretical and Experimental Investigations of Dynamic Delamination in Carbon Fibre/Epoxy Composites* (The Aeronautical Research Institute of Sweden FFA TN)

[8] Huntley J M, Whitworth M B, Field J E, Benckert L R, Sjödahl M, Thesken J C and Henriksson A 1993 High resolution moiré photography: application to impact and dynamic fracture of polymers and composites *ESIS Symposium on Impact and Dynamic Fracture of Polymers and Composites (Sardinia)*

[9] Whitworth M B and Huntley J M 1994 Dynamic stress analysis by high-resolution moiré photography *Opt. Eng.* **33** 924–31

[10] Arakawa K, Ishiguma M and Takahashi K 1994 Study of mode I interlaminar fracture in CFRP laminates by moiré interferometry *Int. J. Fracture* **66** 205–12

[11] Thesken J C, Brandt F and Nilsson S 1994 Investigations of delamination growth rates and criticality along heterogeneous interfaces *19th ICAS Congress/AIAA Aircraft Systems Conf. (Anaheim, CA)* (Reston, VA: AIAA)

[12] Razi H and Kobayashi A S 1993 Delamination in cross-ply laminated composite subjected to low-velocity impact *AIAA J.* **31** 1498–502

[13] Jih C J and Sun C T 1993 Prediction of delamination in composite laminates subjected to low velocity impact *J. Composite Mater.* **27** 684–701

[14] Lammerant L and Verpoest I 1994 The interaction between matrix cracks and delaminations during quasi-static impact of composites *Composites Sci. Technol.* **51** 505–16

[15] Doxsee L E, Rubbrecht P, Li L, Verpoest I and Scholle M 1993 Delamination growth in composite plates subjected to transverse loads *J. Composite Mater.* **27** 764–81

[16] Andersson B, Babuska I, von Petersdorff T and Falk U 1992 *Reliable Stress and Fracture Mechanics Analysis of Complex Aircraft Components using a h-p Version of the Finite Element Method* (The Aeronautical Research Institute of Sweden FFA TN) pp 1992–17

[17] Andersson B, Babuska I and Falk U 1990 Accurate and reliable determination of edge and vertex stress intensity factors in three-dimensional elastomechanies *ICAS-90-4.9.2, Proc. 17th Congress of the International Council of Aeronautical Sciences* (Stockholm: ICAS) pp 1730–46

[18] Melin L G, Thesken J C, Nilsson S and Benckert L R 1995 Tensile impact delamination of a cross-ply interface studied by moiré photography *Fatigue Fracture Eng. Mater. Structures* **18** 1101–14

[19] Melin L G 1995 Moiré photography and its application in high speed photography to measure deformation fields in fibre composites *Luleå University of Technology Research Report* p 39

[20] Melin L G 1998 High speed moiré photography for studying dynamic properties in carbon fibre composites *Opt. Eng.* **37** 642–9

[21] Burch M and Forno C 1975 A high sensitivity moiré grid technique for studying deformation in large objects *Opt. Eng.* **14** 178–85

[22] Huntley J M and Field J E 1989 High resolution moiré photography: application to dynamic stress analysis *Opt. Eng.* **28** 926–33

[23] Hutley M C 1982 *Diffraction Gratings* (New York: Academic)

[24] nicholas t 1985 split-hopkinson Bar in Tension *Metals Handbook* vol 8, 9th edn (Metals Park, OH: American Society of Metals) pp 212–14

[25] Li M, Wang R and Han M-B 1993 A Kolsky bar: tension tension–tension *Exp. Mech.* **33** 7–14

[26] Kujawinska M 1993 Spatial phase measurement methods *Interferogram Analysis, Digital Fringe Pattern Measurement Techniques* ed D W Robinson and G T Reid (Bristol: Institute of Physics Publishing)

[27] Huntley J M 1989 Noise-immune phase unwrapping algorithm *Appl. Opt.* **28** 3268–70

[28] Melin L G, Goldrein H T, Huntley J M, Nilsson S and Palmer S J P 1998 A study of mode I loaded delamination cracks by high magnification moiré interferometry *Composites Sci. Technol.* **58** 515–25

Chapter 12.3

Real-time moiré image processing

M Bruynooghe
University Louis Pasteur of Strasbourg, France

12.3.1 Introduction

Moiré methods are optical measurement methods [12, 25, 26, 31], that are based on the effect of the superposition of grating lines and have been widely used to contour surfaces and applied to form measurement.

The superposition moiré is a well-known phenomenon which occurs when periodic structures such as line gratings are superposed. The moiré effect consists of visible fringes which are clearly observed at the superposition, although they do not appear in any of the original line gratings [23].

In industrial quality control applications by standard moiré methods, the following computations—pre-processing by image filtering, fringe skeletonizing and fringe numbering—have to be performed for each test object before comparison between the numerically reconstructed test object shape and its computed aided design (CAD) model. The use of these algorithms for real-time quality control is not possible without specific implementation on costly VLSI chips.

We have developed original methods to overcome these difficulties and to get a real-time defect detection system. Image-processing algorithms that can be optically implementable as well as robust statistical methods have been used to achieve this objective.

Up to now, optical correlation has not been used to perform a global analysis of moiré fringe patterns for industrial quality control. In order to get a real-time quality control system by moiré image analysis, we propose hereafter detect defects directly from moiré image data by using an optoelectronic correlator system that is associated with a digital signal processor for real-time image smoothing and subsampling.

We present hereafter three original algorithmic approaches that are respectively based (1) on skeletonizing moiré fringes, (2) fringe shape conformity by a hit/miss transform [2] and (3) defect detection by multi-dimensional statistical analysis.

The two first approaches have been applied successfully to inverse moiré images of test objects that are supposed to be exactly positioned. The third algorithmic approach is an industrial application of the phase-shifting moiré technique and of optical correlation that are intended for the automatic 3D inspection of manufactured objects that are approximately positioned. The practical objective of the third approach was to design and to test a real-time optoelectronic defect detection system that would be applicable even when the object under test is supposed to be only approximately positioned, due to possible lateral or transverse deviations, as would be the case in the context of industrial applications.

These three algorithmic approaches are applicable to real-time defect detection by a global analysis of moiré image data. With these methods, it is not necessary to reconstruct 3D shapes to detect defective objects. The choice of a particular algorithmic approach depends on the respective costs of optical and digital devices and on the required performance of the defect detection system.

12.3.2   Experimental set-up for real-time defect detection

The optoelectronic correlator system for real-time defect recognition and quality control by moiré analysis consists of (i) a $4f$ Van der Lugt optical correlator, with two Fourier lenses L1 and L2, (ii) a first 480×640 liquid-crystal television screen (LCTV) that is used to transfer the moiré image from the microcomputer to the input plane of the optical correlator, (iii) a second LCTV that is used as a real-time quantized phase-only matched filter, (iv) a video CCD camera that acquires the result of each optical correlation, (v) a digital signal processor that can be associated to a vector-coprocessor and (vi) a standard micro-computer that controls optical and digital devices.

Coherent optical recognizers using correlation criterion for image processing are very attractive due to their very high speed in processing image data. The basic principles of coherent optical correlator construction are firmly established, bottlenecks due to optical/digital interfaces are well known and specifications of components are well defined. The development and the evaluation of coherent optical correlators of different types have been the subject of recent research, including the classical Van der Lugt correlator [2, 10, 11, 29], the joint transform correlator [16, 35], and hybrid devices [20].

In our experiments, we have used the Van der Lugt optical correlator of the National Optics Institute (NOI) of Québec [2]. This optical correlator is integrated into a hybrid information processing system that allows optoelectronic thresholding [3]. The illumination source is a linearly polarized laser diode beam ($\lambda = 688$ nm), spatially filtered through a 15 μm pinhole and collimated by a lens. The test phase moiré image $\phi(x, y)$ is injected into the optical correlator via an LCTV that can be used for real-time operation at video rate. The incident collimated beam is spatially modulated by this first LCTV and forwarded to the first Fourier lens L1, that Fourier transforms the input image $I_p(x, y)$. Then, the transformed data are multiplied by the quantized phase-only representation of a conform phase moiré image by use of a second LCTV, that is located in the so-called filter plane of the optical correlator, The light transmitted through this second LCTV is Fourier transformed by the second Fourier lens L2. The ouput is the two-dimensional correlation image corr(x, y), that is given by

$$\text{corr}(x, y) = \int_{\infty}^{\infty} \int_{\infty}^{\infty} I_p(\zeta, \eta) h^*(\zeta - x, \eta - y) \, d\zeta \, d\eta \qquad (12.3.1)$$

where $h(x, y)$ represents the implulse response $h(x, y)$ of the phase filter that is used in the filter plane. The output correlation image corr(x, y) is forwarded to a CCD camera that performs image acquisition at video rate. The NOI-packaged portable correlator [2] is based on spatial light modulators that can handle images of 480×640 pixel, with VGA standard resolution. The main characteristics are the following: NTSC (video compatible) input and output signals, 256 amplitude levels in the input plane and 256 phase levels in the filter plane. The frame rate is 30 frames/s and the contrast ratio is 1000:1. The size of this packaged portable correlator is 30 cm $\times$ 20 cm $\times$ 10 cm.

12.3.3 Moiré image data

12.3.3.1 Inverse moiré image data

In shape measurement, there are two classical methods for implementing the moiré technique: the shadow moiré method and the projection moiré method. In the shadow moiré method, fringes are formed between a grating and its own shadow. In the projection moiré method, a grating is projected onto the test object and moiré fringes are formed by imaging the projected fringes on another grating. In industrial quality control by these classical moiré methods, preprocessing by image filtering, fringe skeletonizing and fringe numbering have to be performed for each test object, before comparison between the numerically reconstructed test object shape and its CAD model.

In order to reduce the computing time required by the preceding computations, the *inverse moiré method* [14] uses a pre-computed specific grating, instead of a grating made of parallel straight lines. This specific grating is formed of curved lines such that the moiré pattern is composed of parallel straight fringes if the test object shape conforms to its CAD model. Defects in the test object shape are then characterized by a deformation and a curvature of these parallel fringes.

The inverse moiré technique is the first method that we have used to get moiré images, in the case of objects that are supposed to be exactly positioned.

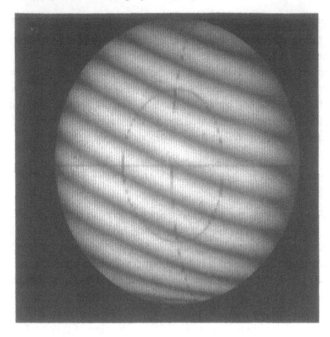

Figure 12.3.1.   Inverse moiré image I_1 related to a conform cylinder, without any deformation.

Five inverse moiré images have been analysed in our experiments. The first inverse moiré image (figure 12.3.1) corresponds to a cylinder without any deformation and the other ones (2, 3, 4, 5) are characterized by an increasing deformation of the observed cylinder. Figures 12.3.2 and 12.3.3 respectively correspond to a cylinder with a medium and a very strong deformation.

12.3.3.2   Moiré data acquisition by the phase shifting technique

The phase-shifting technique has also been used to acquire moiré image data in the case of approximately positioned objects.

The phase-shifting technique is based on the use of a line grating to generate projected type moiré fringes whose relative phases are directly related to the surface height of the object under test. Moiré interferometry has made great advances by adopting the effective phase-shifting technique, in which the relative phases of moiré fringes can be precisely computed even though there exists a significant level of errors in phase shifting due to miscalibration and external vibration [15]. The phase-shifting technique allows much enhanced measuring resolution for practical applications, since the ambiguity problem that characterizes the fringe-following technique is inherently eliminated.

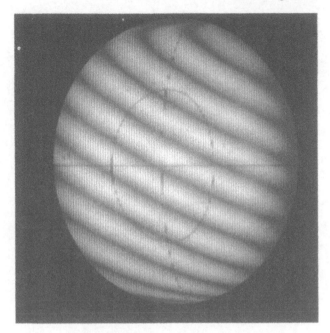

Figure 12.3.2. Inverse moiré image I_3 related to a medium deformation of the observed cylinder.

The moiré data acquisition system is specially configured to implement the phase-shifting technique. In order to induce phase shifts, the grating is allowed to move in the direction normal to the grating itself. Therefore, the intensity of a moiré fringe pattern varies in a sinusoidal form and can be approximated by the following equation if we neglect high-frequency patterns:

$$I_k(r, c) = a(r, c) + b(r, c) \cos 2\pi n(\phi(r, c) + \delta_k). \tag{12.3.2}$$

In this equation, $I_k(r, c)$ denotes the intensity level of moiré fringes that are captured by the CCD camera and that corresponds to the kth phase shift. The array I_k is a digital image whose grey level is $I_k(r, c)$ for the pixel that is located at the intersection of line r and column c. Arrays $a(r, c)$ and $b(r, c)$ respectively represent the mean intensity and the modulation amplitude, $\phi(r, c)$ is the relative phase and δ_k is the phase shift that has been introduced.

In the process of fringe pattern analysis, the intensity $I(r, c)$ is the only measurable quantity. A single fringe pattern does not contain enough information to allow phase extraction without any ambiguity. This is because the sign of $\phi(r, c)$ cannot be determined from a single measurement of $I(r, c)$ without supplementary *a priori* knowledge.

In order to solve the previous ambiguities, the phase-shifting technique steps the phase a known amount between intensity measurements to get N intensity

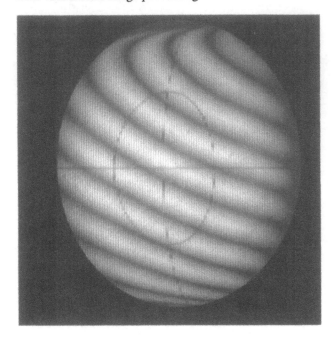

Figure 12.3.3. Inverse moiré image I_5 related to a very strong deformation of the observed cylinder.

values $I_k(r, c), k = 1, 2, \ldots, N$ that are associated with phase shift values $\delta_k(r, c), k = 1, 2, \ldots, N$. The most common algorithms for determining the phase, use three phase shifts and three intensity measurements to determine the three unknowns of the fringe pattern equation and are:

- the three-phase step technique with a $2\pi/3$ phase shift per exposure

$$\phi = \arctan \sqrt{3} \frac{I_1 - I_3}{2I_2 - I_1 - I_3} \tag{12.3.3}$$

- the three-phase step technique with a $\pi/2$ phase shift per exposure

$$\phi = \arctan \frac{I_3 - I_2}{I_1 - I_2}. \tag{12.3.4}$$

With the well-known Carré algorithm [9], there is no need for the phase-shift calibration and the phase-shift modulo π can be computed as follows:

$$\phi = \arctan \frac{\sqrt{((I_1 - I_4) + (I_2 - I_3))(3(I_2 - I_3) + (I_1 - I_4))}}{(I_2 + I_3) - (I_1 + I_4)}. \tag{12.3.5}$$

With this algorithm, it is possible to compute the phase $\phi(r, c)$ at each point (r, c) without needing to worry about the errors that could result from differences in phase calibrations across the sampled area [24].

With a larger number N of phase shifts ($N \geq 4$), phase estimation by least-square fitting allows one to get a higher accuracy, without any restriction on the number and spacing of phase shifts. Furthermore, by using least-square fitting, phase shifts can also be considered as parameters whose values can be estimated from the intensity information without prior calibration. Then, fringe phase estimation and phase shift estimation can be performed iteratively until convergence, as suggested by Kim *et al* [15]. This iterative technique has been found very efficient even though the phase shift values are not known precisely. With this iterative least-square fitting, it is not necessary to move the grating very precisely.

12.3.4   Fringe skeletonizing

In this section, we present examples showing that standard fringe extraction by automatic thresholding is not that easy. To overcome this difficulty, we propose a four-stage algorithmical approach that allows fringe detection in inverse moiré images with high sensitivity and specificity.

First, we used the well-known image-processing technique called unsharp masking to enhance the moiré image and to emphasize the low-contrast fringes.

The second step is to extract bright fringes by image segmentation and constrained contour modelling. An original textural image segmentation algorithm has been used to perform this task, together with contour modelling subjected to geometrical constraints. After deletion of these bright fringes inside the zone of interest of the moiré image, we get the thick skeleton comprised of the adjacent background and dark fringes.

The third step is to skeletonize this thick skeleton using morphological thinning which is applicable to well-composed sets. This algorithm ensures that each fringe skeleton will be one pixel thick in contrast to standard thinning techniques.

The fourth step is to apply a graph technique to isolate the individual dark fringes. When all these four steps have been followed, one is left with a binary image showing the dark fringe pattern skeleton.

Then, for industrial quality control applications, the detection of defects in test object shapes can be performed by a neural network or by the nearest-neighbour method which can be used for supervised classification of fringe curvature features. Hence, the detection of defects can be performed directly from the fringe pattern data and quality control no longer requires the reconstruction of the 3D shape of the test object.

Experimental results have shown the robustness of this algorithmical approach for the analysis of noisy inverse moiré images.

Figure 12.3.4. Filtering of original inverse moiré image I_5.

12.3.4.1 Moiré image enhancement and subsampling

We have used the well-known image-processing technique called unsharp masking, in which one smooths the image heavily and then substracts the smoothed image from the original moiré image. The result is that low-contrast fringes are emphasized in the enhanced moiré image. The wavelet decomposition and a multi-resolution approach [21], which are optically implementable [30], have been used to perform this task. The background has been heavily smoothed by zeroing details generated by the wavelet decomposition and by interpolating the approximation image, obtained at low resolution, using a quadrature mirror filter. We have verified that the proposed unsharp masking approach is faster than the classical one based on Gaussian filtering.

An example of an original moiré image, which corresponds to a cylinder with a deformation of 4 mm is given in figure 12.3.3. The filtered original moiré image is presented in figure 12.3.4.

The objectives of this moiré image pre-processing are twofold. First, by reducing 512×512 initial moiré images to smoothed and subsampled 128×128 smaller images, it will be possible to use spatial multiplexing for optical correlation experiments. This will reduce the overall processing time, especially when the test objets are approximately positioned and when the number of possible locations is greater than 16. Second, iterative smoothing by a QMF filter

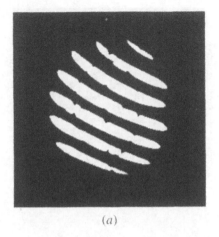

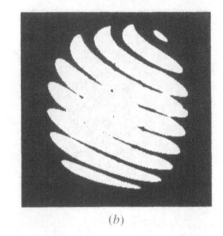

(a)   (b)

Figure 12.3.5. Examples of manual thresholdings of the filtered inverse moiré image I_5. These two examples illustrate the difficulty of the choice of a pertinent threshold to extract bright fringes with high sensitivity and specificity.

and subsampling eliminates the noise and small details that have been introduced by the moiré image acquisition system. This will improve the robustness of the defect detection system as it will not be sensible to small variations in moiré image data.

12.3.4.2 Extraction of bright fringes by image segmentation and constrained contour modelling

The second step is to segment the enhanced moiré image by thresholding. Examples of thresholded moiré images are given in figures 12.3.5–12.3.7. These examples show that manual and automatic thresholding by the moment-preserving method [34] are not that easy. If the threshold has too high a value, then some bright fringes are not detected. In contrast, a low thresholding value falsely aggregates adjacent bright fringes. This results in a lack of sensitivity and specificity of fringe detection by standard thresholding. To overcome this difficulty, we propose, hereafter, an approach for bright fringe detection that is based on image segmentation by pixel clustering and on constrained contour modelling.

12.3.4.2.1 Image segmentation by pixel clustering

We have segmented the enhanced moiré image by recursive thresholding using a factorial analysis whose first factor automatically discriminates pixels of interest from image background. This fast image segmentation technique is a simplified variant of a more general image segmentation approach by multi-dimensional pixel clustering, that uses the *reducible neighbourhood clustering algorithm* [4].

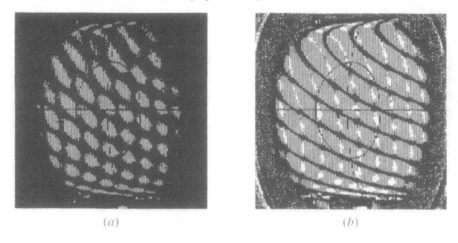

(a) (b)

Figure 12.3.6. Examples of manual thresholdings of detail images that have been obtained by a wavelet decomposition of inverse moiré image I_5. These two examples show that it is necessary to smooth the original moiré image for the extraction of bright fringes.

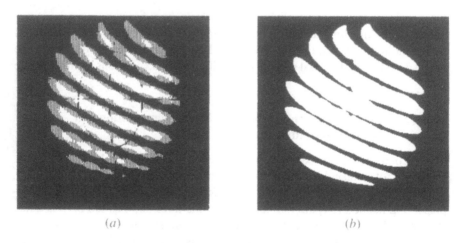

(a) (b)

Figure 12.3.7. (a) Automatic thresholding of the original inverse moiré image I_5, by the moment preserving method. (b) Automatic thresholding of the filtered inverse moiré image I_5, by the moment preserving method. These two examples show that bright fringes are not correctly extracted by the moment preserving method.

The theoretical justification for this fast hierarchical algorithm is the so-called reducibility property that lays down the condition under which the exact clustering hierarchy may be constructed locally by carrying out aggregations in restricted regions of the representation space. Consider the aggregation of clusters or objects, a and b, into the cluster $a \cup b$. Let us also consider some other cluster or object t. If $\delta(a, b)$ represents the dissimilarity between clusters a and b, the

Bruynooghe reducibility property [4] is:

$$\delta(a, b) \leq \inf\{\delta(a, t)\delta(b, t)\} \implies \inf\{\delta(a, t)\delta(b, t)\} \leq \delta(a \cup b, t).$$

It can be shown that the thresholding results of the proposed clustering approach are invariant with respect to monotonic transformations of the scale of image intensities. This is a guarantee for the robustness of the proposed method of segmentation by recursive thresholding.

12.3.4.2.2 *Constrained fringe contour modelling*

Deformable contour models, often referred to as *snakes* or *active contour models*, were proposed first by Kass *et al* [17, 32]. These contour models turn boundary localization into an optimization problem and seek to minimize a global energy $E = \alpha E_{int} + \beta E_{ext}$ where the internal energy E_{int} imposes continuity and contour smoothness and the external energy E_{ext} attracts the deformable contour to image features. The weighting coefficients α and β are regularization parametres whose estimation largely vary from one dataset to another one. Furthermore, if a deformable model is not initialized near the contour of interest, it can converge to a point or a line and fail to localize the true contour of the object of interest.

To overcome these drawbacks, we have used our image segmentation algorithm by pixel clustering to extract the thick skeleton of bright fringes in inverse moiré images. Our approach is based on the following principles:

- The initial contour models are automatically generated by our unsupervised segmentation algorithm. The first originality of the proposed approach is that this initial contour is a close approximation of the contour of the object to be extracted.
- The external energy function to be minimized is constrained by nonlinear geometrical constraints in order to take tolerances about maximal contour deformation and maximal model dissimilarity into account. The advantage of such an approach is to avoid the introduction of multiple Lagrange parameters or weighting coefficients that are difficult to choose *a priori*.

After extracting the thick skeleton of bright fringes by image segmentation and constrained contour modelling (figure 12.3.8), we delete small and compact connected components that are not interpretable as bright fringes. The obtained results clearly show that this fringe detection approach has a better sensitivity and a better specificity than standard thresholding by the moment-preserving method. A morphological filter is used to smooth the contours of the detected bright fringes. Then, we determine the thick skeleton of dark fringes and adjacent image background of the inverse moiré image by deleting the detected bright fringes inside the zone of interest of the moiré image. The remaining domain of interest (figure 12.3.9) is then thinned to get a thin skeleton of dark fringes (figure 12.3.11), as explained herafter.

Figure 12.3.8. Thick skeleton of bright fringes of inverse moiré image I_5. The automatic thresholding of the filtered inverse moiré image I_5 has been performed by our image segmentation algorithm.

12.3.4.3 The skeletonizing of dark fringes and background

In classical moiré image analysis algorithms [28], fringe skeletonizing is generally based on the previous extraction of thick fringe skeletons that have been obtained by thresholding the Laplacian or gradient of the moiré image. Then, morphological thinning is used to process this binary thick skeleton pattern. This standard approach presents a major drawback: it is necessary to select a threshold to generate the binary thick skeleton pattern and the automatic selection of such a threshold is not very robust. Furthermore, spikes that have been introduced by the morphological thinning have to be removed. A length threshold has then to be defined to eliminate false branches or spikes whose length is lower than the pre-specified length threshold. This can result in the elimination of low-contrast fringe extremities.

To overcome these difficulties, a parallel morphological thinning algorithm, applicable to well-composed sets, has been used. By definition, a binary image is well composed if it does not contain *critical configurations*, i.e. four corner-adjacent pixels such that two diagonally adjacent pixels are black and the other two are white. *Well-composed sets*, recently introduced by Latecki *et al* [19], have very nice topological properties: the Jordan curve theorem holds for them, their

Figure 12.3.9. Thick skeleton of dark fringes and adjacent image background of inverse moiré image I_5 (original thick bright fringes are now black).

Euler characteristic is locally computable and they have only one connectedness relation, since 4 and 8 connectedness are equivalent. Morphological thinning is an internal operation on well-composed sets. Furthermore, the skeletons obtained have a graph structure and are *one point thick*. Thus, the problem with irreducible thick skeletons disappears.

In order to benefit of the advantages offered by thinning of well-composed sets, we have transformed the thick skeleton of background and dark fringes into a well-composed binary image by adding a few white pixels whose positions can be identified by the detection of critical configurations.

A variant of the Rosenfeld four-phase thinning algorithm [27] has been used to process the grey-level image, whose domain of definition is restricted to the thick skeleton of the background and dark fringes and whose grey values are the complement of the filtered values. The result of this first morphological thinning has then been binarized and further thinned by the four-phase thinning algorithm. As a result, the skeleton fringes are one pixel thick, in contrast to standard thinning techniques.

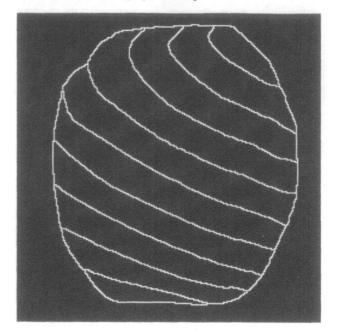

Figure 12.3.10. Thin skeleton of dark fringes and adjacent background of inverse moiré image I_5.

12.3.4.4   Extraction of dark fringes by a graph technique

A graph technique has been used to isolate the individual fringes as links between nodes located at the boundary of the domain of definition of the inverse moiré images. The selected links which, are interpreted as dark fringes, are those that are not adjacent to the moiré image background. This specific algorithmical technique presents the advantage of producing fringe skeletons without any spike and does not require the definition of a length threshold, in contrast to standard skeletoning algorithms that have to eliminate false branches or spikes whose length is lower than a pre-specified length threshold.

To illustrate the proposed method, a thick skeleton of dark fringes and the adjacent background, the result of their morphological thinning and the thin skeleton of dark fringes are, respectively, presented in figures 12.3.9–12.3.11.

12.3.4.5   Experimental results

The first experimental results have been very promising, even though they were obtained on a relatively small number of cases related to various deformations of a cylinder. With the proposed algorithmical method, moiré fringes have been systematically detected with perfect accuracy, even in the case of faint fringes.

The main improvements to the existing methods are:

Figure 12.3.11. Extraction of the thin skeleton of dark fringes of inverse moiré image I_5.

- that the proposed method is able to deal efficiently with noisy inverse moiré images;
- that no user interaction is required to extract moiré fringes;
- that low-contrast fringes are detected with a high sensitivity;
- that the skeleton fringes are one pixel thick in contrast to standard thinning techniques; and
- that the fringe extraction is robust and reproducible.

12.3.5 Defect detection by hit/miss transform

In this section, we present an optoelectronic hit/miss morphological transform for real-time quality control by image moiré analysis that integrates a Van der Lugt optical correlator and a fast digital signal processor (DSP) that is associated with a vector co-processor.

The procedure for real-time defect detection is a three-stage process. The first step is to enhance the moiré image by unsharp masking using wavelet decomposition and a multi-resolution approach. The second step is to segment the enhanced moiré image automatically, using the moment-preserving thresholding algorithm. These two steps are implementable in real time on an optical correlator and DSP. The third step is to apply the morphological hit/miss transform directly to recognize moiré images that correspond to defective objects. This new procedure of defect detection by global analysis of moiré image data has been compared to another new technique that is based on multi-dimensional supervised classification of optical correlations between the test object moiré image and reference moiré images.

The experimental results have demonstrated the robustness of this algorithmic approach for the analysis of noisy inverse moiré images.

12.3.5.1   Optical and digital methods for real-time processing of moiré images by morphological hit/miss transform

Our algorithmic procedure for supervised defect recognition and for real-time quality control by moiré analysis is a three-stage process. First, we used the well-known image-processing technique called unsharp masking that is optically and digitally implementable to emphasize low-contrast fringes and to enhance moiré images. Wavelet decomposition and a multi-resolution approach have been used to perform this unsharp masking. The second step is to segment the enhanced moiré image automatically, using the moment-preserving thresholding algorithm. These two steps are implementable in real time on a DSP that is associated to a vector co-processor. The third step is to apply the morphological hit/miss transform to detect moiré images that correspond to defective objects. This last step is implementable in real time on an optical thresholding correlator.

The morphological hit/miss transform (HMT) is a basic tool for shape detection [13]. The HMT has been used to detect any bright or dark fringes in the thresholded test moiré image that are roughly similar in shape to reference fringes of the conform moiré image. The HMT can be decomposed into three elementary operations. The first operation, called the *hit operation*, detects all regions in the thresholded test moiré image where bright fringes are similar to bright fringes of the reference moiré image. This hit operation is performed by morphological erosion of the thresholded test moiré image A with a *hit-structuring element H* that is formed from pixels associated with conform bright fringes of the reference moiré image. Recall that the erosion $A \ominus H$ of A by H is the set of locations of the origin of H such that H is completely contained in A. The second operation, called the *miss operation*, detects all regions in the thresholded test moiré image where dark fringes are similar to dark fringes of the reference moiré image. This miss operation is performed by morphological erosion of the complement of the thresholded test moiré image A^c with a *miss-structuring element M* that corresponds to conform dark fringes and to their local background in the reference moiré image. Then, the set of locations for which H exactly fits inside A and for which M exactly fits inside the complementary of A is the intersection of the erosion of A by H and the erosion of A^c by M.

In other words, the HMT transform $A \odot (H, M)$ contains the locations at which, simultaneously, H found a match (*hit*) in A and M found a match (*miss*) in A^c:

$$A \odot (H, M) = (A \ominus H) \cap (A^c \ominus M). \qquad (12.3.6)$$

Morphological erosion $A \ominus H = T_t(A \star H)$ can be optically implemented as a correlation of the image A with the structuring element H, followed by a thresholding operation: $A \ominus H = T_t(A \star H)$, where $\star$ stands for correlation and $T_t(.)$ is a thresholding operator that returns the value 1 if its argument is greater than the threshold t and 0 otherwise. The threshold level t has been set to the peak correlation value: $t = \max\{A \star B\}$. The HMT transform has been optically implemented by first correlating image A with the hit structuring element, H, and thresholding the result at the peak correlation level. Then A^c is correlated with the miss structuring element, M, and the resulting miss correlation image is similarly thresholded. Finally, the two binary eroded images are multiplied to get the HMT of A by the pair (H, M) of structuring elements:

$$A \odot (H, M) = T_h(A \star H) \times T_m(A^c \star M) \tag{12.3.7}$$

The standard hit/miss transform will detect conform fringes only if they exactly match the shape of the bright and dark reference fringes that respectively define the hit- and miss-structuring elements. However, in the analysis of moiré images, bright and dark fringes that are related to conform objects may vary slightly in shape. These small deformations of conform fringes have been taken into account by performing an erosion of the hit- and miss-structuring elements with a small disc and also by slightly reducing the peak correlation values that are used to threshold hit and miss correlations.

12.3.5.2 Experimental results

The five inverse moiré images, that have been presented in section 12.3.3, have been analysed by digital image enhancement and segmentation and then by optical correlation.

The hit-structuring element H_1 and the miss-structuring element M_1, that have been obtained as explained herabove by analysis of the conform inverse moiré image I_1, are illustrated in figure 12.3.12.

Experimental results obtained with the hit/miss transform are twofold. First, the match is perfect for the reference image since the distance between the correlation peaks is null. Second, the intersection of thresholded hit and miss correlations is always empty when the test moiré images correspond to the defective objects. For example, figure 12.3.13 shows that the two hit and miss correlation peaks are separated by more than 15 pixels in the case of the moiré image related to a medium deformation of the observed object. These experimental results show that the defect detection has always been correct.

For comparison purposes, we have also performed supervised recognition experiments and we have applied a standard test technique that bipartitions the ensemble of moiré images into a set of reference images and a set of test images. The nearest-neighbour classification rule has been used to perform supervised recognition. The set of reference images always contains the first image that correspond to a normal object and two or three images that correspond to objects

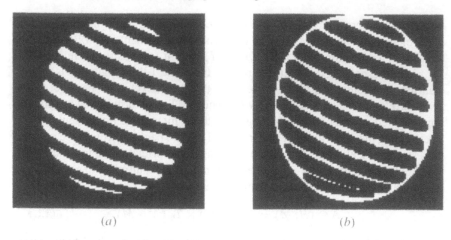

(a)   (b)

Figure 12.3.12. Structuring elements related to the conform inverse moiré image I_1. The (a) hit and (b) miss structuring elements H_1 and M_1 have been obtained by automatic thresholding and by morphological erosion of the enhanced and subsampled conform inverse moiré image.

(a)   (b)

Figure 12.3.13. Thresholded correlations in the case of inverse moiré image I_3. (a) Thresholded *hit* correlation, (b) thresholded *miss* correlation. The intersection of these thresholded correlation images is empty. This shows that the moiré image I_3 corresponds to a defectuous object. Thus, defect detection has been performed correctly by the hit/miss transform.

with a median and/or a high deformation respectively. For example with reference images I_1, I_3 and I_5, the result of supervised recognition is the following: images I_2 and I_4 are respectively similar to images I_3 and I_5. Thus, the observed object is classified correctly as defective for each of these two test cases. Experimental results have shown that defect classification is always correct when an example of an object with a medium or a high deformation is included in the set of reference moiré images.

The advantage of the hit/miss transform compared to supervised detection is that it is not necessary to define a learning set of moiré images and that it is only necessary to perform two correlations with the two hit- and miss-structuring elements.

Using a $4f$ optical correlator, we have experimentally demonstrated a proof of principle hit/miss transform for the detection of defects by direct moiré image data analysis. The main advantage of the proposed quality control system is that the detection of defects is performed directly from the fringe pattern data and that quality control no longer requires the numbering of fringes and the reconstruction of the 3D shape of the test object.

12.3.6 Defect detection by multi-dimensional statistical analysis

We present hereafter an industrial application of the phase-shifting moiré technique and of optical correlation that are intended for the automatic 3D inspection of manufactured objects that are approximately positioned. The practical objective was to design and test an optoelectronic defect detection system that would be applicable even when the object under test is supposed to be approximately positioned, due to possible lateral or transverse deviations, as would be the case in the context of industrial applications.

12.3.6.1 Multi-dimensional statistical analysis of the results of optical correlation experiments

In this section, we present fast and robust algorithms for real-time defect detection by global analysis of moiré image data. The procedure for real-time defect detection is a two-stage process. The first step is to pre-process the moiré image, by smoothing the phase moiré image with a quadrature mirror filter [21] and by subsampling the resulting smoothed image to get smaller phase images for subsequent optical correlation. This first step is implementable in real time on a DSP that is associated with a vector co-processor. Then, for industrial quality control applications, the detection of defects in test object shapes is performed directly from smoothed and subsampled phase image data by an optoelectronic system that integrates a Van der Lugt optical correlator and a DSP. The real-time defect detection approach that has been evaluated is based on Benzécri factorial correspondence analysis [1] and on the definition of decision areas in the factorial space for supervised defect detection.

The object under test is supposed to be approximately positioned, due to possible lateral or transverse deviations, as would be the case in the context of industrial applications. In order to allow the analysis of any defect type and to get a more flexible defect detection system, we have chosen not to include examples of defective objects in the learning image dataset L so it only contains examples of phase images of conform objects with a lateral deviation x and a transverse deviation z: $L = \{\Phi_{x,z}(r, c) | x \in X, z \in Z\}$.

Supervised detection of defects for quality control has been performed by characterizing phase moiré images $\Phi_{T,x,z}(r, c)$ (with defect type T) by their correlation peak values with the phase moiré images $\Phi_{x,z}(r, c)$ of the learning set L and by using the nearest-neighbour recognition method. Thus, phase moiré images of objects under test are represented by feature vectors in the multi-dimensional space whose coordinates are their correlation peak values with the phase moiré images of the learning set L. In this feature space, the χ^2 metric that verifies the *distributional equivalence principle* has been choosen to get a robust method for moiré image supervised classification and Benzécri correspondence analysis has been used to obtain a low dimension factorial representation space where χ^2 distances are approximated by Euclidean distances.

12.3.6.2   Factorial correspondence analysis

The feature vectors that have been previously defined can be considered as the lines of the data table $t_{I,J} = \{t(i, j) | i \in I, j \in J\}$, where $t(i, j)$ represents the normalized correlation peak value between the phase image $i \in I = L$ and the conform phase image $j \in J = L$. Benzécri *factorial correspondence analysis* [1] is particularly suited to the analysis of the table $t_{I,J}$ representing correlation peak profiles. This factorial analysis will allow the dimension of the feature space to be reduced and experimental noise eliminated.

The correspondence analysis allows the similarities and dissimilarities between row profiles to be described by the χ^2 distance which takes the weight differences between the rows into account. The χ^2 distance between rows i and i' is defined as follows:

$$d(i, i') = \sum_{j \in J} \frac{1}{t(j)} \left(\frac{t(i, j)}{t(i)} - \frac{t(i', j)}{t(i')} \right)^2 \qquad (12.3.8)$$

where $t(i) = \sum_{j \in J} t(i, j)$ and $t(j) = \sum_{i \in I} t(i, j)$.

As correspondence analysis is based on the χ^2 distance, it satisfies the distributional equivalence principle which has a stabilizing influence on the results of the analysis. According to this principle, the results of the analysis remain the same after aggregating rows which have identical profiles.

The aim of Benzécri correspondence analysis is to represent multi-dimensional data by pairs of factors $(F_1, G_1), (F_2, G_2), \ldots, (F_k, G_k)$ which are defined on I and J respectively and which verify the so-called *transition formula*:

$$F_k(i) = \frac{1}{\sqrt{\lambda_k}} \sum_{j \in J} \frac{t(i, j)}{t(i)} G_k(j) \qquad (12.3.9)$$

and

$$G_k(i) = \frac{1}{\sqrt{\lambda_k}} \sum_{i \in I} \frac{t(i, j)}{t(j)} F_k(i) \qquad (12.3.10)$$

where λ_k is the common variance of factors F_k and G_k.

To determine factors $(F_1, G_1), (F_2, G_2), \ldots, (F_k, G_k), \ldots,$ Benzécri correspondence analysis first computes the following inertia matrix:

$$s(j, j') = \sum_{i \in I} \frac{t(i, j)t(i, j')}{t(i)\sqrt{t(j)t(j')}}. \qquad (12.3.11)$$

Then, the eigenvalues, λ_k, and eigenvectors, u_k, of this inertia matrix are computed:

$$\sum_{j' \in J} s(j, j')u_k(j') = \lambda_k u_k(j) \qquad (12.3.12)$$

and the eigenvectors are sorted by decreasing order of their corresponding eigenvalues.

The factors that correspond to the eigenvector u_k are defined by the pair of functions (F_k, G_k) that are defined on I and J respectively by the following equations:

$$G_k(j) = \frac{u_k(j)}{\sqrt{t(j)/t}} \qquad (12.3.13)$$

$$F_k(i) = \frac{1}{\sqrt{\lambda_k}} \sum_{j \in J} \frac{t(i, j)}{t(i)} G_k(j) \qquad (12.3.14)$$

where $t = \sum_{i \in I, j \in J} t(i, j)$. The first factor corresponds to the eigenvalue that is equal to 1. This first factor is eliminated because it is trivial.

12.3.6.3 Analysis of phase moiré images by correspondence analysis

As a result of Benzécri correspondence analysis, the phase moiré images, $\Phi_{x,z}$, of the learning set, L, are represented by points in the factorial space whose coordinates are denoted as follows: $F_1(x, z), F_2(x, z), F_3(x, z), \ldots, F_K(x, z)$, where K is the number of selected factors.

Furthermore, the dissimilarity between a supplementary test object and each reference object of the learning image database is equal to the Euclidean distance between points that represent corresponding phase images in the factorial space. With such a dissimilarity definition, the conformity decision will be based on the whole set of correlation peak values between a test object image and various phase images of the conform object with different locations.

Moiré images of objects under test may then be classified by dividing the feature space, i.e. the factorial space, into decision areas and by assigning a test moiré image to a given class when its feature vector belongs to the corresponding decision area.

12.3.6.4 Rules and decision areas for supervised defect detection

For supervised classification experiments, we have systematically applied two defect recognition rules that are based on the results of Benzécri correspondence analysis.

First, the nearest-neighbour classification rule has been used to perform supervised recognition. Moiré phase images $\Phi_{x,z}$ of the learning dataset, which correspond to different lateral and transverse positioning of the conform object, are represented by points in the factorial space. The decision area that characterize conform objects in this factorial space is the union of balls $B_{x,z}$ associated with the different conform examples of the learning image dataset. The ball $B_{x,z}$ is centred on the point $F_1(x, z), F_2(x, z), F_3(x, z), \ldots, F_K(x, z)$ that represents the conform moiré phase image $\Phi_{x,z}$ in the factorial space. The radius of the ball $B_{x,z}$ is equal to the distance, in the factorial space, that separates the moiré phase image $\Phi_{x,z}$ from its nearest conform neighbour. An object under test is then classified as defective when its feature vector v in the factorial space is such that $d(v, nn(v)) > \rho(nn(v))$, where $nn(v)$ denotes the nearest neighbour of v in the learning dataset, $\rho(nn(v))$ is the radius of the ball centred on $nn(v)$ and $d(v, nn(v))$ is the distance in the factorial space between v and its nearest conform neighbour $nn(v)$.

The second recognition rule is based on a local correspondence analysis of correlation peaks between phase images. The nearest neighbour of the object under test is adjacent to, at most, four patches of the learning dataset in the physical plane spanned by the two orthogonal directions along which lateral and transverse deviations are measured. Benzécri correspondence analysis of these elementary patches allows the phase images to be represented by points in the local factorial space spanned by the first three factors (F_1, F_2, F_3). Then it is possible to classify as defective those objects with factorial components $(F_1(v), F_2(v), F_3(v))$ which cannot be interpreted as the result of an interpolated displacement of the conform object, i.e. when the factorial components $(F_1(v), F_2(v), F_3(v))$ do not belong to the convex hull of conform objects in the local factorial space.

12.3.6.5 Experimental results

Two sets of moiré images that correspond to conform or defective cans have been used for optical correlation experiments when objects under test are not exactly positioned, as it would be the case in the context of industrial applications. The first set of 18 conform moiré images is related to various lateral and/or transverse displacements of the conform can. Six lateral deviations in position that vary between 0 and 10 mm, in steps of 2 mm, have been considered together with three transverse deviations in position that vary from 0 to 4 mm, in steps of 2 mm. The second set of 17 abnormal moiré images is related to various lateral and/or transverse displacements of defective cans, that are characterized

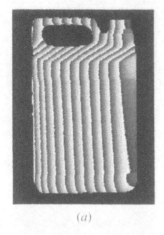

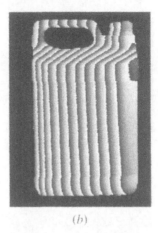

(a)

(b)

Figure 12.3.14. Phase moiré images related to a conform can, without any deformation. (a) Phase image $\phi_{0,0}$, (b) phase image $\phi_{8,0}$.

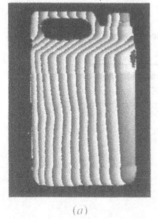

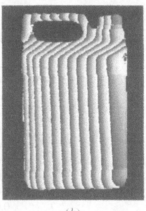

(a)

(b)

Figure 12.3.15. Moiré phase images related to the first example of application of the second defect recognition rule. (a) Phase image $\phi_{C,8,0}$, (b) nearest conform neighbour $\phi_{2,4}$.

by one of the following defect types: A, important deformation; B, mean deformation; C, one small defect; D, two small defects. Figure 12.3.14 shows phase moiré images $\phi_{0,0}$ and $\phi_{8,0}$ that are related to a conform object, without any deformation. Figure 12.3.15 and 12.3.16 illustrate the application of the second defect recognition rule in the case of phase moiré images $\phi_{C,8,0}$ and $\phi_{D,8,4}$ that are respectively characterized by one or two small defects.

Optical correlation experiments have been realized with phase images that have been generated by the phase-shifting technique. The correlation peaks between smoothed and subsampled phase images, that have been measured in volts in the correlation plane, vary between 0.30 and 1.85 V. A non-symmetric

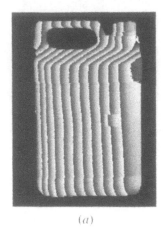
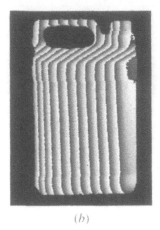

(*a*)   (*b*)

Figure 12.3.16. Moiré phase images related to the second example of application of the second defect recognition rule. (*a*) Phase image $\phi_{D,8,4}$, (*b*) nearest conform neighbour $\phi_{6,4}$.

Table 12.3.1. Percentages of variances explained by the first factors of the Benzécri correspondance analysis.

Factor k	Variance ratio τ_k	Cumulative variance ratio s_k
1	0.7043	0.7043
2	0.1039	0.8081
3	0.0820	0.8901
4	0.0434	0.9336
5	0.0183	0.9518
6	0.0150	0.9669
7	0.0121	0.9790
8	0.0076	0.9866
9	0.0042	0.9908
10	0.0033	0.9941

normalized correlation matrix has been obtained for two reasons. First, the use of a quantized phase-only matched filter introduces a dissymmetry in the estimation of the correlation coefficient. Second, the nonlinear response of the spatial light modulator has not been taken into account to synthesize the quantized phase-only matched filters that correspond to reference moiré images.

The percentages of variances explained by the first factors of the Benzécri correspondence analysis are presented in table 12.3.1. χ^2 distances have been approximated by euclidean distances in the factorial space spanned by the first eight factors of the Benzécri correspondence analysis.

Table 12.3.2. Maximal admissible distances defined by χ^2 distances between conform moiré phase images and their nearest conform neighbours. Note: χ^2 distances have been approximated by Euclidean distances in the factorial space spanned by the first eight factors of the Benzécri correspondence analysis.

Phase image of the deviated conform object	Phase image of its nearest neighbour	χ^2 distance to its nearest neighbour
$\Phi_{0,0}$	$\Phi_{2,0}$	0.055 96
$\Phi_{0,2}$	$\Phi_{0,0}$	0.070 34
$\Phi_{0,4}$	$\Phi_{2,2}$	0.059 96
$\Phi_{2,0}$	$\Phi_{4,0}$	0.055 07
$\Phi_{2,2}$	$\Phi_{0,4}$	0.059 96
$\Phi_{2,4}$	$\Phi_{4,2}$	0.046 79
$\Phi_{4,0}$	$\Phi_{4,2}$	0.041 15
$\Phi_{4,2}$	$\Phi_{4,0}$	0.041 15
$\Phi_{4,4}$	$\Phi_{6,2}$	0.042 10
$\Phi_{6,0}$	$\Phi_{4,0}$	0.074 91
$\Phi_{6,2}$	$\Phi_{4,4}$	0.042 10
$\Phi_{6,4}$	$\Phi_{8,2}$	0.067 39
$\Phi_{8,0}$	$\Phi_{8,2}$	0.070 08
$\Phi_{8,2}$	$\Phi_{6,4}$	0.067 39
$\Phi_{8,4}$	$\Phi_{8,2}$	0.099 68

By applying of the first defect recognition rule, 14 objects under test have been correctly classified, since the χ^2 distance to their nearest conform neighbour is greater than the corresponding maximal admissible distance, as shown by tables 12.3.2 and 12.3.3. The objects under test whose phase images are $\Phi_{C,4,0}$, $\Phi_{C,8,0}$, $\Phi_{D,8,0}$ and $\Phi_{D,8,4}$ that have not been correctly classified by the first defect recognition rule, will later be correctly classified by the second defect recognition rule. For example, the object under test with phase image $\Phi_{C,8,0}$ is correctly classified as defective as shown hereafter. First, note that the nearest conform neighbour of $\Phi_{C,8,0}$ is $\Phi_{2,4}$. Thus, the two adjacent patches in the plane (x, z) that share the node $(2, 4)$ are: $(0, 2)$, $(2, 2)$, $(0, 4)$, $(2, 4)$ and $(2, 2)$, $(4, 2)$, $(2, 4)$, $(4, 4)$. The results of local Benzécri correspondence analysis do not support the hypothesis according to which the phase image of the object under test could be interpreted as the result of an interpolated deviation within each of these patches. In fact, tables 12.3.4 and 12.3.5 show that the representative point of $\Phi_{C,8,0}$ in each of the two local factorial planes does not belong to the convex hull of the corresponding patch nodes. Consequently, $\Phi_{C,8,0}$ is correctly classified as defective. Tables 12.3.6 and 12.3.7 illustrate another example, in the case of the phase moiré image $\Phi_{D,8,4}$, for which the third factor has too high a

Table 12.3.3. Defect detection by the nearest-neighbour method. If the χ^2 distance of the object under test to its nearest conform neighbour is greater than the corresponding maximal admissible distance, then the considered object is classified as defective. The objects under test whose phase images are $\Phi_{C,4,0}$, $\Phi_{C,8,0}$, $\Phi_{D,8,0}$ and $\Phi_{D,8,4}$ will later be correctly classified by the second defect recognition rule.

Phase image of the object under test	Phase image of its nearest conform neighbour	χ^2 distance to its nearest conform neighbour	Maximal admissible χ^2 distance	Object classification
$\Phi_{A,0,0}$	$\Phi_{0,4}$	0.297 31	0.059 96	Defective
$\Phi_{A,2,4}$	$\Phi_{0,4}$	0.194 53	0.059 96	Defective
$\Phi_{A,4,0}$	$\Phi_{0,4}$	0.224 27	0.059 96	Defective
$\Phi_{A,8,0}$	$\Phi_{4,4}$	0.140 85	0.042 10	Defective
$\Phi_{A,8,4}$	$\Phi_{8,0}$	0.163 48	0.070 08	Defective
$\Phi_{B,2,4}$	$\Phi_{0,4}$	0.135 91	0.059 96	Defective
$\Phi_{B,4,0}$	$\Phi_{0,0}$	0.070 56	0.055 96	Defective
$\Phi_{B,8,0}$	$\Phi_{2,4}$	0.062 08	0.046 79	Defective
$\Phi_{B,8,4}$	$\Phi_{6,4}$	0.190 52	0.067 39	Defective
$\Phi_{C,2,4}$	$\Phi_{4,0}$	0.063 06	0.041 15	Defective
$\Phi_{C,4,0}$	$\Phi_{0,0}$	0.038 43	0.055 96	$\ldots$
$\Phi_{C,8,0}$	$\Phi_{2,4}$	0.037 85	0.046 79	$\ldots$
$\Phi_{C,8,4}$	$\Phi_{6,4}$	0.086 79	0.067 39	Defective
$\Phi_{D,0,0}$	$\Phi_{8,4}$	0.157 02	0.099 68	Defective
$\Phi_{D,2,4}$	$\Phi_{2,2}$	0.064 85	0.059 96	Defective
$\Phi_{D,8,0}$	$\Phi_{2,4}$	0.030 15	0.046 79	$\ldots$
$\Phi_{D,8,4}$	$\Phi_{6,4}$	0.042 54	0.067 39	$\ldots$

value: consequently, the object under test cannot be classified as conform since its representative point in the two possible local factorial 3D spaces does not belong to the following convex hulls whose patch nodes are, respectively, $\Phi_{4,2}$, $\Phi_{6,2}$, $\Phi_{4,4}$, $\Phi_{6,4}$ and $\Phi_{6,2}$, $\Phi_{8,2}$, $\Phi_{6,4}$, $\Phi_{8,4}$. Experimental results have shown that the second defect recognition rule has correctly classified the four objects that were not classifed as defective by the first recognition rule.

In conclusion, the objects under test have always been correctly classified as defective, by using the two proposed defect recognition rules. Furthermore, we have chosen to process phase images, instead of original intensity moiré images, because of their higher and more uniform contrast. The practical result is that it is no longer necessary to enhance low-contrast fringes by unsharp masking and this results in a lower processing time.

Table 12.3.4. Example of application of the second defect recognition rule. The object under test with phase moiré image $\Phi_{C,8,0}$ cannot be classified as conform since its representative point in the local factorial plane does not belong to the convex hull of patch nodes $\Phi_{0,2}$, $\Phi_{2,2}$, $\Phi_{0,4}$ and $\Phi_{2,4}$ (the factor F_2 has too high an absolute value)

Phase image	Factor F_1	Factor F_2
$\Phi_{0,2}$	−1.05	−0.54
$\Phi_{2,2}$	0.26	1.33
$\Phi_{0,4}$	−0.25	0.03
$\Phi_{2,4}$	0.99	−0.90
$\Phi_{C,8,0}$	0.77	−2.69

Table 12.3.5. Example of application of the second defect recognition rule. The object under test with phase moiré image $\Phi_{C,8,0}$ cannot be classified as conform since its representative point in the local factorial plane does not belong to the convex hull of patch nodes $\Phi_{0,2}$, $\Phi_{2,2}$, $\Phi_{0,4}$ and $\Phi_{2,4}$ (the factor F_2 has too high a value)

Phase image	Factor F_1	Factor F_2
$\Phi_{2,2}$	−1.52	−0.58
$\Phi_{4,2}$	0.06	1.86
$\Phi_{2,4}$	0.36	−0.55
$\Phi_{4,4}$	1.05	−0.73
$\Phi_{C,8,0}$	0.88	3.48

Table 12.3.6. Example of application of the second defect recognition rule. The object under test with phase moiré image $\Phi_{D,8,4}$ cannot be classified as conform since its representative point in the local factorial plane does not belong to the convex hull of patch nodes $\Phi_{4,2}$, $\Phi_{6,2}$, $\Phi_{4,4}$ and $\Phi_{6,4}$ (the factor F_3 has too high a value).

Phase image	Factor F_1	Factor F_2	Factor F_3
$\Phi_{4,2}$	0.22	4.26	−0.22
$\Phi_{6,2}$	0.68	−1.88	−1.21
$\Phi_{4,4}$	0.77	−0.94	1.35
$\Phi_{6,4}$	−1.56	−1.31	0.1
$\Phi_{D,8,4}$	−1.01	1.81	3.55

Table 12.3.7. Example of application of the second defect recognition rule. The object under test with phase moiré image $\Phi_{D,8,4}$ cannot be classified as conform since its representative point in the local factorial plane does not belong to the convex hull of patch nodes $\Phi_{6,2}$, $\Phi_{8,2}$, $\Phi_{6,4}$ and $\Phi_{8,4}$ (the factor F_3 has too high an absolute value)

Phase image	Factor F_1	Factor F_2	Factor F_3
$\Phi_{6,2}$	2.78	−2.04	−0.23
$\Phi_{8,2}$	−0.81	4.04	−0.61
$\Phi_{6,4}$	0.04	2.08	0.92
$\Phi_{8,4}$	−2.18	−4.46	−0.05
$\Phi_{D,8,4}$	−0.14	1.02	−4.22

12.3.7 Conclusion

The main advantage of the proposed quality control system is that the detection of defects is performed directly from the fringe pattern data and that quality control does not require the numbering of fringes and the reconstruction of the 3D shape of the object under test. Using a $4f$ optical correlator and a fast DSP, we have experimentally proved the principle of the skeletonizing, hit/miss transform and statistical approach for the detection of defects by the global analysis of moiré image data.

In fact, by using a DSP TMS320C80 associated with a SHARP vector co-processor, initial image smoothing and subsampling can be performed in 0.1 s and optical correlation requires, at most, two video cycles for correlation of the input image with selected matched filters that correspond to smoothed and subsampled phase images of conform objects. With pipeline processing, the computing time required by optoelectronic correlation is less than 0.1 s. Thus, the total computing time is less than 0.2 s per test object.

The use of a very fast electrically addressable spatial light modulator (frame rate > 1000 Hz) [18] and of a very fast CCD camera or the use of a high-resolution multi-channel system (frame rate 30 Hz) would later allow a faster analysis of a larger number of object positions for the realization of an even more efficient real-time quality control system by moiré image analysis.

References

[1] Benzécri J P 1973 L'Analyse des Données, Tome I: La Taxinomie *Dunod* Paris
[2] Bergeron A 2002 Low-light level recognition using COTS optical correlator *Proc. SPIE* **4734** 65–72
[3] Bergeron A 1997 Optodigital neural network classifier *Opt. Eng.* **36** 3134–9

[4] Bruynooghe M 1993 Maximal theoretical complexity of fast hierarchical clustering algorithms based on the reducibility property *Int. J. Pattern Recognition and Artificial Intelligence* **7** 541–71

[5] Bruynooghe M, Sadki M, Harthong J and Becker A 1996 Fast algorithms for automatic moiré fringe analysis. Application to non-contact measurements for quality control of industrial components. *Int. Symp. Lasers, Optics and Vision for Productivity in Manufacturing Proc. SPIE* **2786** 54–67

[6] Bruynooghe M and Bergeron A 1997 Optoelectronic hit/miss transform for real-time defect detection by moiré image analysis *Euro. Symp. Lasers, Optics and Vision for Productivity in Manufacturing Proc. SPIE* **3101** 30–7

[7] Bruynooghe M and Bergeron A 1997 New digital/optical methods for real-time moiré fringe processing *Fringe'97. Proc 3rd Int. Symp. Automatic Processing of Fringe Patterns (Akademie Verlag Series in Optical Metrology)* ed W Jüptner and W Osten (Bremen: Akademie) pp 411–5

[8] Bruynooghe M, Bergeron A and Colon E 1997 Real-time digital/optical system for quality control by moiré image processing *SPIE's Int. Symp. Intelligent Systems and Advanced Manufacturing* Pittsburgh, PA, October 13–17 *Proc. SPIE 3208, Intelligent Robots and Computer Vision XVI: Algorithms, Techniques, Active Vision and Material Handling* ed D P Casasent (Bellingham, WA: SPIE) pp 445–54

[9] Carre P 1966 Installation et utilization du comparateur photoélectrique et interférentiel du Bureau International des Poids et Mesures *Metrologia* **2** 13–23

[10] Chen X and Cher Z 1995 Amplitude-modulated circular-harmonic filter for pattern recognition *Appl. Opt.* **34** 879–85

[11] Curtis K and Psaltis D 1994 Three-dimensional disk-based optical correlator *Opt. Eng.* **33** 4051–4

[12] J L Doty 1983 Projection moiré for remote contour analysis *J. Opt. Soc. Am.* **73** 366–72

[13] Gonzalez R C and Woods R E 1993 *Digital Image Processing* (Reading, MA: Addison-Wesley) pp 528–30

[14] Harthong J, Sahli H, Poinsignon R and Meyrueis P 1991 Analyse de formes par moiré *J. Phys. III* **1** 69–84

[15] Kim S W, Choi Y B and Oh J T 1997 Three-dimensional profile measurement of fine objects by phase-shifting shadow moiré interferometry *Three-dimensional Imaging and Laser-Based Systems for Metrology and Inspection II (Proc. SPIE 2909)* ed K G Harding and D J Svetkoff (Bellingham, WA: SPIE) pp 28–36

[16] Javidi B, Li J, Fazlollali A M and Horner J 1995 Binary nonlinear joint transform correlator performance with different thresholding methods under unknown illumination consitions *Appl. Opt.* **34** 886–96

[17] Kass M, Witkin A and Terzopoulos D 1987 Snakes: active contour models *Int. J. Computer Vision* **1** 321–31

[18] McKnight D J, Johnson K M and Serati R A 1994 *Appl. Opt.* **33** 2775

[19] Latecki L, Eckhardt U and Rosenfeld U 1995 Well-composed sets *Computer Vision and Image Understanding* **61** 70–83

[20] Li J, Wang Y and Hu J 1994 Experimental investigation of real-time nonlinear joint transform correlator *Opt. Eng.* **33** 3302–6

[21] Mallat S 1989 Multiresolution approximation and wavelet orthonormal bases of $L^2(R)$ *Trans. Am. Math. Soc.* **315** 69–87

[22] Kalms M K, Jüptner W and Osten W 1997 Automatic adaption of projected fringe patterns using a programmable LCD-projector *Euro. Symp. Lasers, Optics and Vision for Productivity in Manufacturing, Proc. SPIE* **3101**

[23] Oster G 1969 The Science of Moiré Patterns (Tonawanda, NY: Edmund Scientific) 2nd edn

[24] Patorski K and Kujawińska M 1993 *Handbook of The Moiré Fringe Technique* (Amsterdam: Elsevier)

[25] Reid G T 1984 Moiré fringes in metrology *Opt. Lasers Eng.* **5** 63–93

[26] Robinson D W and Reid G T 1993 *Interferogram Analysis. Digital Fringe Pattern Measurement Techniques* (Bristol: Institute of Physics Publishing)

[27] Rosenfeld A 1975 A characterization of parallel thinning algorithms *Information and Control* **29** 286–91

[28] Sahli H 1991 Perception et modélisation d'objets tridimensionnels par la méthode du moiré *Thèse de Doctorat de l'Université Louis Pasteur de Strasbourg* (in French)

[29] Stefanov V A, Tsibulkin L M, Tsurikov Y L and Samoilov V P 1996 Holographic recognizer with disk memory unit *Int. Symp. Lasers, Optics and Vision for Productivity in Manufacturing, Proc. SPIE* **2785** 300–3

[30] Stollfuss A 1994 Optische wavelet transformation *Diplomarbeit* Universität Karlsruhe, Fakultät für Informatik (in German)

[31] Takeda N 1982 Fringe formula for projection type moiré topography *Opt. Lasers Eng.* **3** 45–52

[32] Terzopoulos D and Metaxas D 1991 Dynamic 3D models with local and global deformations: Deformable superquadrics *IEEE PAMI* **13** 703–14

[33] Schönleber M and Tiziani H J 1997 Computer-generated holograms from 3D objects written on twisted-nematic liquid crystal displays *Opt. Commun.* **140** 299–308

[34] Tsai W T 1985 Moment-preserving thresholding: a new approach *Computer Vision, Graphics, and Image Processing* **29** 377–93

[35] Wang W, Jin G, Yan Y and Wu M 1995 Joint wavelet-transform correlator for image feature extraction *Appl. Opt.* **34** 370–6

[36] Young R C, Chatwin C R and Scott B F 1993 High-speed hybrid optical/digital correlator system *Opt. Eng.* **32** 2608–14

Appendix A

Micro-scale strain measurement: limitations due to restrictions inherent in optical systems and phase-stepping routines

James McKelvie
University of Strathclyde, Glasgow, UK

A.1   Introduction

Optical methods, in general, and moiré interferometry, in particular, have been hugely successful in the determination of strain fields and the validation of theoretical and computational models. In recent years, their capabilities have been extended by the application of computational processing of optically delivered information.

This combination has been particularly powerful as a means for revealing increased detail in studying strain fields. However, the matter of how far we may proceed in this quest for detail (or the minimum gauge-length that we may consider) is something that we must address if we are to avoid the equivalent of what, in ordinary optical terms, would be called 'empty magnification', with consequent erroneous conclusions.

We are concerned with cases where the strain fluctuates rapidly—for example in slip fields or where we have material discontinuities (e.g. at bi-material interfaces) or in composite materials (cases where we might like to have strain gauges perhaps several micrometres long).

Three aspects are to be considered:

(1) the fringe structure, i.e. the extent to which low-frequency fringes carry high-frequency information;
(2) the limits inherent in the optical system, i.e. the degree of detail that the system can actually see and the distortions it might it introduce; and
(3) the restrictions imposed by the recording and interrogation system.

A.2   The fringe structure

In fringe-forming techniques, the mathematical description of the light intensity at the specimen is identical to that for a frequency-modulated radio signal, with a d.c. term added, i.e.

$$I(x) = I_0 + a \cos k(2\pi f_c x + \phi(x) + \delta) \tag{A.1}$$

where the desired information (the 'message') is contained in the function $\phi(x)$, k is the 'sensitivity' factor which includes, for example, the pitch of a grating, f_c is the 'carrier' frequency and δ is an unknown initial phase. The amplitude 'a' $\leq I_0/2$. (The treatment is restricted to a one-dimensional description.)

The question has been raised [1] as to whether such a signal can be properly demodulated to yield the 'message' if the latter contains frequencies that are higher than the 'carrier' frequency f_c (which is the average fringe frequency and is typically quite low—a few tens of cycles over the field of view). The manifestation of such a condition is a 'rippled' fringe profile, as in figure A.1. There is some evidence [2], using simulation with random $\phi(x)$, that when the frequency content of $\phi(x)$ becomes large in relation to f_c then distortion is introduced. However, the author has constructed numerous simulated ideal intensity distributions with high-frequency $\phi(x)$ and they always demodulate exactly using the 'perfect' demodulation algorithm—the simple rearrangement of equation (A.1)—introduced originally in [3]:

$$\phi(x) = \frac{1}{k} \cos^{-1} \left[\frac{I - I_0}{a} \right] - 2\pi f_c x - \delta. \tag{A.2}$$

The unknown δ is a problem in photoelasticity but not in displacement-measuring techniques where it disappears in the subtraction of two neighbouring $\phi(x)$ values in the calculation of strain. 'a' is readily measurable and f_c can be measured prior to applying the test load. (Any change in average frequency then appears as part of $\phi(x)$.) The ambivalent arcos is readily dealt with by the usual 'unwrapping' process.

The problem arises when we are presented, not with ideal, but with real intensity profiles. We then get

$$I(x) = I_0(x) + a(x) \cos k(2\pi f_c x + \phi(x) + \delta). \tag{A.3}$$

The d.c. level and the fringe amplitude vary, as a result of non-uniformity in the illumination, reflectivity (or transmissivity) or diffraction efficiency. This gives a signal that is modulated in amplitude as well as frequency. Rihaczek and Bedrosian [4] have analysed such a signal and the conclusion is that it cannot be reliably demodulated to give $\phi(x)$. There is no way of discriminating between a change due to a modulation of $a(x)$ and one of $\phi(x)$.

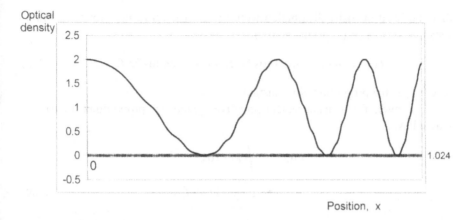

Figure A.1. Intensity profile of a low-frequency moiré with a high-frequency strain component.

A.2.1 Phase-shifting

The techniques of phase-shifting, or phase-stepping, are available to the photomechanician (where they are not normally for the radio engineer). By capturing more than one image, in each of which an increment in the fringe phase is introduced, these ambiguities can be eliminated. This is equivalent to sending the same signal several times with a phase-shift between them; (if practical, this would resolve the problem for Rihaczek).

This stratagem provides several independent values for $I(x)$ in equation (A.1) and the unknowns $I_0(x)$, $a(x)$ and δ can then be eliminated and $\phi(x)$ determined. A thorough review is provided in [5].

These techniques are robust against additive noise (affecting $I_0(x)$) and multiplicative noise (affecting $a(x)$). However, any noise affecting the argument inside the cosine—such as decollimation of the incident beams or non-uniformity in the grating frequency—remains unaffected. Also, of course, if $a(x)$ is locally zero (due to small holes in the grating for example), then phase-shifting will achieve no improvement. There are techniques described in [5] to arrive at best estimates for such 'bad pixels'.

A.2.2 Displacement *versus* strain sensitivity

Displacement-sensitive systems have a distinct disadvantage in the fringe structure vis-à-vis strain-sensitive systems:

If we consider a sinusoidal strain variation in frequency f_ε,

$$\varepsilon(x) = \varepsilon_0 \cos 2\pi f_\varepsilon x \tag{A.4}$$

then the ideal intensity distribution would have the following form for a strain-sensitive system (ϕ proportional to ε):

$$I(x) = I_0(x) + a(x)\cos k(2\pi f_c x + \kappa_1 \varepsilon_0 \cos 2\pi f_\varepsilon + \delta) \quad \text{(A.5)}$$

where κ_1 is the proportionality constant

However, for a displacement-sensitive system (ϕ proportional to u), we would have:

$$u(x) = \int \varepsilon(x)\,dx$$

$$= \frac{\varepsilon_0}{2\pi f_\varepsilon} \sin 2\pi f_\varepsilon x \quad \text{(A.6)}$$

giving the intensity distribution

$$I(x) = I_0(x) + a(x)\cos k\left(2\pi f_c x + \kappa_2 \frac{\varepsilon_0 \sin 2\pi f_\varepsilon x}{2\pi f_\varepsilon x} + \delta\right) \quad \text{(A.7)}$$

where κ_2 is the proportionality constant

It is clear the modulation achieved in (A.7) diminishes with increasing f_ε; i.e. it is harder to 'see' the effect of the strain. In principle, this is not a problem but it has clear implications in practice for the detectability of high-frequency variations.

A.3 The optical limits

We are concerned here with problems that arise when we use a lens at the image plane to reproduce the intensity distribution that propagates from the specimen.

There are several approaches to be considered in the matter of the optical limits and they are closely linked: the Airy disc; the Rayleigh criterion; the maximum detectable spatial frequency; the modulation and phase transfer functions; misfocus; off-axis effects; and second-order diffraction effects.

A.3.1 Under ideal conditions

A.3.1.1 The Airy disc

This is also known as the 'circle of confusion'. For a lens of diameter D with the object at distance U, the disc extends over a diameter, referred to as the object space, of

$$d_a = \frac{1.22\lambda}{N.A.} \quad \text{(A.8)}$$

where $N.A.$ is the numerical aperture $=$ sin(acceptance half-angle) $=$ sin{arctan($D/2U$)}. Within this dimension, it is sometimes inferred that there is no information.

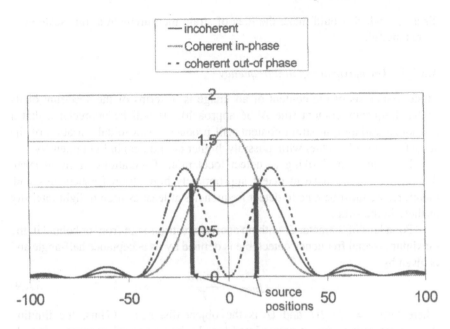

Figure A.2. Variations in the intensity profile from two point sources at the 'Rayleigh criterion' separation. (For the two complex amplitudes of the light, the sinc distributions, as for a lens with rectangular aperture, have been employed, rather than the more computationally demanding Bessel function distributions corresponding to a circular lens. The results, though, would show very similar behaviour.)

A.3.1.2   The Rayleigh criterion

This distance is (arbitrarily) defined to be 'the limit of resolution' of two point sources and is half the dimension of the Airy disc (which implies that there is, in fact, some information within the Airy disc). However, at this level we are not in the measurement business at all. What we are saying is merely "it looks like there are two point sources here". Furthermore, for the same distance between the sources, the intensity profile changes and the distance between the two maxima varies, according to whether the sources are incoherent, coherent in phase or coherent out-of-phase, as shown in figure A.2 (obtained by the appropriate addition or subtraction of the complex amplitudes of two Airy-type distributions—see [6] and caption of figure A.2). The phase relationship of the light emanating from two neighbouring points on a moiré interferometry specimen will depend on how perfectly symmetrically arranged, and collimated, and matched for intensity, the illuminating beams are.

(It is also interesting to note that even with the simpler case of incoherent light, as two point sources positioned at the Rayleigh distance move slightly apart or together, the two maxima of the distribution do not quite move by the same distance—which would make the idea of strain measurement at this scale even more fanciful.)

A.3.1.3   The maximum spatial frequency

Modern analysis of the content of an image is in terms of the spectrum of its spatial frequency content (the Abbé approach). It will be appreciated that a sinusoidal variation in strain content will produce a sinusoidal variation of the same frequency (together with, possibly, higher harmonics) in light intensity.

If we are now looking to detect local rapid fluctuations in strain level, this means that we have to detect appropriately high strain frequencies—and, therefore, we must be able to detect frequencies at least as high in light intensity as those in the strain.

For a lens of diameter D with no aperture restriction in front or behind it, the maximum spatial frequency detectable is defined by its acceptance half-angle and is given by

$$f_{max} = \frac{2 \sin \theta}{\lambda} \qquad (A.9)$$

where $\tan \theta = D/2U$ and U is the object distance. (Thus, the limiting f_{max} is determined for an unrestricted lens by its numerical aperture—since U cannot be less than the focal length for the formation of a real image. With aperture restriction, a lower value pertains and the common practice of inserting an aperture plate in the rear focal plane—while having the desirable effect of reducing optical noise—also has the highly undesirable feature of greatly reducing f_{max}.)

The crucial point, now, is that the Uncertainty Principle [7] states that there is no information within a dimension $1/f_{max}$.

In displacement-sensitive systems, such as moiré interferometry, the maximum spatial frequency criterion is the one that may be most readily applied to quantify the ultimate limits. Since, for any gauge-length over which we might consider making a measurement, there will, at either end, be a dimension $1/f_{max}$ over which there will be complete uncertainty on the position to be ascribed to each displacement value, the minimum possible gauge-length is, therefore, seen to be

$$g_{min} = \frac{100}{E f_{max}} \qquad (A.10)$$

where E is the percentage (plus or minus) uncertainty margin permitted in the strain value.

This is readily appreciated in the simplest case when the pre-load condition gives a zero-fringe moiré pattern and we merely wish to measure the distance P between two fringe maxima so as to derive the mean strain between them. Since the position of a maximum is totally uncertain within $1/f_{max}$, we then have an absolute uncertainty of $\pm 1/f_{max}$ on the measurement of P. Since P is the gauge-length, equation (A.10) follows. For the case when there is a significant initial carrier f_c which changes due to the strain, the matter is rather worse, for we are then in the ever-undesirable situation of subtracting two quite similar values to get the desired answer.

A.3.1.4   An example

Using an unrestricted microscope objective with an F-number $(F/D) = 1.5$ operating at large magnification ($U \approx$ focal length) in red light, say $\lambda = 0.6 \ \mu m$, and with a $\pm 10\%$ allowable uncertainty, we obtain

$$f_{max} = 1054 \ \text{mm}^{-1} \qquad 1/f_{max} = 0.94 \ \mu m$$

and the minimum gauge-length would be

$$g_{min} = 9.4 \ \mu m.$$

This dimension is at the specimen not at the image.

This estimation is based on the assumption that there may be strain variations at frequencies above f_{max} that are of sufficient amplitude that they could affect the result if it were possible somehow to include them. If one could be certain that there are no such strain frequencies, then we would be justified to argue as follows.

If the maximum frequency that is known to occur is f_k, then we could say

$$g_{min} = \frac{100}{E f_k}.$$

However, it is not clear to the author how one might suggest a value for f_k.

The Airy diameter for such a lens is 2.3 μm (in object space) and were we to take that as the 'no information dimension', then we would have, for $\pm 10\%$ allowable uncertainty,

$$g_{min} = 23 \ \mu m.$$

However, the f_{max} approach appears to be the more rigorous—and is actually the more 'encouraging' in that it yields a smaller value for g_{min} (as we can, in fact, glean information on dimensions smaller than the Airy disc).

A.3.1.5   Multiple-lens imaging

The previous treatment presumes a single-lens imaging system. It can be very convenient to use a collecting lens positioned at less than one focal length from the specimen and to have a second lens system to form the final real image. (The work of the author and his colleagues mostly used such a system. It avoids the problem of dispersion referred to later.) In such circumstances, the acceptance half-angle may approach 90° and the value of f_{max} would approach $2/\lambda$, the limiting value. However, that would be an incorrect calculation: the image formed by such a collecting lens is a large virtual image positioned behind the specimen and a proper calculation would involve the use of the second lens's aperture and the distance to the virtual image, with a correction for the fact that the now-observed 'specimen'—the virtual image—is enlarged compared to the actual specimen and the 'spatial frequencies' are, therefore, correspondingly altered. That calculation will require considerable care.

A.3.2   With non-ideal conditions

A.3.2.1   *The modulation and phase transfer functions*

With incoherent illumination (as is used in photoelasticity) the lens' modulation and phase transfer functions (MTF and PTF) result in an attenuation of the higher spatial frequencies, and the phases of the spectrum may be altered or even reversed (i.e. black becomes white). Unless these effects are accounted for in the analysis, completely misleading results will be obtained.

In coherent illumination, the MTF is, ideally, unity up to f_{max} and phases are preserved but, in practice, the following effects will operate.

A.3.2.2   *Off-axis effects*

In the previous treatment, the field of view is considered small compared to the dimension of the lens. This will not normally be the case, and there are two effects to be considered:

(a) It is well recognized, in standard two-beam coherent collimated illumination for moiré interferometry, that, as the mean strain becomes higher, so the angle between the two normally emerging diffracted orders increases. What is not so obvious is that as the level of strain variation increases so each order itself becomes more divergent—or dispersed. This is dealt with in [8]. Both of these effects, which are additive, mean that the permissible field of view has to be smaller than the lens or the conditions near the periphery of the field will not be properly represented in the image. The required margin will not be very significant unless it is suspected that the frequency of strain variation is likely to approach or exceed 100 per mm (when the divergent half-angle is 3.6°)—but then the whole concern of this discussion is when spatial frequencies might be very high. We should always have the lens significantly larger than the field of view, especially for longer stand-off distances.

(b) The MTF, most typically, alters as the region of interest moves off the optical axis. It is easily seen, for instance, that for a lens positioned one diameter from the specimen, the half-angle subtended from the optical axis is 30° but at half a diameter off-axis it is 22.5°: this materially reduces the cut-off frequency. The PTF is also affected and, in the coherent moiré interferometry set-up, it is unlikely that off-axis viewing does not affect the PTF.

A.3.2.3   Misfocus

Loss of focus results in deterioration of the MTF and in phase changes. With high-resolution lenses, there is a very small depth of focus, e.g. $\lambda/(4 \times N.A.^2)$, and practitioners may have observed the black-to-white changes that occur at small imperfections as the focus is altered. This must have the most serious implications for processes that involve the interpretation of intensity at high magnification. 'Stopping down' the lens to give a greater depth of field is not an option: it reduces the maximum spatial frequency, thus limiting the ability to look for fine-scale detail.

A.3.2.4   Effects of the second diffracted order

If we are using typical coherent-optical methods, then another complication arises if the strain function contains very high frequencies. It is possible that there will be overlap between the dispersion angles of the first and second diffracted orders. This creates a problem for which there seems to be no immediately obvious solution. This was analysed in [8].

   However, if we are using photoelasticity, then we have a different problem. Any sinusoidal variation in refractive index represents a diffraction grating. This will typically have more than one diffracted order and if we are to reproduce the intensity in the image plane at the specimen, then we are required to capture all the diffracted orders. This means that if we wish to look at variations in frequency f_ε, then we have, in principle, to have a lens that will accept frequencies that are a multiple of f_ε. This requires further quantification but it appears likely that the contribution of the higher orders will be negligible, except possibly for the second. This appears to contrast with moiré interferometry in which it is necessary to exclude the higher orders, not capture them. However, the orders concerned are of different sources: in moiré interferometry they are arise from the applied grating, whereas in photoelasticity they are caused by the variations in the strain. There are also such orders in moiré emissions (referred to as 'mini-orders' in [8]) and they too must be captured.

A.3.3   How then to proceed?

If we wish to carry out a detailed investigation, what is necessary is that we first calibrate the optical system for MTF, PTF and out-of focus, using standard black-and white gratings up to the highest frequency available. This should be done under the experimental conditions and using a field of appropriate size. (It is easy to see how to calibrate for out-of-focus but it is not so obvious how to estimate the extent of any out-of-focus occurring in the experiment. We should at least alter the focus slightly as part of the experiment and see whether it affects the results.) This information can then be used to correct the intensity data that are recorded, so that we can have an improved estimate of the emission distribution at the specimen, rather than as it is collected in the image.

A.4   The recording and analysis systems

We are considering pixel-by-pixel analysis using an ideal phase-shifting or phase-stepping system.

First, it is important to know the 'contrast index' (gamma) of the camera. It should preferably be unity, so that the output signal is proportional to the incident intensity. If it is not, then it will effectively introduce higher harmonics in the presumed expression for the fringe structure in equation (A.3), i.e.

$$I(x) = I_0(x) + a(x)\cos k(2\pi f_c x + \phi(x) + \delta_1)$$
$$+ b(x)\cos 2k(2\pi f_c x + \phi(x) + \delta_2) + \text{etc.} \qquad \text{(A.11)}$$

(Reference [5] describes algorithms that can minimize this effect.)

Of course, there are limits to the linearity, most notably at the saturation point. There will also be some variations in gamma from pixel to pixel to be accommodated. These will typically be specified to be within 5% of each other.

With improvement in technology, the number of pixels in a typical field is continuing to increase. However, what has been quite static for many years is the number of grey levels—at 256. This is a major limitation in the recording and analysis. Saturation must be avoided at all costs and the 'comfortable' number is, therefore, about 230. Allowing for typical amplitude and d.c. variations, even in good quality low-noise fringes, some cycles will have less than 200 useful grey level peak-to-peak. This restriction on the precision of the discretization represents a form of digital noise.

A.4.1   The sampling theorem

Since the image is sampled at pixel spacing, d_{pix}, the Sampling Theorem states that the maximum frequency that can be recorded without distortion is

$$f_{max} = \frac{1}{2d_{pix}} \qquad \text{(A.12)}$$

and there is, therefore, no information within a length $2d_{pix}$. (However, in fact, the situation is worse: any frequencies that do exist above f_{max} will be aliased to appear as very low frequencies.) This gives us another criterion for minimum gauge-length: analogously to the earlier example, for a $\pm 10\%$ uncertainty level,

$$g_{min} = 20 \text{ pixel spacings} \qquad \text{(A.13)}$$

or, more correctly, whatever that may translate as in mm with the magnification used. This is quite a discouraging result: we would probably tend to opt to accept a greater uncertainty or to argue that there are no significant strain frequencies above f_{max}.

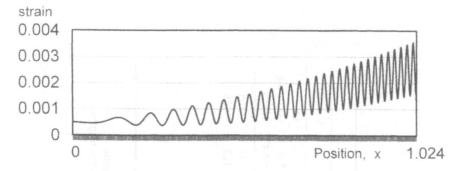

Figure A.3. Strain used for simulation (it corresponds to the moiré in figure A.1): $\varepsilon(x) = 10^{-3}[0.5 + 2x^2 - x\cos(10^4\pi x^2/180)]$.

A.4.2 A simulation exercise

Figure A.3 shows a strain distribution that the author has used to simulate various effects. Fast Fourier transformation shows that the significant spatial frequencies go approximately from zero to 60 per unit of x—confirming the impression from the graph. If x is in mm, that would correspond to about 1 cycle per 16 μm, which is well within what we would consider to be the optical limit for good lenses. At the same time, it could represent a 10× magnification onto a CCD array 1 cm across. The corresponding moiré pattern is that shown in figure A.1, for which FFT shows that there are no significant frequencies above 10 per unit of x; this is confirmation of the attenuation effect of the moiré process on the higher frequencies, as indicated in the discussion at equations (A.4)–(A.7).

The conditions simulated were:

(1) grating pitch 1/1200 mm,
(2) 250 grey levels throughout,
(3) perfect contrast characteristics (i.e. gamma = 1 for all pixels),
(4) random variation in gamma (maximum ±2.5%) from pixel to pixel,
(5) gamma = 1.2 for all pixels,
(6) the introduction of a first harmonic with 10% of the amplitude of the fundamental (i.e. $b(x) = 0.1$ in equation (A.11)),
(7) the strain is calculated on a gauge-length of (a) 1 , (b) 3 or (c) 9 pixel-lengths
(8) four-bucket algorithms was standard but a three-bucket one was explored,
(9) A 1024 pixel camera was assumed; the pixel fill factor was not considered relevant to this study.

A.4.3 Discussion of the simulation results

(1) Albeit the simulated strains go as high as 3000 microstrain, the errors indicated, especially for the single-pixel gauge-length, are surprisingly high.

Table A.1. Results of the simulation exercise.

Gamma	Bucket	Gauge length	Pixel-length					
			1		3		9	
			S.d. error[a]	Max. occurring error[a]	S.d. error[a]	Max. occurring error[a]	S.d. error[a]	Max. occurring error[a]
1 for all pixels	4	'averaging'[b]	217	−599	54	165	21	65
		'point'[c]	217	−599	79	−251	128	−458
	3	'averaging'[b]	178	−548	45	125		
±2.5% random	4	'averaging'[b]	281	−1150	72	−265		
1.2	4	'averaging'[b]	187	857	108	638		
With 10% first harmonic	4	'averaging'[b]	215	724	117	411		

[a] The errors are quoted in microstrain.

[b] These errors were calculated by comparison with the true average strain over the gauge lengths.

[c] These errors were calculated by comparison with the true strain at the centre point of the gauge lengths.

To be indicating errors of 500–600 $\mu\varepsilon$ even under these ideal conditions is, on a subjective basis, quite disappointing.

(2) The three-bucket algorithm is, on this limited evidence, slightly superior to the four-bucket. This is somewhat counterintuitive and requires further investigation.

(3) The errors are due to the lack of discrimination in the 250 grey levels. Consequently (although this is not inferable from the table) the largest errors occur around the points where the moiré in figure A.1 maximizes or minimizes, i.e. where there is the slowest rate of change in the intensity profile and the use of 256 grey levels is insufficient. This gives an insight as to why the three-bucket algorithm may be superior to the four-bucket one: since the four-buckets are at 0, $\pi/2$, π, $3\pi/2$, wherever the 0-bucket, say, occurs near a maximum or minimum, the π-bucket too will occur near a minimum or maximum, so that the algorithm will use two buckets from the least discriminatory areas (similarly where the $\pi/2$-bucket is near a maximum or minimum): however, the three-buckets are at 0, $\pi/3$, $2\pi/3$, so that only one bucket in the algorithm can be near a maximum or minimum.

(4) The use of a three-pixel gauge-length improves the s.d. errors but there are still quite high individual errors occurring.

(5) The nine-pixel gauge length gives a marked improvement in the 'averaging' error but worsens the 'point' error. This is exactly what would be expected, as the rapid changes will be smeared out by the averaging. However, the whole point of the process is to elicit the detail, and such an averaging process is, in principle, inconsistent with that objective. Somewhere about a three-pixel gauge-length seems to give the optimum compromise between errors due to discretization and those due to averaging. The best solution would be increased grey-level discrimination.

(6) A small random variation in gamma, although typically within standard product tolerances, causes a substantial increase in the errors.

(7) It is not shown in table A.1 but, in fact, using the 256 grey levels, the errors are considerably larger if the arcosine formula (equation (A.2)) is used, rather than the phase-stepping algorithm. That is because near the maxima and minima the arcosine is using a single value of $I(x)$ to demodulate, whereas the phase-stepping algorithm includes values from points well away from the turning points. (The arcosine does, of course, give error-free results if grey levels are not restricted.)

(8) For the one-pixel values with gamma $= 1.2$, we appear to have a better result than for gamma $= 1$. This could be attributable to the fact that the higher exponent will make the moiré more 'peaky', so that the problem of discrimination in the regions of maxima and minima could be reduced. This reasoning is reinforced by the deterioration in the gamma $= 1.2$ results once we go to the three-pixel gauge-length (relative, that is, to the other three-pixel gauge-lengths).

(9) The inclusion of 10% of the first harmonic introduces errors of approximately the same scale as for gamma = 1.2

A.5 Conclusions

When seeking to look at strain variations in great detail using photomechanics, the use of sophisticated optical systems and methods of recording and analysis has to be considered with the greatest care.

Methods have been described for evaluating the optical and recording systems, based on what spatial frequencies are of interest and what the system is capable of seeing.

The implication of the Sampling Theorem is that if a 10% error be specified for the strain, the minimum gauge-length is 20 camera pixels—whatever that may represent in the specimen.

In the phase-stepping recording and analysing system, the 256 grey-level restriction has serious implications. A superior specification should be sought on both the gamma value, which should be unity and specified to better than the ±2.5% typical standard (or, alternatively, each individual pixel should be calibrated).

Any results obtained should be considered in those contexts and subjected to appropriate final filtering to eliminate what appear to be very rapid changes in strain or very sharp changes of direction of a strain contour, that imply spatial frequencies that can only be artefacts of the system, or of noise, and, therefore, cannot be attributed to the actual strain variation.

References

[1] McKelvie J 1986 On the limits to the information obtainable from a moiré fringe pattern *Proc. 1986 SEM Spring Conf. on Experimental Mechanics* (New Orleans, LA: SEM) pp 971–80
[2] Green R J, Walker J G and Robinson D W 1988 Investigation of the Fourier transform method of fringe pattern analysis *Opt. Laser Eng.* **8** 29–44
[3] Sciamarella C A 1965 Basic optical law in the interpretation of moiré patterns applied to the analysis of strains—part I *Exp. Mech.* **5** 154–60
[4] Rihaczek A W and Bedrosian E 1966 Hilbert transforms and the complex representation of real signals *Proc. IRE* **54** 434–4
[5] Huntley J M 1998 Automated fringe pattern analysis in experimental mechanics: a review *J. Strain Anal.* **33** 105–26
[6] Born M and Wolf E 1975 *Principles of Optics* 5th edn (Oxford: Pergamon) ch 8.6, equation (51)
[7] Yu F T S 1976 *Optics and Information Theory* (New York: Wiley)
[8] McKelvie J 1990 On moiré interferometry and the level of detail that it may legitimately reveal *Opt. Laser Eng.* **12** 81–99

Index

additive noise, 487
Airy disc, 488
algorithm, 276, 403
alignment, 363
Almansi, 6, 11
annealing, 275, 277–8, 284, 312, 315
aperture restriction, 490
array, 43, 128, 145, 251
Au–Si eutectic, 160

(BGA) ball grid array package, 128, 135, 136, 154, 171, 387
bimaterial, 287
bias fringes, 403
bolt, 66, 361
buckling, 375

calibration, 185, 192, 403, 433
carrier frequency, 486
caustic, 342
CBGA package, 135, 145, 154
CCGA (ceramic column grid array), 145
clustering algorithm, 463
composite, 230
composite, carbon fibre, 79
composite, epoxy-carbon, 230
compression, 221, 223
concrete, 66
contact, 342, 357
contact stresses, 357
correlator, 455
correspondence analysis, 474
crack, 443

crack closure, 53, 110
crack growth, 91
crack growth criteria, 91
crack-tip, 51, 56, 92, 110, 261
creep, 104, 108
creep relaxation, 104
cross grating, 325
cyclic plastic zone, 62
cylindrical, 207
cylindrical surface, 195, 199, 207

defect, 455
deflection, 378
delamination, 443
demodulation algorithm, 486
dovetail joint, 257

enhancement, 462
epoxy, 172, 181, 361, 446
expansion, 250

factorial correspondence analysis, 474
failure, 451
fasteners, 361
fatigue, 56, 243
fatigue crack, 244
finite element, 92, 156, 264
finite element modelling, 60, 156
flatness, 397, 404
flip-chip, 29
Fourier processing, 306
Fourier transform, 9, 327, 447
fracture mechanics, 261
fracture, 323, 450

processes, 51, 70
fringe analysis, 106
fringe contours, 382–3
fringe pattern, 5
fringe shifting, 23, 380, 390

gauge length, 187, 197
gerbil, 433
gerbil eardrum, 433
grain boundaries, 35
graph technique, 468
grating, 312
grating application, 172, 181, 288, 322
grating diffraction, 417
grating frequency, 367
grating photography, 299
grating replication, 209, 213
grid method, 321
grid transfer, 367
grilles, 375

heat resistant gratings, 303, 313
heatsink, 153
high-frequency information, 485
high temperature, 299
hit/miss transform, 456, 469
hole, 207
hole-drilling, 240
hole expansion, 239
Hopkinson, 323
    bar, 323, 445
    split bar, 323, 445
HRR fields: Hutchison, Rice and Rosengren crack-tip stress fields, 531
hybrid, 72, 295, 343
hybrid mechanics, 230, 255, 261, 263

image processing, 431
impact, 443
in-plane, 275, 343, 448
interface, 287

interferometer, 19
interlaminar, 195
interlaminar deformation, 198
inverse moiré, 457, 461

J-integral, 75, 89, 103, 118
joint, 27, 361

laminated, 207
laminates, 221, 228, 361
leadless chip carrier (LCC), 27
ligament, 332
linerboard, 69
load, 385
low-frequency fringes, 485
lug, 239–40

maximum spatial frequency, 490
metal-matrix, 42
micromechanical deformation, 47
microscope, 11, 322, 419
microscopic, 5
microscopic moiré, 18, 25, 158
minimum possible gauge length, 490
misfocus, 493
mixed-mode, 62, 75
modulation and phase transfer functions, 492
moiré photography, 443
multiple lens imaging, 491
multiplicative noise, 487

nonlinear analysis, 5
nonlinearity, 342
Normal Stress Ratio, 90

off-axis effects, 492
optical correlation, 455
optical/digital fringe multiplication (O/DFM), 23, 42, 379
orthotropic, 75

packaging, 123, 134
paper, 66

phase grating, 414
phase moiré, 475
phase-shifting, 487
phase-stepping, phase-shifting, 276, 381, 395, 458
photoelastic, 294
plastic, 51, 242, 250, 329
plastic ball grid array (PBGA), 128, 381
plated through holes, 33
ply, 207
polarized shearing, 348
printed circuit board (PCB), 126, 145
printed wirer board (PVB), 394
profilometry, 408
projection, 401
projection moiré, 399

railway rail, 272
Rayleigh criterion, 489
ray tracing, 408
real-time, 455
reference grating, 380, 385, 390
residual stress, 56, 160, 239
resolution, 433
roller expansion, 250

sampling theorem, 494
second diffracted order, effects of, 493
segmentation algorithm, 465
shadow moiré, 375, 379
shear strains, 28
shear, 179, 182, 227, 231
simulated ideal intensity distributions, 486
skeletonizing, 456, 461, 466
solder bump, 29, 171, 174
solder column connection, 145
solder, 135, 154
specimen, 446
splines, 264
statistical analysis, 456

strain, 157
strain gauge, 179, 185
strain sensor, 421
stress distribution, 292
stress intensity factor, 56, 75, 89
stress relaxation, 104
submicrovoids, 339

T^* fracture parameter, 104
T^* integral, 103
tangential strain, 200
thermal, 145, 271
thermal cycle, 145, 148
thermal deformation, 42
thermal expansion, 125
thermal expansion coefficient (CTE), 123, 161, 166
thermal strain, 25
thermal strain analysis, 140, 145, 153
thermal stress, 134, 271
thermal warpage, 391
thermo mechanical, 171
thin small outline package (TSOP), 26, 125
three dimensional, 195, 271, 294
transverse, 201
transverse strain, 201
tympanic membrane, 433, 437

validation, 98
van der Lugt, 456
van der Lugt correlator, 456
verification, 217
video, 427
video moiré, 427, 437
void, 322
von Mises, 53, 106
von Mises yield criterion, 51

wood, 66

yield, 51